Improved Reservoir Models and Production Forecasting Techniques for Multi-Stage Fractured Hydrocarbon Wells

Improved Reservoir Models and Production Forecasting Techniques for Multi-Stage Fractured Hydrocarbon Wells

Special Issue Editors

Ruud Weijermars
Wei Yu
Aadi Khanal

MDPI • Basel • Beijing • Wuhan • Barcelona • Belgrade

MDPI

Special Issue Editors
Ruud Weijermars
Texas A&M University
USA

Wei Yu
Texas A&M University
USA

Aadi Khanal
Texas A&M University
USA

Editorial Office
MDPI
St. Alban-Anlage 66
4052 Basel, Switzerland

This is a reprint of articles from the Special Issue published online in the open access journal *Energies* (ISSN 1996-1073) from 2018 to 2019 (available at: https://www.mdpi.com/journal/energies/special_issues/Reservoir_Models_and_Production_Forecasting).

For citation purposes, cite each article independently as indicated on the article page online and as indicated below:

LastName, A.A.; LastName, B.B.; LastName, C.C. Article Title. *Journal Name* **Year**, *Article Number*, Page Range.

ISBN 978-3-03921-892-9 (Pbk)
ISBN 978-3-03921-893-6 (PDF)

Contents

About the Special Issue Editors

Ruud Weijermars is Professor at the Harold Vance Department of Petroleum Engineering at Texas A&M University. His current research interests include the development of practical and fast methods for modeling fracture propagation and reservoir drainage simulations at high resolution. Weijermars holds a BS (1977) and MS (1981) in Geoscience and Mathematics from the University of Amsterdam, and a PhD in Geodynamics from the University of Uppsala (1987). His first academic position was as Associate Professor in 1990 at Uppsala University, from which he moved on to teaching and research activities in Switzerland (ETH Zurich), the United States (Bureau of Economic Geology, UT Texas at Austin), Saudi Arabia (King Fahd University of Petroleum and Minerals), the Netherlands (Delft University of Technology), and the United States (Texas A&M University). Weijermars has worked extensively as a retained consultant for industry via Alboran Energy Strategy Consultants.

Wei Yu is currently Research Associate both at Harold Vance Department of Petroleum Engineering, Texas A&M University, USA, as well as at the Hildebrand Department of Petroleum and Geosystem Engineering, University of Texas at Austin, USA. He holds a PhD degree in petroleum engineering from the University of Texas at Austin. He is Associate Editor of *SPE Journal and Journal of Petroleum Science and Engineering*. His research interests include EDFM (embedded discrete fracture model) technology for handling any complex fractures, unconventional reservoir simulation, and automatic history matching. He has more than 100 technical papers and holds five patents and one book (Shale Gas and Tight Oil Reservoir Simulation).

Aadi Khanal is Postdoctoral Researcher at Harold Vance department of Petroleum Engineering, at Texas A&M university. He graduated with a BS from Lafayette College, USA, MS, and PhD from the University of Houston, USA, all in Chemical Engineering. His research interests include reservoir simulation, production forecasting, analytical modeling, and data driven modeling for unconventional reservoirs. He has authored several scientific papers in international refereed journals and conferences.

Preface to "Improved Reservoir Models and Production Forecasting Techniques for Multi-Stage Fractured Hydrocarbon Wells"

Petroleum engineering has drastically evolved in the past few decades, especially with the current focus on the development of unconventional reservoirs requiring new approaches. We think this is an opportune moment to dedicate an entire Special Issue to reservoir models and production forecasting in unconventional reservoirs. The papers collated here cover a wide variety of topics, with the common theme being how to improve the accuracy of performance models for unconventional reservoirs. In particular, production forecasts can still be improved from the current state to offer more accurate results, based on models that better capture the physics of the complex subsurface flow in hydraulically fractured shales.

This Special Issue aims to showcase systematic studies with a focus on unconventional reservoirs, with particular emphasis on presenting recent developments in the field of reservoir modeling and production forecasting. We received numerous submissions from researchers all over the world on a wide range of topics, making our task of selecting the best articles for this issue both a challenging and rewarding one. This issue features the application of analytical modeling, numerical modeling, statistics, decline curve analysis, history matching, and other techniques. Applications based on real world reservoirs are given, using data from a variety of active shale plays in the United States (Eagle Ford, Permian Basin, Barnett, Marcellus, and Bakken) and China (Changning Shale Gas Field). As a result of such diverse topics and case studies, this Special Issue is truly interdisciplinary.

This Special Issue will serve as a valuable resource to anyone interested in reservoir modeling and production forecasting of unconventional reservoirs. We hope that this issue will not only provide answers to some of the relevant current questions in this topic, but also encourage readers to identify new problems which may further advance the field of reservoir modeling and production forecasting of unconventional reservoirs. Finally, we would like to thank all the authors, publishers, industry partners, scientific community, and the reviewers without whom this Special Issue would not have been possible.

Ruud Weijermars, Wei Yu, Aadi Khanal
Special Issue Editors

![energies logo] *energies*

MDPI

Article

Impact on Drained Rock Volume (DRV) of Storativity and Enhanced Permeability in Naturally Fractured Reservoirs: Upscaled Field Case from Hydraulic Fracturing Test Site (HFTS), Wolfcamp Formation, Midland Basin, West Texas

Kiran Nandlal * and Ruud Weijermars

Harold Vance Department of Petroleum Engineering, Texas A&M University, 3116 TAMU, College Station, TX 77843-3116, USA; r.weijermars@tamu.edu
* Correspondence: knandlal@tamu.edu

Received: 7 August 2019; Accepted: 2 October 2019; Published: 11 October 2019

Abstract: Hydraulic fracturing for economic production from unconventional reservoirs is subject to many subsurface uncertainties. One such uncertainty is the impact of natural fractures in the vicinity of hydraulic fractures in the reservoir on flow and thus the actual drained rock volume (DRV). We delineate three fundamental processes by which natural fractures can impact flow. Two of these mechanisms are due to the possibility of natural fracture networks to possess (i) enhanced permeability and (ii) enhanced storativity. A systematic approach was used to model the effects of these two mechanisms on flow patterns and drained regions in the reservoir. A third mechanism by which natural fractures may impact reservoir flow is by the reactivation of natural fractures that become extensions of the hydraulic fracture network. The DRV for all three mechanisms can be modeled in flow simulations based on Complex Analysis Methods (CAM), which offer infinite resolution down to a micro-fracture scale, and is thus complementary to numerical simulation methods. In addition to synthetic models, reservoir and natural fracture data from the Hydraulic Fracturing Test Site (Wolfcamp Formation, Midland Basin) were used to determine the real-world impact of natural fractures on drainage patterns in the reservoir. The spatial location and variability in the DRV was more influenced by the natural fracture enhanced permeability than enhanced storativity (related to enhanced porosity). A Carman–Kozeny correlation was used to relate porosity and permeability in the natural fractures. Our study introduces a groundbreaking upscaling procedure for flows with a high number of natural fractures, by combining object-based and flow-based upscaling methods. A key insight is that channeling of flow through natural fractures left undrained areas in the matrix between the fractures. The flow models presented in this study can be implemented to make quick and informed decisions regarding where any undrained volume occurs, which can then be targeted for refracturing. With the method outlined in our study, one can determine the impact and influence of natural fracture sets on the actual drained volume and where the drainage is focused. The DRV analysis of naturally fractured reservoirs will help to better determine the optimum hydraulic fracture design and well spacing to achieve the most efficient recovery rates.

Keywords: naturally fractured reservoirs; time of flight; particle paths; enhanced permeability; flow modeling; natural fractures; hydraulic fractures; drained rock volume; fracture porosity; hydraulic fracturing; hydraulic fracturing test site; wolfcamp formation; midland basin

1. Introduction

Production from unconventional reservoirs can only be economically achieved via the creation of high permeability conduits in the form of man-made hydraulic fractures. The modeling of these hydraulic fractures is difficult due to very little direct data about the true nature of the hydraulic fracture orientation and hydraulic fracture properties in the subsurface. Further complications arise from the uncertainty in other subsurface properties such as reservoir anisotropy and heterogeneity and the intricacies of the flow and recovery process due to the interactions of gravity, capillary and phase behavior [1]. Reservoir heterogeneity in shale formations may be further complicated by the presence of pre-existing natural fractures.

The interaction of natural fracture sets with newly created hydraulic fractures may cause various effects that bear on the drainage and production from reservoirs. Natural fractures introduce added complexity that affect the flow of hydrocarbons in the reservoir, which may directly affect well performance [2]. The interaction between natural and hydraulic fractures and the modeling of this behavior is relevant for understanding variations in production due to varying the fracture treatment design [3]. Even when natural fractures are non-conductive, they can still influence sweep patterns due to the local blockage and deflection of waterfloods [4]. If conductive, such natural fractures may have an even greater effect on flow regions near hydraulic fractures [5]. Due to the potential impact that natural fractures may have on flow paths in the subsurface, accurate production forecasting and simulation models need to account for the critical fracture attributes (e.g., natural fracture orientation, distribution, connectivity, and interaction with the hydraulic fractures) [6,7]. Natural fractures have been found to interact with hydraulic fractures and influence production via three major mechanisms and each is explained in detail subsequently (Section 2.1).

Hydraulic fracture design is currently based on the estimation of geomechanical rock properties in conjunction with fracture propagation models utilized in numerous hydraulic fracturing simulators. Comparisons of the various simulators started with the work of Warpinski et al. [8] but to date, no consensus has been reached on the relative merits of the different fracture propagation codes. What most commercial codes have in common is that fracture propagation is based on poro-elastic models that assume homogenous rock properties which favor the formation of sub-parallel, planar hydraulic fractures [9]. The question arises of how realistic this assumption of sub-parallel, planar hydraulic fractures is. Current fracture diagnostics cannot resolve the detailed nature of created hydraulic fractures in the subsurface [10]. Nonetheless, a large amount of empirical evidence has suggested that deviations from single planar fracture geometry are common. For example, cores sampled from a hydraulically fractured reservoir showed more fractures than the number of perforation clusters [11]. The idea that pre-existing natural fractures are necessary for the creation of complex fracture networks is no longer valid as fracture complexity in terms of branching, deflection and offset is possible due to local reservoir heterogeneities such as bedding planes. The presence of natural fractures can lead to even further complexity when interacting with hydraulic fractures due to reactivation of the natural fractures and other planes of weakness present [12].

Further evidence for the complexity of created hydraulic fractures is indicated by data from microseismic surveys [13,14]. The assumption of planar, subparallel fractures rarely matches such generated microseismic data. Furthermore, natural fractures in rocks due to hydrothermal intrusions can be considered as analogs to the hydraulic fracturing process in terms of fracture geometry. An example of this phenomenon is found in hydrothermally fractured rocks from the Aravalli Supergroup in the state of Rajasthan, India [15–17]. The fracture patterns found in these rocks are highly complex and dendritic (Figure 1) and not planar in nature.

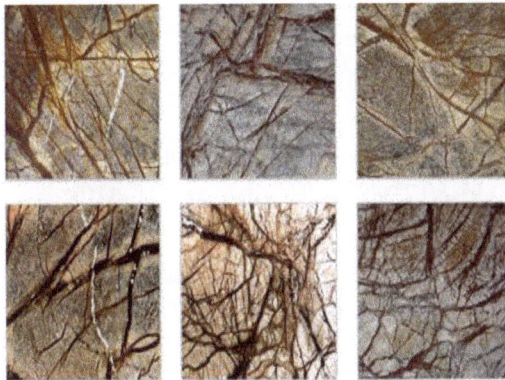

Figure 1. Polished rock slab examples from Bidasar showing bifurcating hydraulic injection veins. Image dimensions about 1 square meter (courtesy Dewan Group).

Beyond natural field examples, mineback and laboratory experiments by Huang and Kim [18] confirmed that man-made hydraulic fractures do not always propagate linearly perpendicularly to the direction of minimum far-field stress and are not always planar in shape. This evidence provides support for the idea that hydraulic fractures are complex branching structures and models that do not honor this geometry may give incorrect results. The assumption of planar hydraulic fractures brings with it the creation of stagnation zones between individual fractures due to flow interference [19,20]. Such dead zones are detrimental to reservoir drainage and affect well productivity. If these hydraulic fractures are not planar, then the influence of stagnation zones changes [21]. It is now obvious how crucial it is to accurately model fracture geometry to ensure the optimization of wells under production.

The present study looks at the flow interaction in a reservoir between natural fracture sets and hydraulic fractures, with an emphasis on how these fracture networks influence the development of the drained rock volume (DRV). A series of methodical simulations allows us to understand how the natural fractures impact the DRV evolution. The effect on the DRV of complex fracture branching has been modeled elsewhere [21]. The present study uses closed-form analytical solutions based on complex analysis methods (CAM) to model flow in a 2D model of both the natural and hydraulic fractures. In this method, the hydraulic fractures are represented as line sinks and the natural fractures are modeled as an infinite number of line doublets using a new algorithm [22,23]. The interaction of the natural fractures and the hydraulic fractures is modeled in CAM to determine the flow response and pressure changes in the reservoir. Based on these responses, Eulerian particle tracking can then quantify the impact of natural fractures and hydraulic fractures on the DRV. Insights generated from the models can be used to optimize well production and recovery factors in unconventional reservoirs.

This paper is organized as follows. We first highlight how natural fractures can impact flow in the subsurface via three major mechanisms (Section 2). These mechanisms are expounded upon and the ways in which they can affect the drainage area in unconventional reservoirs are discussed. A review of the previous ways natural fractures were modeled (in classical numerical reservoir simulations) together with a brief analysis of the drawbacks of such methods when compared to CAM is next presented. Following this is a discussion about the relation of natural fracture porosity and permeability and the surprising lack of data on these crucial properties. The next section (Section 3) looks at the methodology of using CAM to model natural fractures and hydraulic fractures in 2D. The following section (Section 4) describes the additional methods used to incorporate and model the natural fracture mechanisms that impact subsurface flow in the CAM workflow. Once the methods used are explained, we then move to the results from various flow simulations (Section 5). For the results, modeling begins with simple representative elementary volume (REV) models to show the impact of natural fracture with altered porosity and permeability. Subsequently, the models are extended to synthetic cases of flow effects of natural fractures around hydraulic fractures. A final case study makes use of natural fracture properties from the Hydraulic Fracturing Test Site (HFTS) in the Midland Basin to accurately model the DRV of an unconventional reservoir in the Wolfcamp Formation in West Texas.

2. Natural Fracture and Hydraulic Fracture Models

The importance of natural fractures and their possible effects on fluid flow and well production in unconventional reservoirs has been noted by many authors, the work of whom is outlined (Sections 2.1 and 2.2). We propose three major mechanisms by which natural fracture properties may impact flow and describe each mechanism in detail (Section 2.1). We also look at previous attempts to model natural fractures in reservoir simulations and the drawbacks of the prior attempts (Section 2.2). The relationship between natural fracture porosity and permeability is reviewed (Section 2.3). These properties can affect reservoir drainage patterns, which is why field data were obtained from an unconventional shale reservoir in West Texas (Section 2.3).

2.1. Natural Fracture and Hydraulic Fracture Interaction Mechanisms

Numerous authors have stated that the presence of natural fractures increase production in hydraulically fractured wells in unconventional reservoirs [2,24]. Such a broad statement neglects the intricacies in natural fracture morphology, distribution and its ability to impact production. Gale et al. [25] stated that *"fracture systems in shales are heterogeneous; they can enhance or detract from producibility, augment or reduce rock strength and have the propensity to interact with hydraulic fracture stimulation"*. Of importance is whether the natural fractures are sealing or not, which depends on the degree of cementation in the natural fractures. For natural fractures lacking cement, with no natural proppant (as used in hydraulic fractures), a significant reduction in permeability is possible but will not result in complete closure and thus, permeability in the natural fracture would still be above that for the intact host rock [26]. Another factor to consider is the connectivity of the natural fracture system. Cross-cutting and abutting fracture systems of different ages may not be hydraulically connected, depending on the degree of sealing. Here, it is possible that hydraulic fracturing can be beneficial for the reactivation of these natural fracture systems, which may lead to natural fracture networks becoming connected to the hydraulic fractures for the first time. In this study, we modelled natural fracture systems with an enhanced conductivity, i.e., cementation is not a hindrance to the flow potential within the system.

Although some ambiguity remains on the true nature of natural fractures influence on well production, research using static, object-based permeability suggests that natural fractures would enhance well productivity [2]. Three major mechanisms for the increase in productivity due to natural fractures have been put forward by Weijermars and Khanal [23]. These three production enhancement mechanisms related to natural fractures involve (1) *equivalent permeability enhancement*, (2) *storage effects, due to enhanced porosity in natural fractures*, (3) *connection of hydraulic to natural fractures*. Each mechanism is further discussed below.

(1) *Equivalent permeability enhancement*: The presence of a natural fracture system open to flow (uncemented) with higher permeability than the matrix, would increase the equivalent permeability of the overall reservoir. This enhanced equivalent permeability will result in a corresponding higher flow rate towards the hydraulically fractured well increasing the well productivity.

(2) *Storage effects due to natural fracture enhanced porosity*: Natural fracture porosity may differ from the matrix either on initial formation of the fracture or due to later dissolution of precipitated minerals in the fracture space [25]. Due to size dependent sealing patterns, larger natural fractures are believed to have greater porosity [27] and as such, porosity in natural fractures is thought to be underestimated in most models. A greater porosity in the natural fractures than in the matrix may affect the extent of the drained area because porosity is a major control on time of flight for

particles traveling along streamlines [28]. If the porous fractures are more fluid-filled than the surrounding matrix, storage effects will affect the well productivity. Uncemented fractures with enhanced porosity will allow for storage of hydrocarbons that, when tapped by the hydraulic fractures, will flow readily towards the well.

(3) *Connection of hydraulic fractures to natural fractures*: Hydraulic fractures will propagate preferentially along planes of weakness in the reservoir such as those created by natural fracture systems. If a hydraulic fracture reactivates and connects to the natural fracture system, this connection leads to the natural fractures essentially becoming a direct extension of the hydraulic fracture pressure sink. The connection of both fracture systems correspondingly increases the total fracture surface area that is in contact with the reservoir matrix and improves the production rate of such wells.

2.2. Natural Fracture Modeling

Numerous attempts have been made to properly model fractured reservoirs that can accurately account for flow in such fractured porous media. The earliest attempt was made by Warren and Root [29] by using the dual-porosity model. Irregular natural fractures were modeled by using homogenous matrix blocks that are separated by orthogonal uniform natural fractures with fluid communication between the isotropic and matrix blocks governed by the inter-porosity flow coefficient (λ) and fracture storage capacity ratio (ω). Starting with this model, Kazemi et al. [30] introduced modifications that allowed for multiphase flow as well as the introduction of a new matrix shape factor. In addition to this work, numerous other authors have tried to adapt the Warren and Root [29] model to account for changes in matrix block geometry with new methods moving from double-porosity models to triple-porosity models [31,32]. Drawbacks of dual and multi porosity-based fracture models are that discrete fractures are not included and actual fracture density is not accounted for. Dual-porosity models also do not account for the flow paths followed when the fluid exchange occurs between the matrix and fractures, which can thus lead to inaccurate modeling of complex flow behaviors and can result in the wrong calculation of pressure gradients [33].

Another method to model naturally fractured reservoirs has been the use of Discrete Fracture Networks (DFN). For this model, fluid flow in the medium is represented through a system of connected natural fractures embedded within the rock matrix. This technique was first introduced by Long et al. [34] and has evolved over the years and seen increased use to model flow in conventional and unconventional naturally fractured hydrocarbon reservoirs [35,36]. The DFN method is typically used when (1) simulations done on a small scale where fracture dominance would otherwise result in an invalid upscaled continuum approximation, (2) in simulations on a larger scale where fracture dominance is small and the upscaled continuum model with only the largest fractures accounted for is valid [37]. Drawbacks of DFN modeling comes from the lack of data for the detailed inputs needed for the model such as fracture orientation, length, aperture and transmissibility along the (natural) fractures. The use of field analogs in surface outcrops may help fill these data gaps but there is no consensus on how accurately these measurements from outcrops match the subsurface. To combat this downside, current modeling attempts use a stochastic approach based on probability density functions to determine parameters of interest. This stochastic realization method can be used to create multiple realizations of the natural fracture patterns with fracture lengths following a power-law distribution [38]. The DFN method is also computationally intensive (and therefore expensive) as it requires very fine grids, which is particularly the case for multi-stage wells in unconventional reservoirs with numerous perforation zones per well [33].

2.3. Natural Fracture Porosity and Permeability

The effect of natural fractures on fluid flow is highly dependent on the reservoir type. Four major naturally fractured reservoir types have been identified by Nelson [39] based on the extent that fractures have altered the reservoir characteristics. Type 1 reservoirs have natural fractures that provide the bulk of the reservoir storage capacity and permeability, and typically have very high natural fracture density. In Type 2 reservoirs, permeability is essentially provided by the fractures while the matrix is responsible for the bulk of porosity. For Type 3, the reservoir matrix has high permeability and porosity but the permeability is further enhanced by the natural fracture system and can result in very high flow rates. Type 4 naturally fractured reservoirs have fractures that provide no additional porosity or permeability enhancement due to the fractures being filled with impermeable minerals. Natural fractures in Type 4 reservoirs are actually detrimental to fluid flow as they create significant reservoir anisotropy, which acts as barriers to flow [40].

One could say that Nelson's classification is mostly valid for conventional reservoirs and less applicable to shale reservoirs. Unconventional shale reservoirs have the majority of porosity contained within the rock matrix, while hydraulic fracturing is needed to create high enough permeability pathways for economical production. The majority of shale reservoirs also exhibit a high degree of natural fracturing. Due to the described attributes, unconventional shale reservoirs can be considered to range between Type 1 and Type 2 classification of naturally fractured reservoirs, with an example of Type 2 being the Spraberry reservoir in West Texas [40]. The extent to which the natural fracture systems in shale reservoirs affect hydrocarbon production due to enhanced storage and permeability is yet unclearly defined and remains nebulous [25]. There is consensus that hydraulic fracture propagation needs to take into account the impact of natural fractures on this propagation [41] but the impact that natural fractures have independently on production is not well constrained [25]. This is because core observations tend to show cemented natural fractures giving lower permeability and porosity measurements. However, field tests indicate much higher values for both permeability and porosity of natural fractures. Soeder [42] stated *"typical natural fractures that enhance reservoir permeability to the point of commercial production are probably not obvious lithological features, such as near-wellbore calcite mineralized joints"*. Description of natural fractures in the Barnett shale show completely cemented fractures before hydraulic fracturing that subsequently became open and might demonstrate stress sensitivity [43]. The cited evidence shows that there is a strong possibility of natural fracture systems with enhanced porosity and permeability in shale reservoirs potentially high enough to impact fluid flow, which is crucial to accurately capture in any flow models.

Important characteristics of natural fractures include fracture length, aperture, orientation, density, spacing, porosity and permeability. Values for most of these parameters are difficult to obtain from the subsurface. Outcrops can give some indication of fracture length, density and spacing there is reason to believe that limited outcrop data do not give a proper representation of subsurface features that lie deeper within the earth [25]. What we do know is that many shales exhibit a wide range of fracture sizes and properties. The larger the natural fracture, the greater the porosity because of size-dependent sealing patterns [27] and it is believed that underestimation of natural fracture porosity may have occurred (due to this phenomenon) in some case studies. A value of 2% or less for the porosity of a natural fracture system is considered typical, however, field data from the Monterey shale Formation using samples from highly fractured parts, have shown values as high as 6% for natural fracture porosity [44]. Studies conducted by Weber and Bakker [45] as well as Lee et al. [46] give values of 2% to 7% for natural fracture porosities of the Marcellus shale [25].

The present study makes use of detailed core descriptions from the Hydraulic Fracturing Test Site (HFTS) for accurate natural fracture property and distribution data for our field case model (Sections 5.3 and 5.4). These descriptions come from six cores from a slanted well that sampled the rock volume around a hydraulically fractured well. These cores were located in the Upper and Lower Wolfcamp formation and this data (type based in origin, dip and dip direction of the fractures) was previously used to visualize fracture orientation, types of fracture and perforation clusters by Shrivastava et al. [12]. We make use of this data for a more realistic representation of the natural fracture system present in the subsurface in our flow models to determine the impact of this system on the DRV and its implication for well productivity. An essential corollary of our model is the introduction of a new upscaling method for natural fractures, which reduces the number of fractures to the critical ones, while maintaining the same equivalent permeability as the prototype. The upscaled model still contains discrete fractures to reveal their key impact on the flow. The revolutionary upscaling method makes use of a combination of object-based and flow-based upscaling techniques (Appendix C).

3. CAM Solution for Hydraulic Fractures and Natural Fractures

This work uses the Complex Analysis Method (CAM) to model fluid flow in the reservoir. CAM is an analytical method that was originally limited to the use of integral solutions to model streamlines for steady state flow [47–49]. This approach was expanded upon to include the Eulerian particle tracking of time-dependent flows with applications to natural lava flows and other gravity flows [50,51]. The basic methodology to describe gravity flows was then applied to flow in subsurface reservoirs, properly accounting for permeability, porosity and fluid volume factors, while benchmarking tests of the CAM using numerical streamline simulations proved highly successful [33,52]. We propose the use of CAM because of it being a gridless meshless method that allows for infinite resolution at the fracture scale and also has faster computational times as compared to gridded numerical simulators.

For the hydraulic fractures that are connected to the horizontal wellbore, we use the concept of line sources and sinks. The complex potential along an interval $[a,b]$ is given by Potter [53] with time dependent strength $m(t)$ as follows:

$$\Omega(z,t) = \frac{m(t)}{2\pi(b-a)}[(z-a)\log(z-a) - (z-b)\log(z-b)] \quad [\text{ft}^2/\text{day}] \quad (1)$$

Differentiation of this equation with respect to z (the complex coordinate that represents a point in the reservoir space) gives Equation (2), which represents the corresponding velocity field:

$$V(z,t) = \frac{m(t)}{2\pi(b-a)}[\log(z-a) - \log(z-b)] \quad [\text{ft}/\text{day}] \quad (2)$$

In general terms for multiple line interval sources (k) with time dependent strength $m_k(t)$ (see Appendix A the instantaneous velocity field at time t can be calculated from:

$$V(z,t) = \sum_{k=1}^{N} \frac{m_k(t)}{2\pi L_k} e^{-i\beta_k} \cdot \left(\log\left[e^{-i\beta_k}\left(z - z_{c,k}\right) + 0.5L_k\right] - \log\left[e^{-i\beta_k}\left(z - z_{c,k}\right) - 0.5L_k\right]\right) \quad [\text{ft}/\text{day}] \quad (3)$$

For the line interval source solution, we are also able to determine the corresponding pressure depletion due to fluid withdrawal from the reservoir. The real part of the complex potential can then be evaluated to quantify the pressure change $\Delta P(z,t)$ at location z at a given time t:

$$\Delta P(z,t) = -\frac{\phi(z,t)\mu}{k} \quad [\text{psi}] \quad (4)$$

The potential function represented by $\phi(z,t)$ has its pressure scaled based on fluid viscosity μ and permeability k of the reservoir. The actual pressure can be calculated from the initial pressure (P_0) plus the pressure change:

$$P(z,t) = P_0 + \Delta P(z,t) = P_0 - \frac{\phi(z,t)\mu}{k} \quad [\text{psi}] \quad (5)$$

A new algorithm for the modeling of natural fractures was proposed by Van Harmelen and Weijermars [22] that makes use of a complex potential function created by the superposing of an infinite amount of line doublets and is:

$$\Omega(z,t) = \frac{-i \cdot v(t) \cdot e^{-i\gamma}}{2\pi \cdot h \cdot n_m \cdot L \cdot W}[(z - z_{a2}) \cdot \log(-e^{-i\gamma}(z - z_{a2})) - (z - z_{a1}) \cdot \log(-e^{-i\gamma}(z - z_{a1}))$$
$$+ (z - z_{b1})\log(-e^{-i\gamma}(z - z_{b1})) - (z - z_{b2})\log(-e^{-i\gamma}(z - z_{b2}))] \quad [\text{ft}^2/\text{day}] \quad (6)$$

Similarly to the solution for a line source (Equations (1) and (2)), differentiation of the specific complex potential equation of Equation (6) yields Equation (7), which gives the instantaneous velocity field in the natural fractures at time t.

$$V(z,t) = \frac{-i \cdot v(t) \cdot e^{-i\gamma}}{2\pi \cdot h \cdot n_m \cdot L \cdot W} [\log(-e^{-i\gamma}(z - z_{a2})) - \log(-e^{-i\gamma}(z - z_{a1})) \\ + \log(-e^{-i\gamma}(z - z_{b1})) - \log(-e^{-i\gamma}(z - z_{b2}))] \qquad [\text{ft}/\text{day}] \qquad (7)$$

Here, $v(t)(\text{ft}^4/\text{day})$ is the strength of the natural fracture, which scales the permeability contrast with the matrix (as further detailed in Section 4.1). The height, width and length of the natural fracture are denoted by h, W and L (ft) respectively, n is porosity, γ is the tilt angle of the natural fracture as shown in Figure 2. The variables z_{a1}, z_{a2}, z_{b1}, and z_{b2} give the corner points of the natural fracture domain.

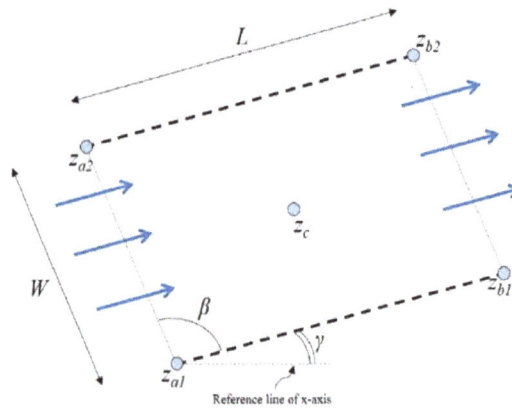

Figure 2. Natural fracture model. L and W are the length and width; z_c is the center; z_{a1}, z_{a2}, z_{b1}, and z_{b2} are the corners; β is the wall angles, while γ is the rotation angle of the natural fracture. The blue arrows give the direction of the flow [22].

As for boundary and initial conditions, CAM can be used to model both steady-state flow as well as transient flow as shown in our models. The initial REV models used in Section 5.1 to demonstrate the fundamental impacts of natural fractures on flow, use constant rate boundary conditions (using a constant far field flow of 2.5 ft/year). For the hydraulic fracture line sink models, we were able to introduce transient flow by the use of a declining flow strength based on the declining rate of the forecasted well production that is allocated back into each hydraulic fracture segment. All of our models were coded using the Matlab programming language.

4. Modeling of Natural Fracture Interaction Mechanisms

The major controls on fluid flow propagation in porous media are the porosity and permeability of the domain. For a naturally fractured reservoir, one may consider two domains for flow, the unfractured rock matrix and the natural fractures present within the reservoir. This assumes that the natural fractures are uncemented and allow for flow. For streamline simulations, the flow paths (FP) and time of flight (TOF) of fluids being transported in porous media due to pressure sources/sinks are calculated by the equation of motion, which is intrinsically dependent on porosity and permeability in the reservoir. A study by Zuo and Weijermars [28] led to the creation of two fundamental rules for FP and TOF in porous media. The first rule shows that an increase in permeability decreased the time of flight, and conversely, an increase in porosity increased the time of flight. The second rule states that the permeability uniquely controls the flight path of fluid flow in porous media and local porosity variations do not affect the streamline path.

Armed with the above rules, we now proceed to explain the three principal mechanisms by which natural fractures may impact fluid flow in the reservoir. Natural fractures may result in localized discrete changes in both permeability (Section 4.1) and porosity or storativity (Section 4.2) in the reservoir domain, creating a direct impact on reservoir drainage patterns and drained areas. The third possibility is the reactivation and connection of natural fractures to the hydraulic fracture network (Section 4.3), which functions as an extension of the hydraulic fracture pressure sink.

4.1. Equivalent Permeability Enhancement

This mechanism is due to the difference of permeability within the natural fracture and the surrounding rock matrix. In unconventional reservoirs, the natural fracture permeability (k_f) is typically greater than that of the rock permeability (k_m). Weijermars and Khanal [23] show via explicit derivations how the permeability ratio (R_k) directly impacts the strength of flow in natural fractures as follows:

$$R_k = \frac{k_f}{k_m} \quad [\text{-}] \tag{8}$$

The fracture hydraulic conductivity (C_f) is determined by the product of its fracture aperture (w_f) and its permeability (k_f):

$$C_f = k_f \cdot w_f \quad [\text{mD·ft}] \tag{9}$$

From this conductivity, we are able to define and scale the strength of the natural fracture segment (v_f) in terms of corresponding permeability contrast with the matrix as follows:

$$v_f = q_f L_f = v_f w_f h_f L_f = \frac{k_f}{k_m} v_m w_f h_f L_f \quad [\text{ft}^4/\text{day}] \tag{10}$$

The length dimensions for the natural fractures (h_f—natural fracture height, L_f—natural fracture length, w_f—fracture aperture) are directly specified in the CAM models and matrix flow velocity (v_m) can be measured near the fracture in the simulation. By fixing the constituent parameters at time t the equation for R_k thus becomes

$$R_k = \frac{v_f(t)}{v_m(t) w_f h_f L_f} \quad [\text{-}] \tag{11}$$

Thus, from the above equation, we can set the permeability ratio using an assigned strength in the natural fractures in our CAM model.

The most important aspect of the permeability enhancement mechanism is that natural fractures do not act as fluid sinks (see also detailed examples in Weijermars and Khanal, [23]). Mechanism 1 assumes the natural fractures are not connected to the hydraulic fractures (unlike mechanism 3). Instead the natural fractures act as zones of flow acceleration and preferentially drain matrix fluid further away from the well at the end of the highly conductive natural fractures rather than from the nearby lower permeability matrix. Change in permeability in the natural fractures impacts both streamline patterns as well as time of flight. This mechanism was thoroughly modeled and investigated by Weijermars and Khanal [23], using a variety of natural fracture parameters and readers are referred to this seminal work for further detail. Although prior studies (e.g., Aguilera, [2]) that use static object-based permeability scaling also give results that natural fractures can enhance well productivity, the method employed by Weijermars and Khanal is based on dynamic flow-based upscaling and is believed to be more accurate. Flow-based upscaling of permeability explicitly shows how for a fractured medium the equivalent permeability increases greatly when compared to similar porous media that are non-fractured. It is this overall increase in equivalent permeability (due to the enhanced permeability of the natural fractures) that leads to a higher flow rate towards the well and thus higher recovery during the economic life of such wells. In this study, we extend this work to investigate the implications of equivalent permeability enhancement due to natural fractures on DRV in conjunction with porosity changes in natural fractures. A new upscaling method for discrete fractures is given in Appendix C, allowing for simultaneous changes to fracture permeability and porosity, which has not been investigated previously.

4.2. Natural Fracture Storativity Effect

In addition to effecting localized permeability changes, natural fractures have the ability to alter porosity. Shale reservoirs tend to exhibit a wide range of fracture sizes. Due to the industries limited data of natural fracture porosity, the effects of this on flow alteration in the subsurface has not been previously studied in any detail. We present a set of high-resolution simulations with altered porosity in the natural fractures to quantify how this parameter affects drainage in the subsurface (see the results in Section 5).

As with the change in permeability, we are now able to define a porosity ratio (R_n) for the porosity change inside of a natural fracture (n_f) compared to the matrix porosity (n_m) surrounding it, given by the following equation:

$$R_n = \frac{n_f}{n_m} \quad [\text{-}] \tag{12}$$

For the CAM analytical solution, natural fracture alignment can be defined in relation to the hydraulic fracture. Equation (7) assumes that the porosity across both the fracture zone and matrix remains the same. If we remove this assumption, based on the evidence presented on porosity differences in natural fractures when compared to the reservoir matrix, Equation (7) can be locally modified to take into account the altered natural fracture porosity as follows:

$$V(z,t) = \frac{-i \cdot v(t) \cdot e^{-i\gamma}}{2\pi \cdot h \cdot n_m \cdot L \cdot W} [\log(-e^{-i\gamma}(z - z_{a2})) - \log(-e^{-i\gamma}(z - z_{a1})) \\ + \log(-e^{-i\gamma}(z - z_{b1})) - \log(-e^{-i\gamma}(z - z_{b2}))] / R_n \quad [\text{ft/day}] \tag{13}$$

This equation will now account for both the altered porosity and permeability within the natural fracture domain. As we manually define the boundaries of the natural fractures, the tracked particles that are displaced based on the time dependent strength of the flow in the reservoir will have velocity increased or decreased based on the porosity and permeability once the fluid particles enter the natural fracture domain. The trajectories of these particles are set by the permeability in the reservoir matrix and natural fractures [28]. Based on Rule 2 for flight paths and time of flight contours in porous media [28]. The time of flight will be slower in natural fractures with a higher porosity than the matrix (the streamline patterns will not be affected). Thus, for a hydraulically fractured well, the presence or absence of natural fractures with different porosities (that may be in situ porosity or increased porosity due to natural fracture reactivation) will affect how far the matrix is drained (i.e., the shape and location of the DRV will be affected), which knowledge is relevant for fracture treatment design and well spacing decisions.

4.3. Natural Fractures as Extension to Hydraulic Fracture Network

The third mechanism that may cause natural fractures to increase well productivity occurs when natural fractures become extensions of the hydraulic fracture system. This can lead to the creation of complex fracture networks, defined as non-planar, branching fracture geometries that are caused by either strong stress shadow effects or by the interactions with natural fractures [54]. Wu and Olsen [54] further state that the efforts to study interaction between natural fractures and hydraulic fractures have taken various forms of theoretical, experimental and numerical work. From this research, they propose three possibilities due to the intersection of natural fractures and man-made hydraulic fractures. The first possibility is that the created hydraulic fracture propagates along its original directions and crosses the natural fracture with no change in orientation. A second possibility is that the hydraulic fracture could be arrested by the natural fracture and then continue to propagate along the natural fracture to finally exit at the tip of the natural fracture. Deflection of the hydraulic fracture into the natural fracture, followed by re-initiating out of the natural fracture at a point of weakness is given as the third possibility [55]. No matter the propagation due to the interaction, the overall effect is that the natural fractures that intersect with the hydraulic fractures become extensions of the pressure sink imposed on the reservoir due to the connection of the fracture network to the wellbore. One way to model these interactions is via the use of fractal theory to replicate the branching fracture geometry that can then be modeled using CAM. The changes and implications on DRV due to such complex branching fracture networks was modeled in detail in Nandlal and Weijermars [21], and as such, is not the subject of focus in the current work. The reader is referred to the previous paper with the major conclusion being that the presence of a complex hydraulic fracture network will increase the initial production rate from the reservoir.

5. Results

Using the CAM approach, we investigated systematically the effects of porosity and permeability alterations within natural fractures on fluid flow using a range of model designs. The changes in these two crucial parameters were studied to determine the effect on the drainage area in the reservoir. Obviously, a proper understanding of the DRV development in naturally fractured reservoirs has implications for production from both conventional and unconventional oil and gas reservoirs.

We adjust the fracture strength and porosity ratio to determine the impact on drained areas in the reservoir. Modeling starts with a simple planar fracture with varying porosity ratios as well as different natural fracture configurations (Section 5.1). The effects of natural fracture storativity and enhanced permeability on DRV are demonstrated and proved (Section 5.2). We investigated the flow patterns near hydraulic fractures (modeled as line sinks using CAM), and how the presence of natural fractures and their corresponding porosity and permeability may change the DRV. It should be noted that the CAM models used in these flow simulations assume hydraulic fractures of infinite conductivity. The creation of hydraulic fracture via geo-mechanical simulations and corresponding coupling to flow simulation is outside the scope of this work.

The results in Sections 5.1 and 5.2 are for synthetic models, intended to systematically demonstrate the effects of natural fractures via the natural fracture interaction mechanisms explained in Section 4. The idealistic representative elementary volumes (REV) and simple fracture models assume porosity changes are independent of any permeability changes. In reality, this may not be true and there are many established correlations that relate increases in porosity with corresponding increases in permeability. We make use of field data for natural fractures to determine the DRV in an actual reservoir (Sections 5.3 and 5.4). Field data obtained from cores in the Hydraulic Fracturing Test Site (HFTS) as well as porosity-permeability correlations are used to determine the impact of natural fractures in the case study. By incorporating real data in our models, we can more accurately determine the impact of natural fractures on the DRV in the field. This is relevant to next proposed methods for optimization of recovery in both highly fractured unconventional and conventional reservoirs.

5.1. Representative Elementary Volume (REV) Models

To properly understand the effects on fluid flow, we started with the modeling of a simple representative elementary volumes (REV) that use a constant far field flow. A representative elementary volume (REV) is defined as a volume over which a measurement can be made that is representative of the whole. Using the REV allows for the understanding of the physics behind any changes in drainage patterns (before moving on to more complex situations). The first model provided is a base case which we use to compare all subsequent models. In this model (Figure 3) we show a reservoir space in 2D with five natural fractures represented by discrete elements that have the same porosity and permeability as the reservoir space. Using Eulerian particle tracking, we determined the flow path based on a constant far field flow. Flight paths (FP) are displayed in blue (Figure 3, left image) with the corresponding time-of-flight contours (TOFC) shown in red (Figure 3, right image). The base model represents a flow time of 30 years with each TOFC representing the fluid displacement after 3 years with reservoir porosity of 5%. Referring back to the two fundamental rules for FP and TOF [28], we observed that with no change in porosity and permeability in the natural fractures, the FP and TOFC remain constant.

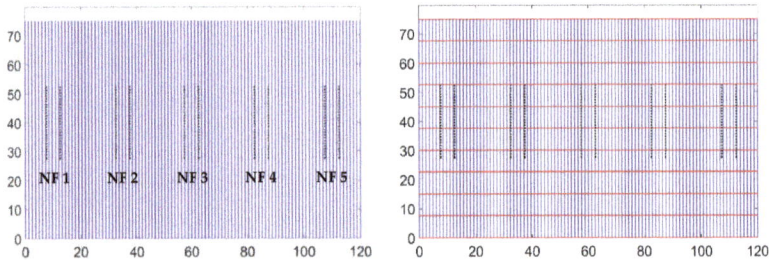

Figure 3. Base case model for homogenous reservoir space with five discrete natural fracture elements all having equal porosity and permeability. Left: Streamlines (blue) for uniform flow from bottom to top through reservoir space and natural fractures (dashed black). Right: Time-of-flight (TOF) contours (red) shown every 3 years during a total simulated time of 30 years.

Porosity effects: The next REV model (Figure 4) highlights the effect of systematically increasing the porosity in the natural fractures. As stated previously, fracture system porosities of 2% or less are considered typical [44], but values as high as 7% for natural fracture porosity in shale formations have been reported [25,46]. With numbers still based on very limited datasets, it is possible that porosity changes in natural fractures can be higher than the values reported thus far. Therefore, we model porosity changes by up to 15% to observe the impact on flow. The initial models decouple the correlation between increased porosity and permeability and such that there is no permeability change in the natural fracture relative to the matrix. When we use the term porosity, we mean connected porosity.

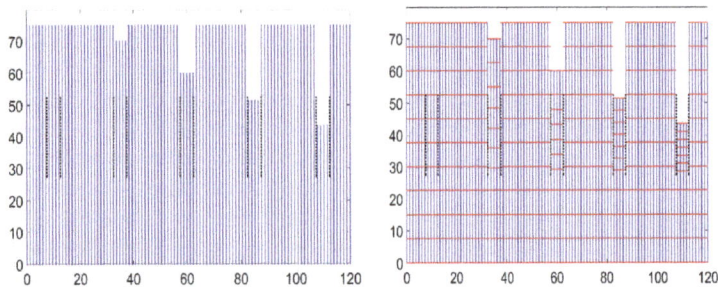

Figure 4. REV model showing impact of different natural fracture (NF) porosity on FP and TOF in a reservoir space of 5% porosity. NF porosity from left to right: NF 1 = 5% (NF 1 porosity same as reservoir), NF 2 = 6%, NF 3 = 8%, NF 4 = 10%, NF 5 = 15%. Streamlines in blue (left side) and TOF in red (right side). Far field flow of 2.5 ft/year scaled by reservoir porosity is used in all REV models.

Figure 4 shows the effects of increasing NF porosity on the FP and TOFC in the reservoir space. The reservoir porosity is kept at 5%, while NF porosity changes incrementally from being equal to the reservoir space to a high of 15%. The results clearly show that the change in porosity within the natural fracture has no effect on the streamline flow paths but does affect the time-of-flight contours. In NF 1, the porosity is the same as the reservoir and as such, there is no slowdown in the TOFC. From NF 2 to NF 5, we progressively increased the porosity to 6%, 8%, 10% and finally, 15%. The model shows that for each successive porosity increase in the natural fractures, the FP stays constant but the

TOF increases. As we are using a constant run time for all models, the increase in TOF results in flow not reaching as far into the reservoir space for the natural fractured with higher porosity. With no porosity change, flow reaches out to approximately 75 ft in the reservoir space. With a porosity change from 5% to 15% in the natural fractures, flow is retarded and reaches only approximately 44 ft out into the reservoir space. Thus, a 10% increase in natural fracture porosity results in a 40% reduction in lateral flow extent. This result can have great implications for accurately determining the DRV in the subsurface when the reservoir rock has a high density of natural fractures with variable porosity.

Permeability effects: The next REV model shows the impact of change in natural fracture permeability on the FP and TOFC after simulation for 30 years. For this model, the porosity in the natural fractures are kept constant with the reservoir to allow for a detailed investigation of the flow effects due to only the permeability change in the fractures. Using CAM, we model a higher permeability in the natural fractures by assigning (scaling with) a particular fracture strength (Equation (10)). As discussed in Section 4.1, an increase in strength can be related back to the natural fracture permeability using the permeability ratio R_k. The REV model (Figure 5) uses a far field flow of 2.5 ft/year (after being scaled by the reservoir space porosity of 5%). The strengths for NF 1 to NF 5 are increased, respectively, from 0.1 ft^4/year to 40, 160, 500, and 1000 ft^4/year.

The results in Figure 5 show that keeping porosity constant in the natural fractures while increasing the natural fracture strength (and thus NF permeability) leads to a change in both the FP and TOF. This is in line with what is expected from the first fundamental rule for FP and TOFC [28]: permeability changes affect the FP and thus, the path of the streamlines is altered. Fluid is seen funneled into the higher permeability natural fractures while the TOF correspondingly decreases. Using the constant run-time of 30 years, this decrease in TOF results in fluid flow reaching further out into the reservoir space. As more of the fluid flow is funneled into the NF due to increasing strength, less of the fluid is transported in the inter-fracture domain (space between the natural fractures). In the space between NF 2 and 3 (though the FP are altered due to the increased NF permeability), fluid still flows in the inter-fracture space as shown by the streamlines. However, in the space between NF 4 and 5 (which are assigned much greater strengths) almost all the fluid flow is funneled into the natural fractures, with most of the inter-fracture space receiving no fluid.

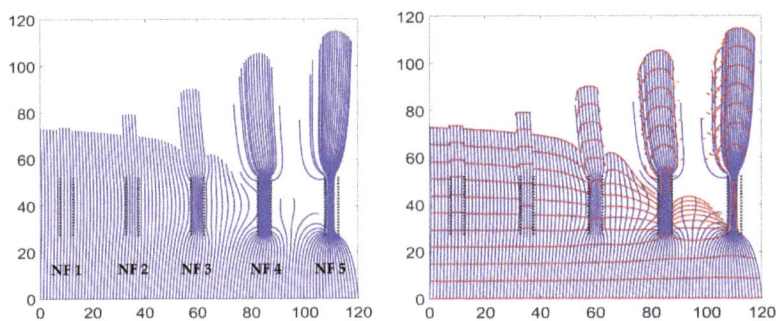

Figure 5. REV model showing impact of different natural fracture (NF) permeability on FP and TOF in a reservoir space of 5% porosity. NF (dashed black lines) strengths from left to right: NF 1 = 0.1 ft^4/year, NF 2 = 40 ft^4/year, NF 3 = 160 ft^4/year, NF 4 = 500 ft^4/year, NF 5 = 1000 ft^4/year. Streamlines in blue (left side) and TOFC in red (right side).

The relationships between the natural fractures input parameters used in Figure 5 and the approximate equivalent natural fracture permeability (based on Equation (11)) are given in Table 1. Fracture input properties used in all subsequent flow models with enhanced natural fracture permeability in this study are included in Table 1.

Table 1. List of natural fracture input properties for models with enhanced permeability.

Figure Number		Natural Fracture Strength (ft⁴/year)	Natural Fracture Length (ft)	Natural Fracture Width (ft)	Natural Fracture Height (ft)	Matrix Flow Rate (ft/year)	Permeability Ratio (R_k)	Natural Fracture Permeability (nD)
	NF1	0.1	25	5	1	2.5	_ ª	_ ª
5	NF2	40	25	5	1	2.5	0.13	12.8 ª
	NF3	160	25	5	1	2.5	0.51	51.2 ª
	NF4	500	25	5	1	2.5	1.60	160
	NF5	1000	25	5	1	2.5	3.20	320
6b		1000	25	5	1	2.5	3.20	320
7		500	25	5	1	2.5	1.60	160
		(ft⁴/day)				(ft/day)		
	a	2500	20	10	60	0.1693	1.23	123.06
9	b	5000	20	10	60	0.1693	2.46	246.11
	c	10,000	20	10	60	0.1693	4.92	492.22
10		5000	20	10	60	0.1693	2.46	246.11
11b		155	20	0.5	60	0.1693	1.53	152.59
12		155	30	0.5	60	0.1693	1.02	101.73

ª R_k formulation gives an approximate natural fracture permeability and does not hold well for low strengths. A matrix permeability of 100 nD is assumed. R_k is calculated from Equation (11) with natural fracture permeability then back-calculated from Equation (8) using the assumed matrix permeability.

Open fracture: A final scenario investigated with the REV model was the effect of a natural fracture with 100% porosity. Theoretically, this can be thought of as an open fracture in the subsurface. Once again, we artificially separate the effects of porosity and permeability to investigate each parameter individually. Figure 6a shows the result for completely open natural fractures set within a reservoir space of 5% porosity. The FP is unchanged but the TOF in the fractures increases dramatically. The fluid drawn from the open fracture does not require long travel paths (due to 100% fluid fill) and drawing the same amount of fluid from the inter-fracture matrix regions requires much longer travel paths in those regions outside the NF.

Figure 6b shows the effect of natural fractures with very high permeability as compared to the reservoir space. The natural fractures in this model have a strength of 1000 ft⁴/year (while porosity is kept the same as that of the reservoir matrix) and fluid flow is simulated for a run-time of 30 years. The marked effect of the change in permeability is seen in the alteration of the FP as well as the decrease in TOF. With such a high fracture strength (high R_k) almost all flow is funneled through the natural fractures, with no fluid being transported via the inter-fracture domain. A similar conclusion was reported in earlier work by our research group [23].

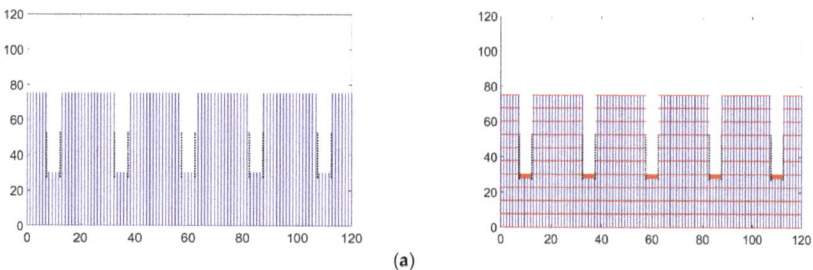

(a)

Figure 6. *Cont.*

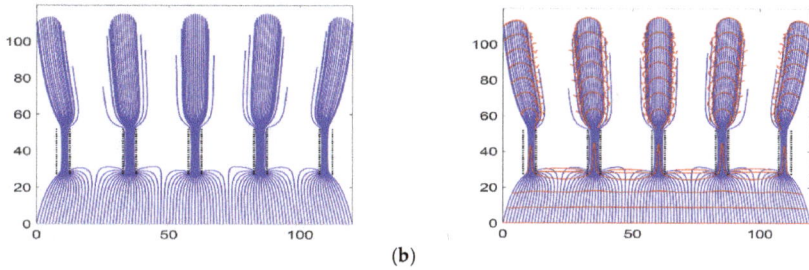

(b)

Figure 6. (a) REV model showing the effect of a natural fracture (NF) porosity of 100% (open fracture) in a reservoir space of 5% porosity with no permeability change. **(b)** Natural fractures with increased strength of 1000 ft⁴/year. Streamlines in blue (left side) and TOFC in red (right side). Natural fractures in black.

All previous REV models have considered the varying effects of porosity and permeability independently of each other. Figure 7 investigates the effect of simultaneous changes of natural fracture porosity on flow, while the permeability contrast with the matrix exists ($R_k > 1$). In this model, we systematically change the porosity within the NF from initially being equal to that of the reservoir space of 5% (Figure 7a) to a high of 30% (Figure 7d), all the while keeping a constant enhanced permeability in the natural fractures.

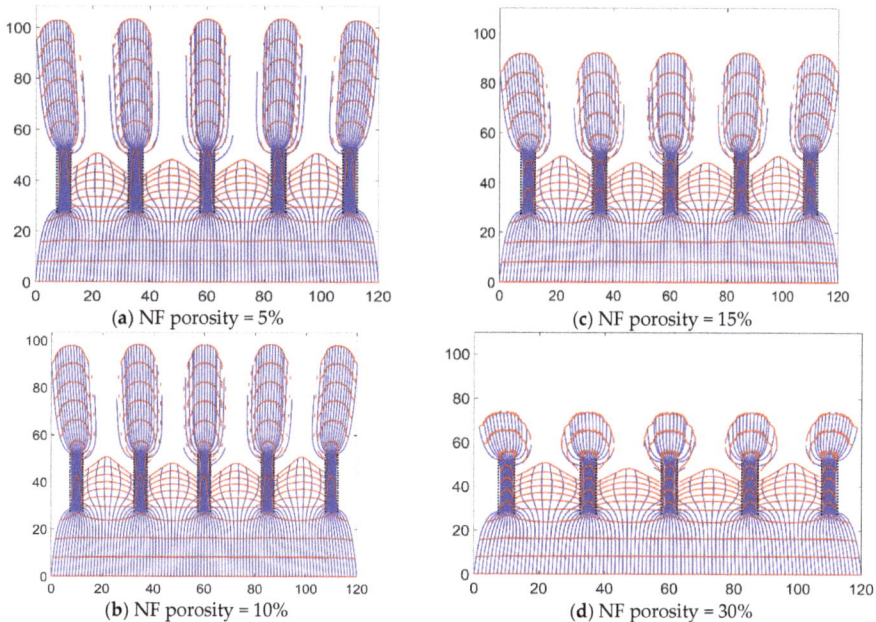

Figure 7. REV model showing effect of various natural fracture (NF) porosity ranging from **(a)** 5% to **(d)** 30% changes in a reservoir space of 5% porosity with enhanced strength in the NF of 500 ft⁴/year. The streamlines are in blue and TOFC are in red. Natural fractures are in black.

The results from the models in Figure 7 show the competing effects between porosity and permeability as defined in the fundamental rules for FP and TOF by Zuo and Weijermars [28]. Figure 7a shows the alteration in FP and decrease in TOF (fast travel times via the fractures) due to the enhanced natural fracture permeability. The successive models (Figure 7b–d) with gradually increasing porosity in the natural fracture conversely increase the TOF and thus reduce the lateral distance reached by the fluid flow in the given run-time. Although the porosity change negates the effect of the enhanced permeability in terms of lateral distance reached, the alteration of the FP by the permeability still occurs. This proves that permeability is responsible for the particle paths while both the permeability and porosity inversely affect the TOF (as stated in Zuo and Weijermars, [28]).

5.2. Synthetic Hydraulic Fracture Models

Using the CAM model, hydraulic fractures can be modeled as either line sinks or as line sources, which is used in this study by applying the principle of flow reversal. Line sinks can show fluid withdrawal contours being forward modeled by line sources (a simple sign reversal in our equations). The effects of fluid flow of enhanced permeability and porosity in natural fractures of an otherwise homogenous reservoir space was modeled in the previous section using a constant far field flow. Models are now presented to demonstrate how natural fracture will alter fluid flow around a single hydraulic fracture. Time-dependent production data from a well completed in the Wolfcamp Formation is used in these models and prorated for fluid allocation produced by a single hydraulic fracture stage (Figure A1). The relatively wide zones (10 ft) of altered permeability and porosity used in these models represent the effect of upscaling numerous smaller individual natural fractures (a detailed upscaling procedure is given in Appendix C). The effect of such altered zones can be clearly demonstrated visually. Each naturally fractured zone has dimensions of 10 ft width by 20 ft in length and the zones are angled at values of 45° and 135° from the hydraulic fracture.

The first model looks at the effect of a synthetic, single hydraulic fracture surrounded by six natural fracture zones having a higher porosity than the reservoir matrix (Figure 8). For this model, the natural fracture zones do not attribute any additional permeability change, only porosity enhancement. Figure 8a,b has a progressively increasing porosity in the natural fracture from left to right, starting with a NF porosity of 10% in 8a and increases to 15% and 20% in Figure 8b,c. The models show that as porosity increases in the natural fracture zone, there is a decrease in the distance drained. In other words, as porosity increases, the time-of-flight also increases. The major observation from these models is that the presence of naturally fractured zones with increased porosity (and assumed fluid storage in those fractures) will decrease the distance drained away from the hydraulic fracture.

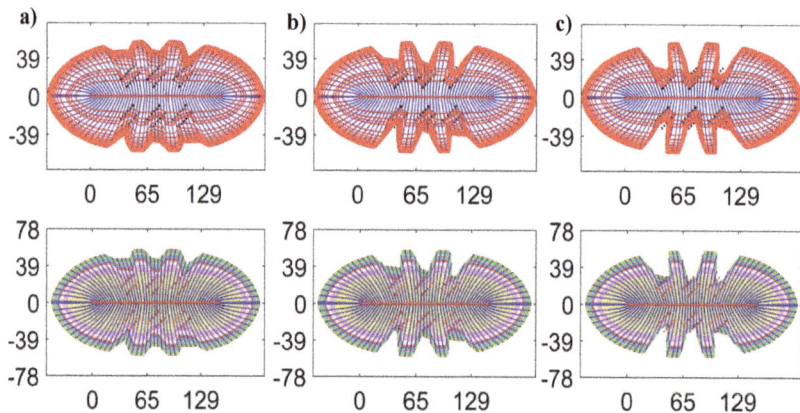

Figure 8. Hydraulic fracture model showing effect of various natural fracture (NF) porosity changes in a reservoir space of 5% porosity with enhanced porosity in the NF of (**a**) 10% (**b**) 15% (**c**) 20%. Streamlines in blue and TOFC in red. Natural fracture zones are shown by dashed lines. The bottom plots use rainbow colors to show drained areas after 3-year time periods.

The next property investigated is the effect of increased permeability (by changing the strength of the natural fractures as compared to rest of the reservoir matrix) (Figure 9). The porosity in the NF zones is kept the same as for the reservoir matrix. Therefore, we can focus solely on the permeability effect. From left to right, the strength in the natural fracture zones is progressively increased from 1000 ft^4/day in Figure 9a to 5000 ft^4/day and 10,000 ft^4/day respectively in Figure 9b,c. Once again, as demonstrated in the REV models of Section 5.1, the streamlines converge into the high permeability zones and lead to larger drainage regions in the direction of the higher permeability zones. One additional point of note is that the direction of the angle of these zones in conjunction with the streamline direction influences how much effect there is on the drainage. If the naturally fractured zones are angled in the same direction as the streamlines, the effect is more pronounced than if they occur at a larger angle to the principal flow direction induced by the hydraulic fracture.

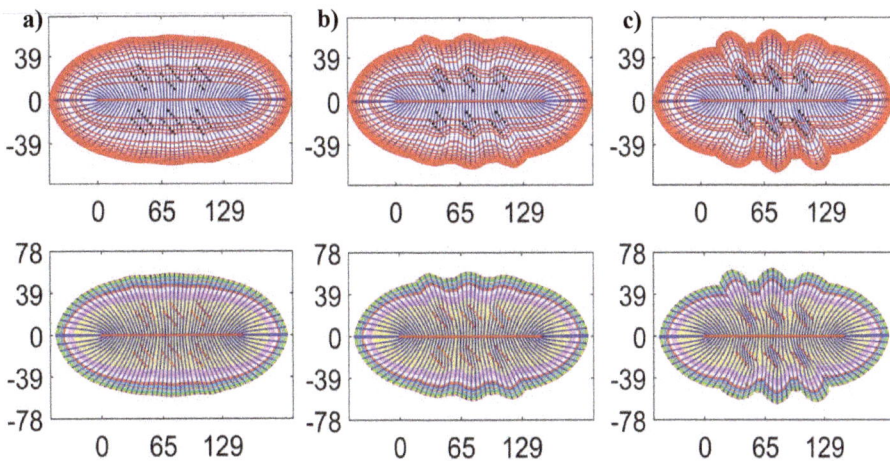

Figure 9. Hydraulic fracture model showing effect of various natural fracture (NF) permeability changes in a reservoir space of 5% porosity with enhanced permeability strengths in the NF of (**a**) 2500 ft^4/day (**b**) 5000 ft^4/day (**c**) 10,000 ft^4/day. Streamlines in blue and TOFC in red. Natural fracture zones in dashed lines and have the same porosity as reservoir. Bottom plots use rainbow colors to show drained areas after 3-year time periods.

The previous models investigated the effect of altered porosity and permeability in naturally fractured zones around a hydraulic fracture independently. Figure 10 looks at the competing effects of altered porosity and permeability together. Figure 10a shows the case with an enhanced permeability in the natural fractured zones, while the porosity is kept the same as the reservoir porosity of 5%. The results show the convergence of the streamlines into these zones, resulting in a lateral extension of the DRV beyond. As we progress from left to right, Figure 10a–c shows the effect of increasing porosity in the NF zones while also having an enhanced permeability. Figure 10b has the same enhanced permeability as in Figure 10a, but now the porosity in the natural fractured zones is increased from 5% (same as the reservoir matrix) to 10%. This model shows that although the streamlines once again converge into the zones of higher permeability, the lateral extent of the DRV is now slightly reduced due to the increased porosity. The enhanced DRV from Figure 10a has now been reduced in Figure 10b to an extent smaller (due to the porosity effect) than if there were no natural fractures. If the natural fracture porosity is increased further to 20%, the extent of the drained area shrinks much further (Figure 10c). The large changes in lateral extent and the spatial location of the DRV due to natural fractures may have significant implications for fracture and well spacing for optimum drainage. The limiting factor for improving models is the lack of fracture diagnostics for field cases (in particular the fracture permeability and porosity values). In the next section, detailed field data abstracted from the Hydraulic Fracture Test Site will be used to constrain fluid withdrawal patterns near the hydraulic fractures that drain the Wolfcamp reservoir space.

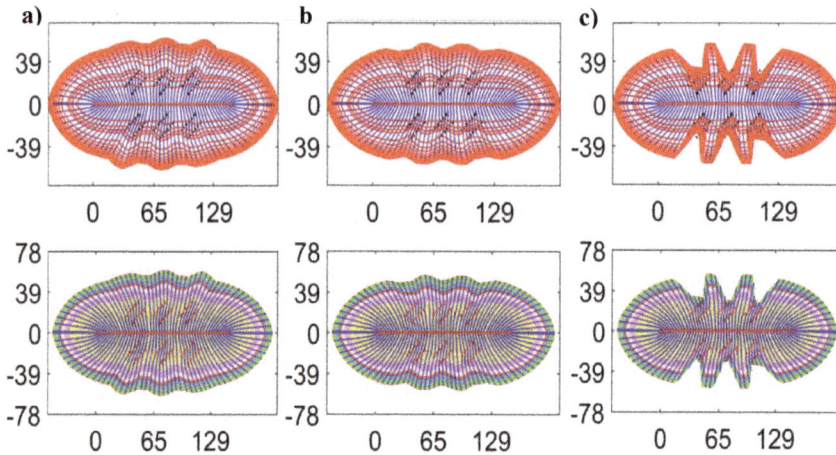

Figure 10. Hydraulic fracture model showing effect of competing changes in natural fracture (NF) porosity and permeability changes in a reservoir space of a 5% porosity. (**a**) NF porosity same as reservoir (5%) and enhanced strength of 5000 ft^4/day (**b**) NF porosity of 10% and enhanced strength of 5000 ft^4/day (**c**) NF porosity of 20% and enhanced strength of 5000 ft^4/day. Streamlines in blue and TOF in red. Natural fracture zones in dashed lines.

5.3. Field Models Using Data from the Hydraulic Fracture Test Site (HFTS)

Data from the Hydraulic Fracture Test Site (HFTS; Midland Basin, West Texas) were used because the natural fracture network present in the subsurface has been characterized in prior studies for this real field case [12,56]. Six cores obtained from the Wolfcamp Formation within the Stimulated Rock Volume (SRV) near to a hydraulically fractured well were studied in detail [12]. One of the aims of the core description was to understand the primary origins of fractures in terms of hydraulic, natural and reactivated natural fractures. The density of the individual types of fractures along the core depths and the dominant orientations of the fractures obtained by Shrivastava et al. [12] were used in our study for a field-based simulation of the impact of natural fractures on the DRV development.

For the present study, certain mean values for natural fracture lengths and aperture were assumed in our models because natural fracture length and aperture values from the HFTS core samples were poorly constrained [12]. In their approximation, the latter authors used a power-law relation to generate a range for natural fracture lengths and the fracture apertures were estimated using a geomechanical fracture propagation simulator. In the present study, we constrain the fracture length to 30 ft (Table 1), corresponding to the maximum value used by Shrivastava et al. [12]. Additionally, the DRV model requires inputs, for every natural fracture, of permeability and porosity. However, almost no data is present in the literature for relating in situ natural fracture porosity with permeability in the subsurface, which is why a Carman–Kozeny (CK) relation was used in our study (Appendix B).

An example of the impact of the Carman–Kozeny porosity–permeability correlation in the natural fractures, but for a still unscaled model, is given in Figure 11. The effect of the enhanced permeability in the natural fractures (Figure 11b) as compared to a single hydraulic fracture without any natural fractures nearby (Figure 11a) is to channel fluid flow faster through these high-speed zones. The effect of the enhanced permeability for this synthetic case completely outweighs any impact of the increased porosity in the natural fracture, which actually increases the time of flight (TOF) and leads to narrowly spaced TOF contours.

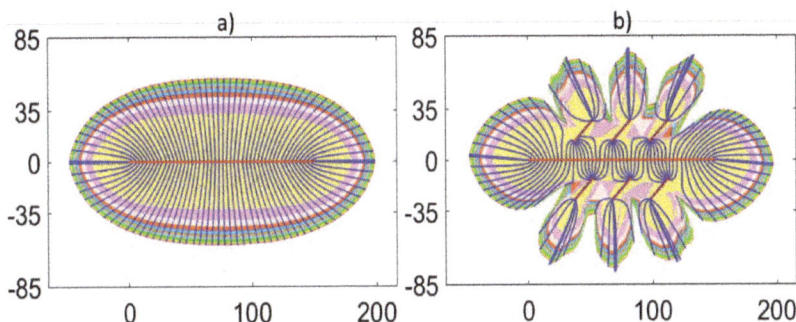

Figure 11. (**a**) DRV around a single hydraulic fracture with no natural fractures around, (**b**) DRV around a single hydraulic fracture with 6 natural fractures with porosity of 8.4% and corresponding strength of 155 ft^4/day from CK correlation after 30 years production. Hydraulic fracture in red, Streamlines in blue, Natural fractures in dashed red lines. Rainbow colored fill shows drained areas after 3-year time periods.

The analysis of the HFTS natural fracture field data suggests that a dense network of natural fractures occurs around the hydraulic fractures [12]. The natural fracture density model based on HFTS field data generated by a discrete fracture network contained over 40,500 individual natural fractures distributed over a domain of 300 m by 300 m [56]. For tractable run times with our smaller model, the number of natural fractures can be reduced by upscaling. A similar approach was used by Kumar et al. [56], whereby the permeability tensor for the entire stimulated rock volume was determined from flowback for input in a discrete fracture network model.

The upscaling method used in the present study sought to reduce the overall number of fractures to be modeled by upscaling the natural fracture widths and fracture permeabilities (strengths) for a dense natural fracture network. Original natural fracture apertures in the subsurface were assumed to be 5 mm (0.2 inches), which follows from core observations that kinematic apertures were estimated to have been more than 1 mm wide [57]. A combination of object-based and flow-based upscaling was developed for this study, with an in-depth discussion of this topic given in Appendix C. The proposed upscaling method was applied to produce field models for DRV around a single hydraulic fracture with a representative, upscaled natural fracture distribution of the HFTS. Using the data input ranges (Table 2) for natural fractures in conjunction with the Carman–Kozeny correlation, the final model was simulated to determine the real-life impact of natural fractures on the DRV.

Table 2. Natural fracture data from HFTS used for model simulations.

Natural fracture orientation (to hydraulic fracture) [a]	−55° and 55°
Natural fracture length [b]	30 ft
Original natural fracture density [c]	0.042 fractures/ft^2
Assumed original natural fracture aperture	0.2 inches
Upscaled natural fracture aperture [d]	6 inches
Number of natural fractures [d]	12
Natural fracture porosity	7.32%
Natural fracture strength	155 ft^4/day

[a] Core data obtained values. [b] Use of maximum value from Shrivastava et al.(2018). [c] From Shrivastava et al. [12].
[d] Values obtained from upscaling (Appendix C).

From the upscaling of the original natural fracture density, the outcome is a model with 12 natural fractures around the single hydraulic fracture. These 12 natural fractures are stochastically placed around the hydraulic fracture using the relevant field data all other parameters needed. Once again, the CK correlation was used to relate natural fracture permeability to porosity. Simulation of this model in CAM gives the representative DRV when affected by natural fractures (Figure 12a).

Figure 12a shows that fluid is preferentially channeled through the natural fractures for the HFTS field case models. The DRV in the upscaled HFTS model is highly convolute (Figure 12a) with numerous undrained matrix zones occurring between the upscaled natural fractures created from field data. Any storativity effects of the enhanced porosity in the natural fractures remain obscured by the enhanced flow due to the enhanced permeability of the natural fractures. For comparison, the pressure plot after 1 month of production was generated using CAM (Figure 12b). Pressure was calculated in CAM by extracting the potential function from the complex potential and normalizing by the ratio of reservoir permeability and fluid viscosity [58]. For the plot presented, the pressure scale was normalized by the maximum pressure present in the reservoir at 1 month production. The lowest pressures occur near the hydraulic fractures. We utilized the process of flow reversal, which means that the highest pressures occur at the hydraulic fractures (which can be simply corrected by flipping the scale in Figure 12b). Anomalous high pressures at the tips of the natural fractures are due to singularities and associated branch cut effects occurring when high permeability contrasts (R_k) are used [59]. The progressive distortion of the pressure field near a hydraulic fracture due to the presence of natural fractures is further discussed in Section 6.2 (see also Figure 15).

The overall pressure field is greatly altered by the presence of natural fractures due to their impact on the flow pattern. The results presented here confirm that the calculated DRV do not conform 1:1 to the pressure field, making the use of pressure plots very poor proxies for reservoir drained areas.

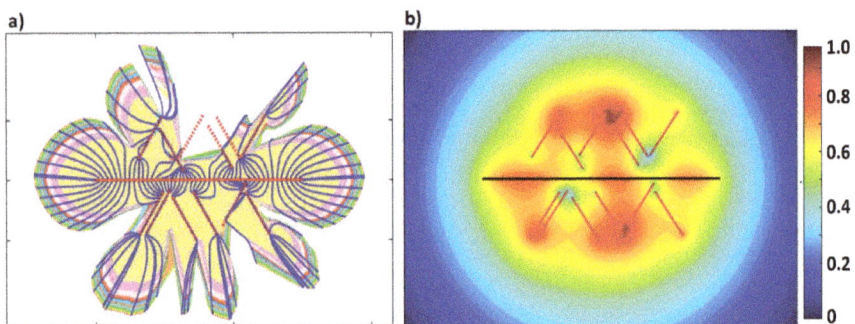

Figure 12. (**a**) DRV generated with upscaled natural fractures using field data from HFTS; hydraulic fracture in red; streamlines in blue; natural fractures in dashed red lines. Rainbow colored fill shows drained areas after 3-year time periods. (**b**) Pressure plot after 1 month production generated from CAM around single hydraulic fracture with HFTS upscaled natural fractures; hydraulic fracture in black; natural fractures in red; pressure scale normalized by highest pressure value.

5.4. HFTS Full Well Model and Implications

The previous section analyzed the impact that natural fractures modeled from field data have on the DRV around an individual hydraulic fracture. This concept is now expanded upon to determine the impact of natural fractures on DRV across multiple fracture stages representative of an entire hydraulically fractured well. The Wolfcamp production used effectively in these models had 22 stages with each stage spanning 300 ft with a total of 131 individual fracture clusters along the entire lateral. Our modeled DRV around a single hydraulic fracture is assumed to be representative of the collated drainage for all the fracture clusters per stage. Each stage has six fracture initiation points (clusters) with 50 ft spacing. The results thus show the total drainage of these six clusters when upscaled to one single hydraulic fracture.

The first model investigates the drainage based on the given 50 ft cluster spacing (corresponding to the stage spacing of 300 ft) with the assumption of a homogenous reservoir with no natural fractures (Figure 13a). Based on this stage spacing and from the DRV calculated, the multi-stage plot shows large undrained areas in between the existing DRV's after 30 years forecasted production. Results indicate that a maximum distance of 50 ft is drained perpendicularly away from the hydraulic fractures, which represents the drainage of all six fracture clusters. The plots (Figure 13a,b) show this stage spacing was sub-optimal due to the large undrained areas that can be targeted for refracs. For comparison, we model the same number of stages but now including the impact of reservoir heterogeneity using the HFTS field data on natural fractures (Figure 13b). When compared to the case with no natural fractures, the maximum area drained perpendicular to the hydraulic fracture increases from 50 ft to approximately 80 ft. Figure 13b shows that even though there is a shift in the spatial location of the DRV due to the natural fractures, this increase in lateral drainage is not enough to efficiently drain in between the fractures at this stage spacing.

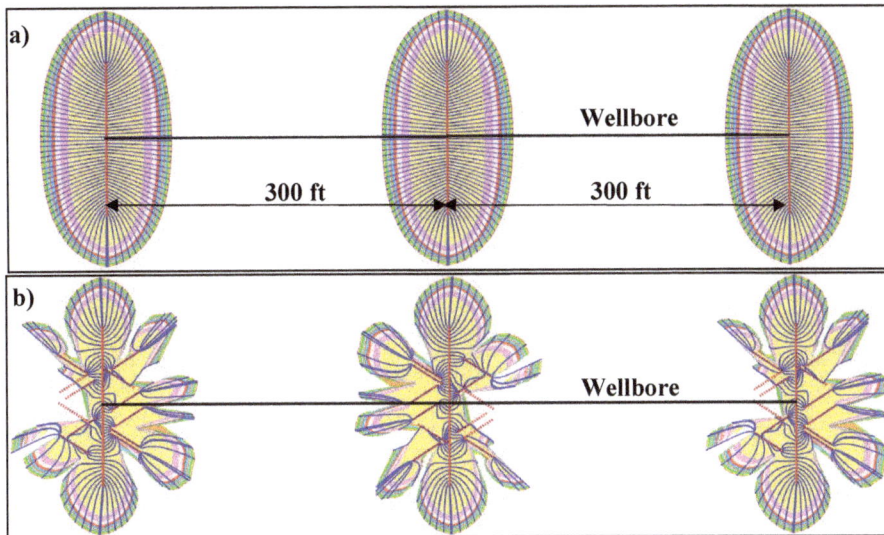

Figure 13. (**a**) Plan view of DRV for modeled well using current stage spacing of 300 ft assuming homogenous reservoir (**b**) Plan view of DRV for multiple stages using current 300 ft spacing with the impact of natural fracture modeled using HFTS data. Hydraulic fracture in red line; natural fractures in dashed red line; streamlines in blue. Rainbow colored fill shows drained areas after 3-year time periods.

Assuming a modified initial fracture cluster spacing of 25 ft, down from 50ft (which corresponds to a stage spacing of 150 ft instead of the field value of 300 ft), the DRV's were modeled using CAM to investigate cases of a homogenous reservoir and heterogeneous reservoir with natural fractures (Figure 14a,b). The first case for a homogenous reservoir (Figure 14a) suggests that the reduction of the cluster spacing based on the upscaled DRV for a single stage, allows for more efficient drainage along the length of the lateral. This decrease in spacing to a more optimal value would lead to enhanced well productivity. Our method visualizes the exact DRV and the new spacing does not create adverse flow interference. In fact, the model shows that the spacing can be further optimized to slightly less than 150 ft per stage due to there still being undrained areas between the hydraulic fractures. The introduction of natural fracture heterogeneity reveals a different finding when the stage spacing is decreased to 150 ft. Natural fractures with enhanced permeability when properly oriented to the hydraulic fracture extend the lateral drained areas as shown in our models (Figure 14b). Although the natural fractures extend the drained areas, at the new stage spacing of 150 ft, there is now nearly an overlapping of the DRVs from each stage (shown by dashed black ellipses in Figure 14b). The proximity of these DRVs implies that reduction of the stage spacing to less than 150 ft will lead to flow interference that will reduce the overall recovery from the well. The conclusion from this is that when natural fractures are present, fracture stage treatment with a spacing of less than 150 ft will now be sub-optimal. These results show the importance of accounting for and properly modeling natural fractures, particularly in flow simulations for unconventional reservoirs.

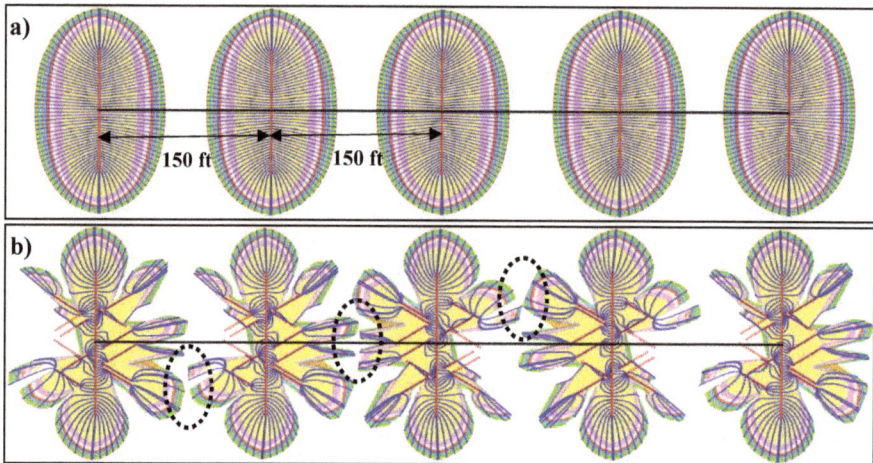

Figure 14. (**a**) Plan view of DRV for modeled well using a possible stage spacing of 150 ft assuming homogenous reservoir (**b**) Plan view of DRV for multiple stages using 150 ft spacing with the impact of natural fracture modeled using HFTS data. Hydraulic fracture in red line; natural fractures in dashed red line; streamlines in blue. Rainbow colored fill shows drained areas after 3-year time periods. Dashed ellipses in black show overlapping of DRV's that can cause unwanted flow interference.

6. Discussion

Proper modeling and forecasting of production from unconventional reservoirs needs to take into account important reservoir heterogeneity such as the presence and the impact of natural fractures. Numerous authors have noted the possible impact that natural fractures can have on production and well performance [2,24], but very few have sought to succinctly delineate and differentiate the ways in which this is possible. The present study puts forward three major mechanisms by which natural fractures can impact well productivity. Natural fractures present in the subsurface can affect well productivity via (1) enhanced permeability, (2) enhanced storativity, and (3) reactivation of natural fractures as extensions to the created hydraulic fracture network. By the use of a simple analytical streamline simulator based on complex analysis methods (CAM), we visualized the drainage patterns around hydraulic fractures by Eulerian particle tracking. The effects of natural fractures, in particular the enhanced permeability and storativity were investigated systematically and the results show that the drainage patterns (DRV) can be greatly altered by the presence of these reservoir heterogeneities.

6.1. Storativity Impact of Natural Fractures

Natural fractures present in the subsurface show a range of measured porosity from 2% to 7% [25] but these measured data sets are very limited in sample size and it is believed that porosity ranges may include even higher values. The altered mineralogy in these natural fractures can lead to a porosity and permeability that is vastly different to that of the unfractured reservoir matrix. With regards to natural fractures present in the Permian Basin, Forand et al. [24] stated that *"despite natural fractures having a calcite fill, the permeability contrast between the fracture and matrix is likely high enough that the healed fractures may be preferential hydrocarbon pathways. Combining this dominant character with the orientation of natural fractures to maximum horizontal principal stress has the potential to affect the efficiency of hydraulic fractures and the size of the total connected and stimulated rock volume."* The change in permeability will also result in an increased porosity, which we see as a cause of enhanced storativity for reservoir fluids.

Enhanced storativity can contribute to better well performance as these naturally fractured regions will have a larger hydrocarbon fluid supply that may last longer [23]. The impact of enhanced storativity in natural fractures on the drainage area around a well is for the first time visualized in our results. Starting with a simple REV model (Figure 4), the effect of increased porosity is seen to slow the time-of-flight (TOF) in the natural fracture as compared to the matrix. Once again, this proves that porosity changes do not affect streamline patterns but only the time-of-flight [28]. When applied to naturally fractured zones around a hydraulic fracture (Figure 8), the increase in the TOF results in a slower expansion of the DRV in the natural fracture zones compared to the rest of the matrix with a lower porosity. This leads to a decrease in the lateral distance drained away from the hydraulic fracture and can thus impact the optimum fracture cluster spacing distance. For a highly naturally fractured reservoir with higher storativity, the well spacing could be decreased compared to a reservoir with no natural fractures, as the drained area laterally would be smaller. This ability to increase the number of wells without introducing interference effects (by draining the same area with multiple hydraulic fractures) will lead to higher recoveries per acreage.

6.2. Enhanced Permeability vs. Enhanced Storativity

For natural fractures with higher permeability, fluid moves preferentially through these high-velocity conduits. REV models for natural fractures with various permeabilities (Figure 5), modeled by individually specified natural fracture strengths in our CAM simulation, show that as fluid moves via the natural fractures, some of the matrix areas between the natural fractures are bypassed or left undrained. When applied to flow around a single hydraulic fracture (Figure 9) the preference for flow through the higher permeability zones creates enhanced lateral drainage in the areas where the drainage plumes near the tips of the natural fractures reach deeper into the lateral reservoir space. Our results show that altered permeability impacts both the streamline patterns (convergence into natural fractures) and TOF. For a greater permeability, the TOF reduces in the natural fractures as compared to the TOF in the matrix. Thus, natural fractures with enhanced permeability can lead to greater lateral drainage with the caveat that there is the possibility of bypassed areas between the natural fractures that can still contain hydrocarbons.

The synthetic models all assumed variations in the porosity being possible independent of permeability changes. In reality, this is not the case as an increase in the effective porosity commonly correlates to an increase in permeability. Nonetheless, the synthetic examples clearly highlight that increased porosity leads to an increase of the TOF (i.e., flow is slowed down in the higher porosity region), whereas increased permeability reduces the TOF (i.e., flow if quickened). The latter also alters the flow paths in the reservoir. This leads to a competing effect of higher porosity reducing the lateral DRV, with greater permeability increasing the lateral DRV assuming otherwise similar production (as used in our models).

The key questions now become: (1) which parameter (permeability vs. porosity) has the more dominant impact on the drainage pattern? and (2) how can one correlate any increases in porosity with permeability, and vice versa? Data for natural fracture porosity values are very limited and any natural fracture permeability values are for typically reactivated fractures that connect directly to the hydraulic fracture. Due to this paucity of data, this paper made use of the commonly used Carman–Kozeny (CK) correlation for determining permeability based on a given natural fracture porosity. The results show (Figure 11) that using this correlation with a limited number of natural fractures, the permeability effect far outweighs the storativity of the enhanced porosity.

The HFTS case (Figure 12), using field data for natural fracture representation (based on natural fracture upscaling), shows that once the CK correlation is used, the impact of the natural fracture enhanced permeability (lateral extension of DRV and undrained matrix between natural fractures), vastly outweighs the storativity effect of said natural fractures. The DRV and pressure field distortion for the HFTS (Figure 12a,b) provide a specific example of what is a generic effect. For example, Figure 15a–d show the pressure field around a single hydraulic fracture without any natural fractures present (Figure 15a) and the stepwise distortion of the associated pressure field due to the presence of one, two and six natural fractures (Figure 15b–d). It should be noted that our models have the highest pressures at the hydraulic fracture due to the flow reversal modeling used (whereby fluid is placed back into the reservoir via the hydraulic fractures at the same rate as that produced).

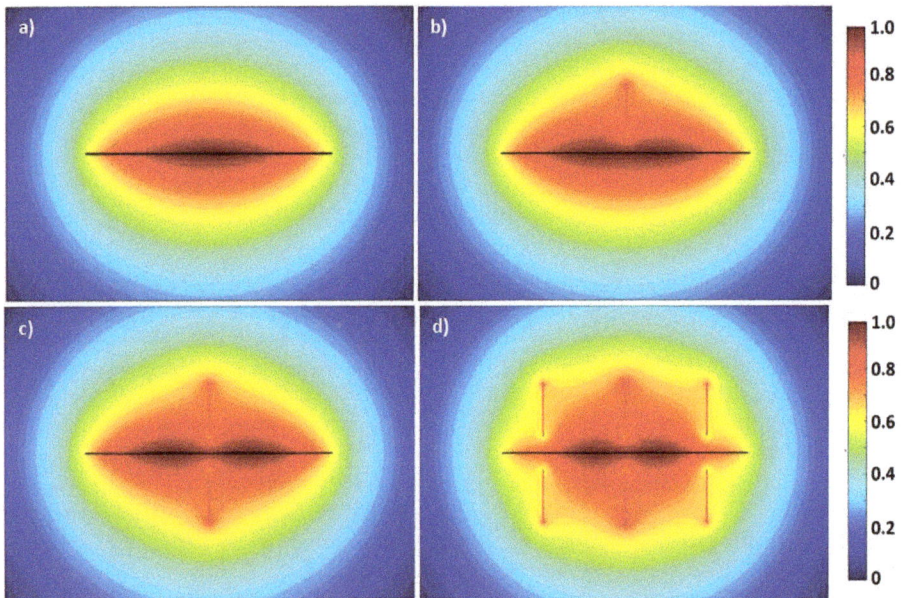

Figure 15. (a) Pressure field around a single hydraulic fracture in a homogenous reservoir with no natural fractures (b) Pressure field with the presence of one natural fracture (c) Pressure field with two natural fractures on either side of hydraulic fracture (d) Pressure field with six natural fracture with three on either side of the hydraulic fracture. Hydraulic fracture in black, natural fractures in dashed red line. The pressure scale was normalized.

6.3. Model Strengths and Limitations

The CAM models presented here are grid-less and meshless, unlike the more often used numerical methods in industry. Due to their being grid-less, CAM is much less computationally intensive than finite-volume/difference numerical methods with the added advantage of high resolution at the scale of the hydraulic and much smaller natural fractures. Other strengths of the CAM model to accurately determine the impact of natural fractures on drained rock volumes comes in the form of this analytical method having closed form solutions as well transparency in all steps of the methodology [23]. The present study is limited to flow in 2D as well as only modeling single phase fluid flow. As the natural fractures are modeled as individual discrete elements, the model would become cumbersome to use and computationally expensive if large scale, stochastically generated natural fracture networks are taken as inputs. This is the rationale behind the use of upscaling methods to represent natural fractures used in the field scale models. In reality, the geometry of both the natural fractures (in terms of inclination angle in 3D) and the hydraulic fractures (as fractal networks instead of simple bi-planar features) are much more complex than that represented here. In spite of these simplifying and reductionist model assumptions (as all other models also have), the CAM tool developed in this paper to include the impact of natural fractures can be used as a quick and simple method to screen optimum hydraulic fracture spacing and to support and direct well spacing decisions in naturally fractured reservoirs. What the 2D studies provide are very valuable systematic insight that will benefit the improvement of 3D model studies as well. Accounting for 3D dimensionality may make for more realistic models, but when coupled with flow, may also disguise some of the systematic effects visualized in our 2D models of flow in hydraulically and naturally fractured reservoirs.

6.4. Practical Implications

The impacts of natural fractures on production in unconventional wells are still debated. However, the interaction of the in-situ stress, hydraulic fractures and natural fractures could be leveraged to optimize well path planning and completions designs [24]. In this study, we distinguished three major mechanisms via which natural fractures may impact flow and, implicitly, acreage productivity. Flow models based on CAM show that enhanced natural fracture permeability and porosity can alter the DRV shapes and spatial location greatly. This can have implications for the spacing of both hydraulic fractures and wells once the nature of the natural fracture network in the subsurface has been accurately characterized. For formations with highly permeable natural fractures, well spacing should be slightly increased to avoid interference as the DRVs would otherwise overlap.

However, this assumes that the spacing is based on DRV modeling. If based on pressure interference models only, our previous work [33,60] argues that such pressure interference occurs for much larger well spacing and fracture spacing. However, such pressure interference should not be used as the sole criterion for well and fracture spacing decisions because of the over one order of magnitude time-lag between the pressure front and the tracer front propagation in ultra-low permeability reservoirs [61].

The models presented emphasize how the spatial orientation, location and lateral extent of the DRV are vastly impacted by the presence of natural fractures. Fluid flows preferentially through the highly conductive natural fractures, altering the shape of the DRV around hydraulic fractures. Any undrained matrix zones that have been bypassed due to flow channeling into the natural fractures with high flow rates can then be preferentially targeted for refracturing. For rock formations where the stress regimes preferentially allow for reactivation of natural fractures to form an extension of the hydraulic fracture, cluster spacing can be decreased to allow for the creation of the largest, most complex fracture network that gives the greatest access to the hydrocarbons trapped in the low permeability reservoir rock [21].

7. Conclusions

Natural fractures present in the subsurface are a major form of heterogeneity in both conventional and unconventional hydrocarbon reservoirs. Highly conductive natural fractures may provide preferential pathways for fluid withdrawal to the production wells, which is why natural fractures are highly crucial for well design decisions (especially in unconventional reservoirs). The major conclusions from our analysis on the impact of natural fractures on subsurface flow are

(1) Natural fractures can affect reservoir flow through three major mechanisms: (i) by enhancing permeability, (ii) by altering the porosity in the fractures, leading to increased storativity, and (iii) by becoming extensions of the hydraulic fracture network due to reactivation.

(2) Enhanced permeability in natural fractures creates high velocity flow zones which preferentially channel fluid flow through them. At high enough permeabilities (or natural fracture strengths as used in our models), this preferential pathway to flow leads to bypassed regions in the matrix blocks between the natural fractures, which are left undrained. These undrained matrix regions

can then be targeted by refracturing to improve recovery factors from hydraulically fractured horizontal wells.

(3)   Altered porosity or enhanced storativity (due to natural fractures with a higher porosity than the reservoir matrix as investigated in synthetic models) leads to a decrease in the lateral extent of the DRV. The impact of both natural fracture storativity and permeability greatly affect the shape and extent of the DRV around the hydraulic fractures.

(4)   The Carman–Kozeny (CK) relation was used to determine the relative impacts of the correlated porosity and permeability in natural fractures on the DRV development. Results based on the CK correlation show that the enhanced flow due to permeability far outweighs any storativity effects (even if natural fractures were to have a higher porosity than the reservoir matrix).

(5)   Use of a hybrid object-based and flow-based method for upscaling allows for the modeling of a high-density natural fracture network. Upscaling is needed to reduce the number of natural fractures modeled while keeping the equivalent permeability the same.

(6)   Field data on in-situ natural fracture characteristics such as porosity and permeability is sparse and lacking in the literature. Industry needs to ensure collection of such data for use in reservoir models to accurately determine subsurface flow and drainage volumes.

(7)   Proper analysis of natural fracture data and the predominant mechanism by which it will affect flow will lead to accurate DRV calculations in the subsurface. From these determined DRV (based on a well type curve), fracture cluster spacing and well spacing could possibly be optimized.

Author Contributions: Conceptualization, K.N. and R.W.; Methodology, K.N. and R.W.; Software, K.N.; Investigation, K.N.; Writing-Original Draft Preparation, K.N. and R.W.; Writing-Review & Editing, K.N. and R.W.; Supervision, R.W.; Funding Acquisition, R.W.

Funding: This project was sponsored by the Crisman-Berg Hughes consortium and startup funds from the Texas A&M Engineering Experiment Station (TEES).

Conflicts of Interest: The authors declare no conflict of interest.

Nomenclature

a	start of interval
b	end of interval
h_k	natural fracture height [ft]
i	imaginary unit
k_f	natural fracture permeability [nD]
k_m	matrix permeability [nD]
k (*subscript*)	summation index
$m(t)$	time dependent flow strength [ft^2/month]
n_f	fracture porosity
n_m	matrix porosity
t	time [month]
v_f	fluid velocity in natural fracture [ft/day]
v_m	fluid velocity in matrix [ft/day]
w_f	natural fracture aperture [ft]
z	complex coordinate
z_{a1}	corner point 1 of natural fracture domain
z_{a2}	corner point 2 of natural fracture domain
z_{b1}	corner point 3 of natural fracture domain
z_{b2}	corner point 4 of natural fracture domain
z_c	center of bounded natural fracture domain
C_f	fracture conductivity [mD.ft]
H_f	hydraulic fracture height [ft]
L_f	natural fracture length [ft]
P	pressure [psi]

P_o	initial pressure [psi]
ΔP	pressure change [psi]
R_k	permeability ratio
R_n	porosity ratio
V	total velocity field in reservoir space [ft/day]
Ω	complex potential [ft^2/day]
Φ	potential function
μ	fluid viscosity [cP]
β	angle between walls of natural fracture [degree]
γ	rotation angle of natural fracture to reference line along x-axis [degree]
υ	natural fracture flow strength [ft^4/day]

Abbreviations

CAM	complex analysis methods
CK	carman kozeny
DRV	drained rock volume
HFTS	hydraulic fracturing test site
REV	representative elementary volume
SRV	stimulated rock volume
TOF	time of flight
TOFC	time of flight contours

Conversion Factors from SI Units to Field Units

1 m	3.28 ft
1 Pa	1.45×10^{-4} psi
1 m^2	1.01×10^{15} md
1 m^3/s	5.434×10^5 STB/day (oil)
1 m^3/s	3049 Mscf/day (gas)
1 Pa-s	1000 cp

Appendix A. Flux Modeling and Production Allocation for Hydraulic Line Sink Models

Flux allocation was proportional to the relative surface areas of each hydraulic fracture. The flux allocation algorithm used is as follows:

$$q_k(t) = Z \cdot S(1 + WOR) \cdot q_{well}(t) \quad [\text{ft}^3/\text{month}] \tag{A1}$$

Z is a conversion factor of 5.61 to convert from barrels to ft^3 as q_{well} is given in bbls/day and q_k in ft^3/day. The WOR though seemingly very high (Table A1) is usual for these Wolfcamp completed wells [9]. S is the prorated factor to scale the total well production, for one hydraulic fracture stage from this well (which had a total of 22 stages):

$$S = (1/22) = 0.0455 \tag{A2}$$

After calculation of the flux using the given algorithm, we next determine the appropriate strength based on the time-dependent flow to use in the velocity and pressure potential equations. This strength is scaled by reservoir properties such as the formation volume factor (B), porosity (n), residual oil saturation (R_0) [60] and hydraulic fracture height (H_k) and is determined as follows:

$$m_k(t) = \frac{B \cdot q_k(t)}{H_k \cdot n_m \cdot (1 - R_0)} \quad [\text{ft}^2/\text{month}] \tag{A3}$$

Production from a well was history matched using the Arp's hyperbolic decline curve method and then forecasted for the 30 year production life of the well (Figure A1). This total well production was then allocated back into a hydraulics fracture representative of a single stage of the well. The production well used has 22 stages with a total of 131 individual fracture clusters. As such our DRV models are representative for the upscaled production from each stage.

Table A1. Reservoir parameters used for modelling.

Matrix Porosity (n_m)	0.05
Matrix Permeability (k_m)	100 nanoDarcy
Water-Oil Ratio (*WOR*)	4.592
Formation Volume Factor (*B*)	1
Viscosity (μ)	1 centipoise
Residual Oil Saturation (*Ro*)	0.20
Hydraulic Fracture Height (*H*)	60 ft
Hydraulic Fracture Length	150 ft

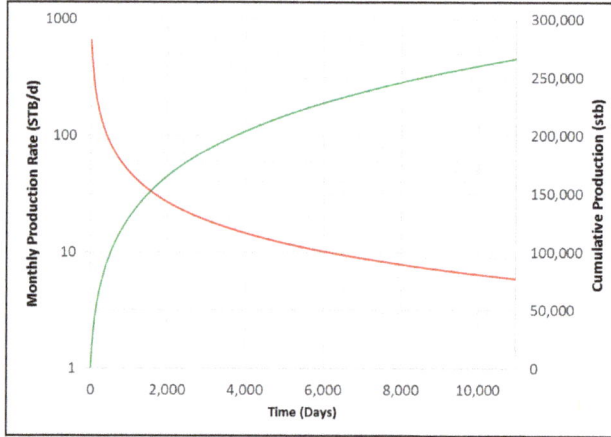

Figure A1. Arps hyperbolic decline curve model used to history match field data to give: (1) Production rate (STB/day, left scale, red curve) and (2) cumulative production (STB, right scale, green curve) for well after forecasted time of 30 years (10,958 days).

Appendix B. Carman–Kozeny Relation for Estimating Natural Fracture Permeability from Porosity

For the field models looking at use of the natural fracture data and its impact on DRV, the Carman–Kozeny correlation was used to determine an effective porosity-permeability relationship. The generic Carman–Kozeny correlation is given by [62]:

$$k = \frac{\phi^3}{\beta T^2 S^2} \tag{A4}$$

This well-known correlation seeks to link the permeability of a porous medium (in our case natural fractures with a predetermined porosity) to the porosity along with other rock properties. β represents the shape factor of the rock and is a constant characteristic for a particular type of granular material, S is known as the specific surface area and is the ratio of the total interstitial surface area to the bulk volume [62]. T is the hydraulic tortuosity defined as by the equation:

$$T = \frac{\langle \lambda \rangle}{\bar{L}} \tag{A5}$$

where $\langle \lambda \rangle$ represents the mean length of fluid particle paths and the variable $\bar{L}$ gives the straight-line distance through the medium in the direction of macroscopic flow. We adopted a T value of 1.41 [62] and a β of 3 for the pore shape coefficient for thin cracks [63]. The specific surface area by volume (S) is calculated from the specific surface area by weight and the average density using data from

Wolfcamp formation samples. Specific surface areas are given by Tinni et al. [64] for various particle sizes in the Wolfcamp formation with an average specific surface area of 9.36 m^2/g. Using this value in conjunction with the average Wolfcamp formation density of 2.73 g/cm^3 [65], S is calculated at 2.55 × 10^7 m^{-1}. Using these values with a given natural fracture porosity, natural fracture permeability is then calculated and converted to the equivalent strength using Equation (11) for use in the CAM models. An example of the correlation is given in Table A2 with the first row values used for Figure 11b.

Table A2. Natural fracture strength from Carman–Kozeny correlation.

Natural Fracture Porosity (%)	Natural Fracture Permeability (nD)	Rk	Matrix Velocity (ft/day)	Natural Fracture Width (ft)	Natural Fracture Length (ft)	Natural Fracture Height (ft)	Natural Fracture Strength (ft^4/day)
8.4	152.6	1.53	0.169	0.5	20	60	155
9.8	246.1	2.46	0.169	0.5	20	60	250

Appendix C. Upscaling for Fractured Porous Media

Upscaling of fractured porous media using an object-based approach is first considered. The object-based upscaling involves no flow simulation and the elements of the equivalent permeability tensor are obtained from the spatial distribution of high permeability zones [23]. Assuming the natural fractures have a uniform width and conductivity simplified expressions for the principal components k_x^* and k_y^* when fractures are parallel to far field flow (Figure A2a), using a 2D Cartesian grid with unit reservoir depth, are given as [23]:

$$k_x^* = \frac{N w_f k_f}{w_{REV}} \tag{A6}$$

$$k_y^* = k_m \frac{w_{REV}}{(N+1) w_m} \tag{A7}$$

With w_{REV} given by the formula:

$$w_{REV} = N w_f + (N+1) w_m \tag{A8}$$

where N is the number of fractures in the pre-determined representative elementary volume (REV), w_f the width of the fracture, k_f permeability of fracture, w_{REV} width of the REV in question, k_m permeability of the matrix and w_m the width of the matrix blocks between the fractures. Normalizing the length scale with respect to w_f and w_m gives:

$$w_{REV}^* = w_f^* + w_m^* = 1 \tag{A9}$$

where

$$w_f^* = \frac{N w_f}{w_{REV}} \tag{A10}$$

and

$$w_m^* = \frac{(N+1) w_m}{w_{REV}} \tag{A11}$$

When the natural fractures are oblique to the far-field flow (Figure A2b), the equivalent permeability tensor can be expressed in terms of the normalized w_f^* as

$$k(\theta) = \begin{bmatrix} w_f^* k_f \cos^2 \theta + k_m \sin^2 \theta & (w_f^* k_f - k_m) \cos \theta \sin \theta \\ (w_f^* k_f - k_m) \cos \theta \sin \theta & w_f^* k \sin^2 \theta + k_m \cos^2 \theta \end{bmatrix} = \begin{bmatrix} k_{xx} & k_{xy} \\ k_{yx} & k_{yy} \end{bmatrix} \tag{A12}$$

It is argued that the object-based method of upscaling cannot accurately capture the physics of flow in fractures porous media and that instead flow-based methods should be used ([23,66]). Chen et al. [67] propose solving the flow problem with a multi-boundary approach which commonly requires the use of numerical simulators. Weijermars and Khanal [23] approached the flow based upscaling by looking at the ratio of the velocity of flow inside and outside of the fracture zones to determine the equivalent permeability for a REV model using CAM. This approach led to the formulation of the 2D equivalent permeability tensor ellipses based on directional flow rates measured in CAM models with the axial ratios (k_x^*, k_y^*) normalized by the matrix permeability k_m:

$$\frac{k_x^*}{k_m} = \frac{\bar{v}_y + \bar{v}_x}{v_{y\_ff}} \tag{A13}$$

$$\frac{k_y^*}{k_m} = \frac{\bar{v}_y - \bar{v}_x}{v_{y\_ff}} \tag{A14}$$

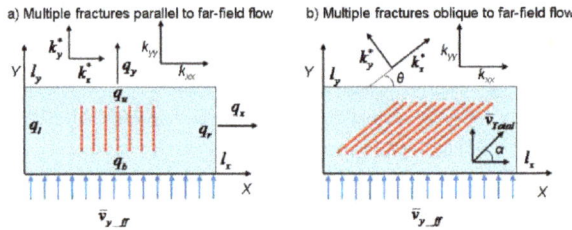

Figure A2. Permeability tensor components for multiple fracture (**a**) parallel and (**b**) oblique to far-field flow (modified from Weijermars and Khanal, [23]). $\bar{v}_y$ is the average velocity in the y direction while $\bar{v}_x$ is the average velocity in the x direction. The variable v_{y_ff} gives the velocity if the far field flow into the REV model.

Our present formulation for upscaling the permeability in fractured porous media is a hybrid between the object-based and flow-based upscaling methods. The object-based upscaling (Equations (A6)–(A12)) is first used to reduce the total number of natural fractures used in the model (essentially decreasing the natural fracture density). Next, the flow-based method (Equations (A13) and (A14)) is used with the upscaled fracture density to ensure the equivalent permeability for the REV of concern remains identical to the prototype.

Object-based upscaling step:

To demonstrate the validity of the proposed procedure we consider two similar REV's (Figure A3).

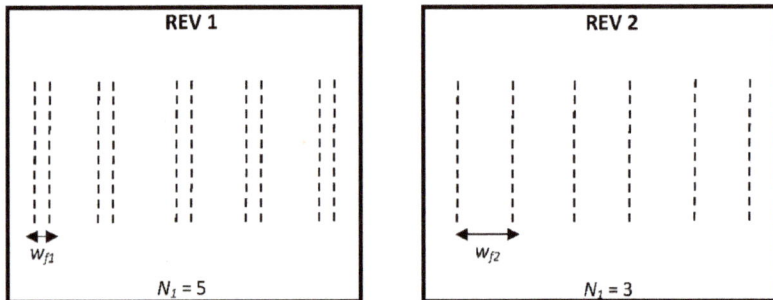

Figure A3. Two equal REV's with different numbers of natural fractures.

For *REV* 1:

$$k_x^* = \frac{N_1 w_{f1} k_f}{w_{REV}} \tag{A15}$$

For *REV* 2:

$$k_x^* = \frac{N_2 w_{f2} k_f}{w_{REV}} \tag{A16}$$

Assuming k_f and w_{REV} are constant and equating the equations for *REV* 1 and *REV* 2 we arrive at;

$$N_1 w_{f1} = N_2 w_{f2}$$

The number of fractures in *REV* 1 can be determined from the natural fracture density and *REV* width and length (L_{REV});

$$N_1 = NF_{density1} \times w_{REV} \times L_{REV}$$

Substituting for N_1;

$$N_2 = \frac{(NF_{density1} \times w_{REV} \times L_{REV}) w_{f1}}{w_{f2}} \tag{A17}$$

Based on a user defined value for a new natural fracture width we can upscale from N_1 fractures to a lower value of N_2 natural fractures which is more practical for use in discrete natural fracture models, including CAM used in our study

Validation of object-based upscaling step:

The proposed object-based upscaling (reduction) of the number of natural fractures in a given reservoir area was validated using the flow-based upscaling method. For the models with N_1 and N_2 fractures, the velocities are calculated in and outside of the natural fractures and the permeability tensor ellipses are generated. To properly account for the reduction of the number of fracture and equivalent upscaling, the assigned natural fracture permeabilities of the original prototype ($v_{f1} w_{f1}$) and upscaled models ($v_{f2} w_{f2}$) needed to maintain the same equivalent permeability are given by;

$$v_{f2}(t) = \frac{v_{f1}(t) w_{f2}}{w_{f1}} \tag{A18}$$

where $v_{f1}(t)$ is the original strength prior to upscaling, and $v_{f2}(t)$ is the new strength (which are proxies to the permeability in our models) to be used after upscaling the number of natural fractures with the corresponding fracture width change. This procedure is demonstrated via the upscaling of natural fractures at an angle of 45° to the far field flow starting with Figures A4 and A5, up to the final upscaled REV in Figure A6.

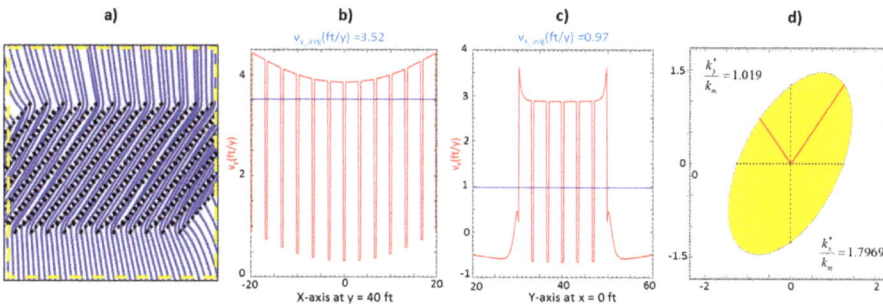

Figure A4. (**a**) Flow in a defined REV space with streamlines in blue and natural fractures (NF) in dashed black, (**b,c**) Velocity profiles along cross-hairs at $y = +40$ and $x = 0$ respectively, (**d**) Equivalent permeability ellipse based on Equations (A13) and (A14). Number of NF = 16; width of NF = 2ft; strength of NF = 120 ft^4/year.

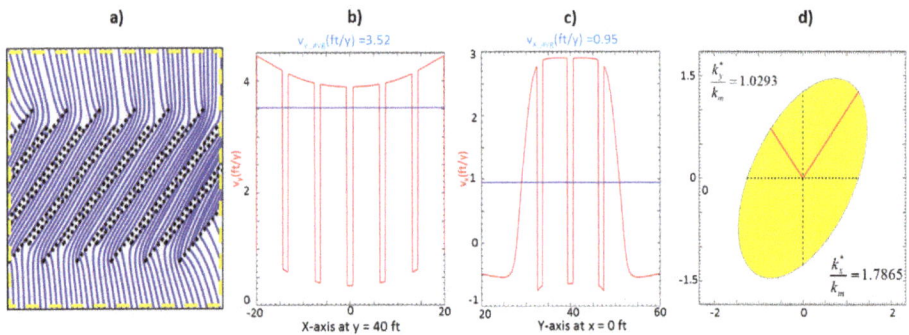

Figure A5. (**a**) Flow in a defined REV space with streamlines in blue and natural fractures (NF) in dashed black, (**b,c**) Velocity profiles along cross-hairs at $y = +40$ and $x = 0$, respectively, (**d**) Equivalent permeability ellipse based on Equations (A13) and (A14). Number of NF = 8; width of NF = 4ft; upscaled strength of NF = 240 ft^4/year.

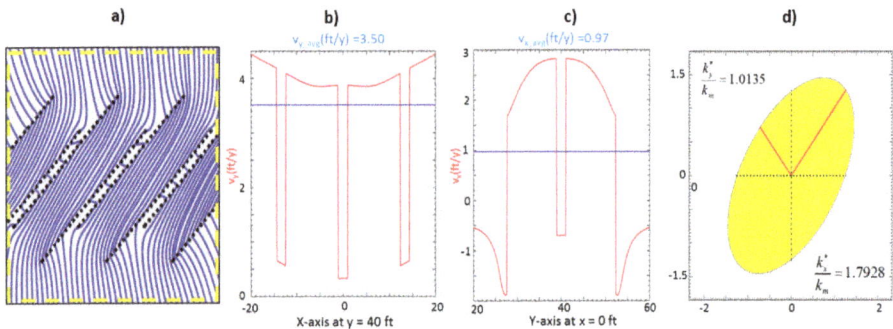

Figure A6. (**a**) Flow in a defined REV space with streamlines in blue and natural fractures (NF) in dashed black, (**b,c**) Velocity profiles along cross-hairs at $y = +40$ and $x = 0$, respectively, (**d**) Equivalent permeability ellipse based on Equations (A13) and (A14). Number of NF = 4; width of NF = 8ft; upscaled strength of NF = 480 ft^4/year.

The above results show that with a reduction in the number of natural fractures by object-based upscaling within a defined REV, using the appropriate upscaling for fracture width and permeability in the natural fractures, the equivalent permeability remains constant. By using this method, we can upscale a realistic fracture density to a manageable number of natural fractures for use in the CAM models for DRV calculations. This upscaling methodology was applied in the next section to field data from the Hydraulic Fracturing Test Site (Midland Basin, West Texas, with completions in the Wolfcamp Formation).

Application of object-based and flow-based upscaling to HFTS field model:

This section makes use of the proposed combination of object-based and flow-based upscaling to reduce the natural fracture density used by Shrivastava et al. [12] in their model to match the data from the HFTS. Selecting a REV located around a hydraulic fracture of 125 ft in length by 45 ft in height above the hydraulic fracture corresponds a true density of 210 natural fractures with an assumed width of 0.2 inches. The 210 fractures are reduced in the proposed upscaling procedure, making use of Equation (A17), and adopting an upscaled natural fracture width of 6 inches (based on object-based upscaling), results in 6 natural fractures of length 30 ft. These 6 natural fractures have fracture centers and angles (kept in range of HFTS data) that are stochastically generated within the specified REV both below and above the hydraulic fracture. This results in a total of 12 upscaled natural fractures that are used in the final HFTS field model (Figure A7). The CK correlation was used with a final upscaled strength of 155 ft^4/year, which gives a corresponding porosity of 7.32% within the natural fractures.

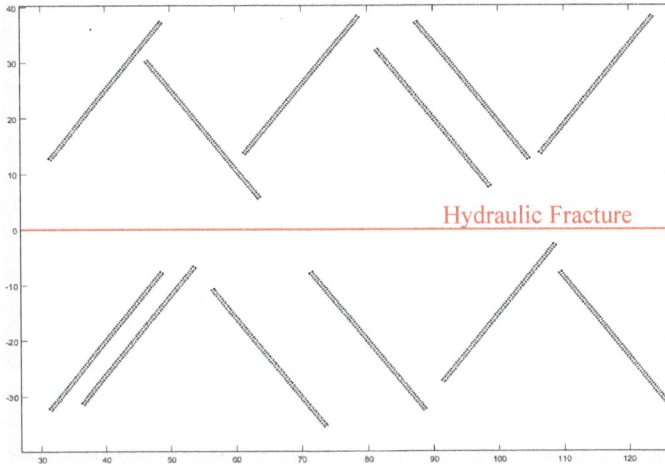

Figure A7. REV near single hydraulic fracture (horizontal red line) with upscaled natural fractures (dashed black lines) based on HFTS field data.

References

1. Kresse, O.; Weng, X.; Gu, H.; Wu, R. Numerical Modeling of Hydraulic Fractures Interaction in Complex Naturally Fractured Formations. *Rock Mech. Rock Eng.* **2013**, *46*, 555–568. [CrossRef]
2. Aguilera, R. Effect of Fracture Dip and Fracture Tortuosity on Petrophysical Evaluation of Naturally Fractured Reservoirs. *Petrol. Soc. Canada* **2008**. [CrossRef]
3. Tutuncu, A.; Bui, B.; Suppachoknirun, T. *An Integrated Study for Hydraulic Fracture and Natural Fracture Interactions and Refracturing in Shale Reservoirs Hydraulic Fracture Modeling*; Gulf Professional Publishing: Oxford, UK, 2018; pp. 323–348. [CrossRef]
4. Weijermars, R.; Van Harmelen, A. Breakdown of doublet recirculation and direct line drives by far-field flow in reservoirs: Implications for geothermal and hydrocarbon well placement. *Geophys. J. Int.* **2016**, *206*, 19–47. [CrossRef]
5. Doe, T.; Lacazette, A.; Dershowitz, W.; Knitter, C. Evaluating the Effect of Natural Fractures on Production from Hydraulically Fractured Wells Using Discrete Fracture Network Models. In Proceedings of the Unconventional Resources Technology Conference, Denver, CO, USA, 12–14 August 2013; pp. 1679–1688.
6. Olson, J.E.; Taleghani, A.D. Modeling simultaneous growth of multiple hydraulic fractures and their interaction with natural fractures. In Proceedings of the SPE Hydraulic Fracturing Technology Conference, Woodlands, TX, USA, 19–21 January 2009.
7. Cipolla, C.L.; Warpinski, N.R.; Mayerhofer, M.J.; Lolon, E.; Vincent, M.C. The Relationship Between Fracture Complexity, Reservoir Properties, and Fracture Treatment Design. *Soc. Pet. Eng.* **2008**. [CrossRef]
8. Warpinski, N.R.; Moschovidis, Z.A.; Parker, C.D.; Abou-Sayed, I.S. Comparison study of hydraulic fracturing models—Test case: GRI staged field Experiment No. 3 (includes associated paper 28158). *SPE Prod. Facil.* **1994**, *9*, 7–16. [CrossRef]
9. Parsegov, S.G.; Nandlal, K.; Schechter, D.S.; Weijermars, R. Physics-Driven Optimization of Drained Rock Volume for Multistage Fracturing: Field Example from the Wolfcamp Formation, Midland Basin. In Proceedings of the Unconventional Resources Technology Conference, Houston, TX, USA, 23–25 July 2018.
10. Popovici, A.M.; Fomel, S.; Grechka, V.; Li, Z.; Howell, R.; Vavrycuk, V. Single-well moment tensor inversion of tensile microseismic events. *SEG Tech. Program Expand. Abstr.* **2017**, *81*, 2746–2751.

11. Raterman, K.T.; Farrell, H.E.; Mora, O.S.; Janssen, A.L.; Gomez, G.A.; Busetti, S.; Warren, M. Sampling a Stimulated Rock Volume: An Eagle Ford Example. In Proceedings of the Unconventional Resources Technology Conference, Austin, TX, USA, 24–26 July 2017. [CrossRef]

12. Shrivastava, K.; Hwang, J.; Sharma, M. Formation of Complex Fracture Networks in the Wolfcamp Shale: Calibrating Model Predictions with Core Measurements from the Hydraulic Fracturing Test Site. *Soc. Pet. Eng.* **2018**. [CrossRef]

13. Fisher, M.; Wright, C.; Davidson, B.; Goodwin, A.; Fielder, E.; Buckler, W.; Steinsberger, N. Integrating Fracture Mapping Technologies to Optimize Stimulations in the Barnett Shale. *SPE Annu. Tech. Conf. Exhib.* **2005**, *20*, 85–93.

14. Maxwell, S.; Urbancic, T.; Steinsberger, N.; Zinno, R. Microseismic Imaging of Hydraulic Fracture Complexity in the Barnett Shale. In Proceedings of the SPE Annual Technical Conference and Exhibition, San Antonio, TX, USA, 29 September–2 October 2002.

15. Pradhan, V.R.; Meert, J.G.; Pandit, M.K.; Kamenov, G.; Mondal, M.E.A. Paleomagnetic and geochronological studies of the mafic dyke swarms of Bundelkhand craton, central India: Implications for the tectonic evolution and paleogeographic reconstructions. *Precambrian Res.* **2012**, *198*, 51–76. [CrossRef]

16. Kilaru, S.; Goud, B.K.; Rao, V.K. Crustal structure of the western Indian shield: Model based on regional gravity and magnetic data. *Geosci. Front.* **2013**, *4*, 717–728. [CrossRef]

17. McKenzie, N.R.; Hughes, N.C.; Myrow, P.M.; Banerjee, D.M.; Deb, M.; Planavsky, N.J. New age constraints for the Proterozoic Aravalli–Delhi successions of India and their implications. *Precambrian Res.* **2013**, *238*, 120–128. [CrossRef]

18. Huang, J.I.; Kim, K. Fracture process zone development during hydraulic fracturing. *Int. J. Rock Mech. Min. Sci. Geéomeéch. Abstr.* **1993**, *30*, 1295–1298. [CrossRef]

19. Weijermars, R.; Van Harmelen, A.; Zuo, L.; Nascentes, I.A.; Yu, W. High-Resolution Visualization of Flow Interference Between Frac Clusters (Part 1): Model Verification and Basic Cases. In Proceedings of the Unconventional Resources Technology Conference, Austin, TX, USA, 24–26 July 2017.

20. Weijermars, R.; Van Harmelen, A.; Zuo, L. Flow Interference Between Frac Clusters (Part 2): Field Example from the Midland Basin (Wolfcamp Formation, Spraberry Trend Field) With Implications for Hydraulic Fracture Design. In Proceedings of the Unconventional Resources Technology Conference, Austin, TX, USA, 24–26 July 2017.

21. Nandlal, K.; Weijermars, R. Drained rock volume around hydraulic fractures in porous media: Planar fractures versus fractal networks. *Pet. Sci.* **2019**, 1–22. [CrossRef]

22. Van Harmelen, A.; Weijermars, R. Complex analytical solutions for flow in hydraulically fractured hydrocarbon reservoirs with and without natural fractures. *Appl. Math. Model.* **2018**, *56*, 137–157. [CrossRef]

23. Weijermars, R.; Khanal, A. High-resolution streamline models of flow in fractured porous media using discrete fractures: Implications for upscaling of permeability anisotropy. *Earth-Science Rev.* **2019**, *194*, 399–448. [CrossRef]

24. Forand, D.; Heesakkers, V.; Schwartz, K. Constraints on Natural Fracture and In-situ Stress Trends of Unconventional Reservoirs in the Permian Basin, USA. In Proceedings of the Unconventional Resources Technology Conference, Austin, TX, USA, 24–26 July 2017. [CrossRef]

25. Gale, J.F.; Laubach, S.E.; Olson, J.E.; Eichhuble, P.; Fall, A. Natural Fractures in shale: A review and new observations. *AAPG Bull.* **2014**, *98*, 2165–2216. [CrossRef]

26. Gutierrez, M.; Øino, L.; Nygard, R. Stress-dependent permeability of a de-mineralised fracture in shale. *Mar. Pet. Geol.* **2000**, *17*, 895–907. [CrossRef]

27. Laubach, S.E. Practical approaches to identifying sealed and open fractures. *AAPG Bull.* **2003**, *87*, 561–579. [CrossRef]

28. Zuo, L.; Weijermars, R. Rules for flight paths and time of flight for flows in heterogeneous isotropic and anisotropic porous media. *Geofluids* **2017**. [CrossRef]

29. Warren, J.; Root, P. The Behavior of Naturally Fractured Reservoirs. *Soc. Pet. Eng. J.* **1963**, *3*, 245–255. [CrossRef]

30. Kazemi, H.; Merrill, L.S.; Porterfield, K.L.; Zeman, P.R. December 1. Numerical Simulation of Water-Oil Flow in Naturally Fractured Reservoirs. *Soc. Pet. Eng.* **1976**. [CrossRef]

31. Huang, T.; Guo, X.; Chen, F. Modeling transient flow behavior of a multiscale triple porosity model for shale gas reservoirs. *J. Nat. Gas Sci. Eng.* **2015**, *23*, 33–46. [CrossRef]

32. Sang, G.; Elsworth, D.; Miao, X.; Mao, X.; Wang, J. Numerical study of a stress dependent triple porosity model for shale gas reservoirs accommodating gas diffusion in kerogen. *J. Nat. Gas Sci. Eng.* **2016**, *32*, 423–438. [CrossRef]

33. Weijermars, R.; Van Harmelen, A. Shale Reservoir Drainage Visualized for a Wolfcamp Well (Midland Basin, West Texas, USA). *Energies* **2018**, *11*, 1665. [CrossRef]

34. Long, J.C.S.; Remer, J.S.; Wilson, C.R.; Witherspoon, P.A. Porous media equivalents for networks of discontinuous fractures. *Water Resour. Res.* **1982**, *18*, 645–658. [CrossRef]

35. Rogers, S.; Elmo, D.; Dunphy, R.; Bearinger, D. Understanding Hydraulic Fracture Geometry and Interactions in the Horn River Basin Through DFN and Numerical Modeling. In Proceedings of the Canadian Unconventional Resources and International Petroleum Conference, Calgary, AB, Canada, 19–21 October 2010.

36. Dershowitz, W.S.; Ambrose, R.; Lim, D.H.; Cottrell, M.G. Hydraulic fracture and natural fracture simulation for improved shale gas development. In Proceedings of the American Association of Petroleum Geologists (AAPG) Annual Conference and Exhibition Houston, Houston, TX, USA, 10–13 April 2011.

37. Jing, L.; Stephansson, O. *Fundamentals of Discrete Element Methods for Rock Engineering. Developments in Geotechnical Engineering*; Elsevier: Amsterdam, The Netherlands, 2007; Volume 85. [CrossRef]

38. Wu, K.; Olson, J.E. Numerical Investigation of Complex Hydraulic-Fracture Development in Naturally Fractured Reservoirs. *Soc. Pet. Eng.* **2016**. [CrossRef]

39. Nelson, R. *Geological Analysis of Naturally Fractured Reservoirs*, 2nd ed.; Gulf Professional Publishing: Oxford, UK, 2001.

40. Tiab, D.; Restrepo, D.P.; Igbokoyi, A.O. Fracture Porosity of Naturally Fractured Reservoirs. *Soc. Pet. Eng.* **2006**. [CrossRef]

41. Zhang, X.; Thiercelin, M.J.; Jeffrey, R.G. Effects of Frictional Geological Discontinuities on Hydraulic Fracture Propagation. In Proceedings of the SPE Hydraulic Fracturing Technology Conference, College Station, TX, USA, 29–31 January 2007; pp. 29–31.

42. Soeder, D. Porosity and Permeability of Eastern Devonian Gas Shale. *SPE Form. Eval.* **1988**, *3*, 116–124. [CrossRef]

43. Gale, J.F.W.; Reed, R.M.; Holder, J. Natural fractures in the Barnett Shale and their importance for hydraulic fracture treatments. *AAPG Bull.* **2007**, *91*, 603–622. [CrossRef]

44. Nelson, R.A. *Geologic Analysis of Naturally Fractured Reservoirs*; Gulf Publishing: Houston, TX, USA, 1985; p. 320.

45. Weber, K.J.; Bakker, M. Fracture and vuggy porosity. In Proceedings of the Society of Petroleum Engineers 56th Annual Fall Technical Conference, San Antonio, TX, USA, 5–7 October 1981; p. 11.

46. Lee, D.S.; Herman, J.D.; Elsworth, D.; Kim, H.T.; Lee, H.S. A critical evaluation of unconventional gas recovery from the marcellus shale, northeastern United States. *KSCE J. Civ. Eng.* **2011**, *15*, 679–687. [CrossRef]

47. Muskat, M. The Theory of Potentiometric Models. *Trans. AIME* **1949**, *179*, 216–221. [CrossRef]

48. Strack, O.D.L. *Groundwater Mechanics*; Prentice-Hall: Englewood Cliffs, NJ, USA, 1989.

49. Sato, K. *Complex Analysis for Practical Engineering*; Springer Science and Business Media LLC: Berlin/Heidelberg, Germany, 2015.

50. Weijermars, R. Visualization of space competition and plume formation with complex potentials for multiple source flows: Some examples and novel application to Chao lava flow (Chile). *J. Geophys. Res. Solid Earth* **2014**, *119*, 2397–2414. [CrossRef]

51. Weijermars, R.; Dooley, T.P.; Jackson, M.P.A.; Hudec, M.R. Rankine models for time-dependent gravity spreading of terrestrial source flows over subplanar slopes. *J. Geophys. Res. Solid Earth* **2014**, *119*, 7353–7388. [CrossRef]

52. Weijermars, R.; Van Harmelen, A.; Zuo, L. Controlling flood displacement fronts using a parallel analytical streamline simulator. *J. Pet. Sci. Eng.* **2016**, *139*, 23–42. [CrossRef]

53. Potter, H.D.P. On Conformal Mappings and Vector Fields. Senior Thesis, Marietta College, Marietta, OH, USA, 2008.

54. Wu, K.; Olson, J.E. Mechanics Analysis of Interaction Between Hydraulic and Natural Fractures in Shale Reservoirs. In Proceedings of the Unconventional Resources Technology Conference, Denver, CO, USA, 25–27 August 2014. [CrossRef]

55. Taleghani, A.D.; Olson, J.E. How Natural Fractures Could Affect Hydraulic-Fracture Geometry. *SPE J.* **2013**, *19*, 161–171. [CrossRef]

56. Kumar, A.; Shrivastava, K.; Manchanda, R.; Sharma, M. An Efficient Method for Modeling Discrete Fracture Networks in Geomechanical Reservoir Simulation. In Proceedings of the Unconventional Resources Technology Conference, Denver, CO, USA, 25 July 2019. [CrossRef]

57. Gale, J.F.W.; Elliott, S.J.; Laubach, S.E. Hydraulic Fractures in Core from Stimulated Reservoirs: Core Fracture Description of HFTS Slant Core. In Proceedings of the Unconventional Resources Technology Conference, Midland Basin, TX, USA, 9 August 2018. [CrossRef]

58. Khanal, A.; Nandlal, K.; Weijermars, R. Impact of Natural Fractures on the Shape and Location of Drained Rock Volumes in Unconventional Reservoirs: Case Studies from the Permian Basin. In Proceedings of the Unconventional Resources Technology Conference, Austin, TX, USA, 31 July 2019. [CrossRef]

59. Khanal, A.; Weijermars, R. Modeling Flow and Pressure Fields in Porous Media with High Conductivity Flow Channels and Smart Placement of Branch Cuts for Variant and Invariant Complex Potentials. *Fluids* **2019**, *4*, 154. [CrossRef]

60. Khanal, A.; Weijermars, R. Pressure depletion and drained rock volume near hydraulically fractured parent and child wells. *J. Pet Sci. Eng.* **2019**, *172*, 607–626. [CrossRef]

61. Weijermars, R.; Nandlal, K.; Khanal, A.; Tugan, M.F. Comparison of pressure front with tracer front advance and principal flow regimes in hydraulically fractured wells in unconventional reservoirs. *J. Pet. Sci. Eng.* **2019**, *183*, 106407. [CrossRef]

62. Duda, A.; Koza, Z.; Matyka, M. Hydraulic tortuosity in arbitrary porous media flow. *Phys. Rev. E* **2011**, *84*, 036319. [CrossRef]

63. Ma, J. Review of permeability evolution model for fractured porous media. *J. Rock Mech. Geotech. Eng.* **2015**, *7*, 351–357. [CrossRef]

64. Tinni, A.; Sondergeld, C.; Rai, C. Particle size effect on porosity and specific surface area measurements of shales. In Proceedings of the International Symposium of the Society of Core Analysts, Avignon, France, 8–11 September 2014.

65. U.S. Energy Information Administration (EIA). *Permian Basin, Wolfcamp Shale Play, Geology Review*; U.S. Energy Information Administration: Washington, DC, USA, 2018.

66. Chen, T.; Clauser, C.; Marquart, G.; Willbrand, K.; Mottaghy, D. A new upscaling method for fractured porous media. *Adv. Water Resour.* **2015**, *80*, 60–68. [CrossRef]

67. Chen, T.; Clauser, C.; Marquart, G.; Willbrand, K.; Büsing, H. Modeling anisotropic flow and heat transport by using mimetic finite differences. *Adv. Water Resour.* **2016**, *94*, 441–456. [CrossRef]

energies

MDPI

Article

Generalized Extreme Value Statistics, Physical Scaling and Forecasts of Oil Production in the Bakken Shale

Wardana Saputra, Wissem Kirati and Tadeusz Patzek *

The Ali I. Al-Naimi Petroleum Engineering Research Center, King Abdullah University of Science and Technology, Thuwal 23955-6900, Saudi Arabia; wardana.saputra@kaust.edu.sa (W.S.); wissem.kirati@kaust.edu.sa (W.K.)
* Correspondence: tadeusz.patzek@kaust.edu.sa

Received: 18 August 2019; Accepted: 16 September 2019; Published: 24 September 2019

Abstract: We aim to replace the current industry-standard empirical forecasts of oil production from hydrofractured horizontal wells in shales with a statistically and physically robust, accurate and precise method of matching historic well performance and predicting well production for up to two more decades. Our Bakken oil forecasting method extends the previous work on predicting fieldwide gas production in the Barnett shale and merges it with our new scaling of oil production in the Bakken. We first divide the existing 14,678 horizontal oil wells in the Bakken into 12 *static* samples in which reservoir quality and completion technologies are similar. For each sample, we use a purely data-driven non-parametric approach to arrive at an appropriate generalized extreme value (GEV) distribution of oil production from that sample's *dynamic* well cohorts with at *least* $1, 2, 3, \ldots$ years on production. From these well cohorts, we stitch together the P_{50}, P_{10}, and P_{90} statistical well prototypes for each sample. These statistical well prototypes are conditioned by well attrition, hydrofracture deterioration, pressure interference, well interference, progress in technology, and so forth. So far, there has been *no* physical scaling. Now we fit the parameters of our physical scaling model to the statistical well prototypes, and obtain a *smooth* extrapolation of oil production that is mechanistic, and not just a decline curve. At late times, we add radial inflow from the outside. By calculating the number of potential wells per square mile of each Bakken region (core and noncore), and scheduling future drilling programs, we stack up the extended well prototypes to obtain the plausible forecasts of oil production in the Bakken. We predict that Bakken will ultimately produce 5 billion barrels of oil from the existing wells, with the possible addition of 2 and 6 billion barrels from core and noncore areas, respectively.

Keywords: EUR; infill wells; (re)fracturing; pressure depletion

1. Introduction

Over the last decade, crude oil and gas production from hydrofractured shales in the United States has accounted for most of the net increase of global crude oil production. Therefore, it is important to have a reliable, quantitative method for delineating the possible futures oil and gas production in the data-rich US shales. The current industry-standard methods of forecasting production from shales are variants of the empirical decline curve analysis (DCA) [1], developed 75 years ago. Lately, some of the more sophisticated methods, for example, Fractional Decline Curve [2] have become popular.

Unlike other analytical and numerical methods that require numerous reservoir parameters and a lengthy calculation or simulation time, DCA only requires production data to predict future production by extrapolating oil or gas rate observed in a boundary-dominated flow regime. Because of DCA's simplicity, most petroleum engineers adopt it for reserve assessment in shales, including Estimated Ultimate Recovery (EUR) predictions from USGS [3,4] and EIA [5].

USGS first split the Bakken region into 6 assessment units (AUs) and defined sweet spots. Then, they calculated the number of wells that could be drilled in each AU by dividing its total area with the average drainage area per well. In parallel, they used DCA to calculate the average EURs for sweet spots and other areas. The 7.4 billion barrels of undiscovered technically recoverable oil was obtained by multiplying the total number of wells that could be drilled in each AU with the corresponding average EUR multiplied by the untested fraction of that AU. EIA used a similar approach by dividing the Bakken region into 5 AUs and refining them by counties to determine the infill well potential. Both USGS and EIA predictions were assessed by Hughes [6–9]. We predict that the undiscovered, technically recoverable oil in the Bakken is 2 billion bbls in the core area and it might be 6 billion bbls in the noncore area (Figure 1) or 8 billion barrels in total.

Most shale wells do not reach the boundary-dominated flow regime for their entire production lives because of the vanishing matrix permeability. Thus, the traditional DCA frequently overestimates EURs of shale wells. To address this issue, many authors have suggested improved DCA methods, specific to shale wells: the Power Law Exponential Decline [10], the Stretched Exponential Decline [11], the Logistic Growth Model [12], the Extended Exponential DCA [13], and the Extended Hyperbolic DCA [14]. To make things worse, the empirical DCA fits of particular wells are ill-suited to forecasting production from a wide area of a given shale play in which reservoir properties vary and uncertainties abound. Therefore, some authors have developed probabilistic models to introduce a range of possible outcomes into their production forecasts [15–20]. The most common assumption is that well productivities in shales are log-normally distributed.

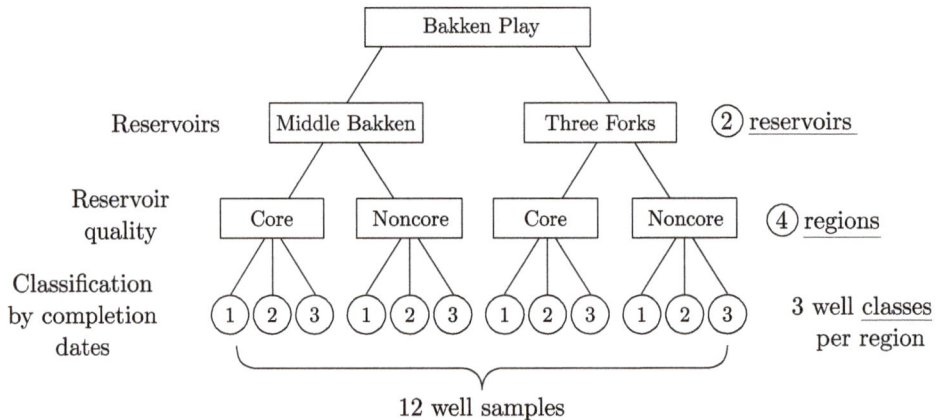

Figure 1. Division of the Bakken shale wells into 12 well samples.

In this paper, we adopt a hybrid data-driven *and* physics-based method of predicting oil or gas production in shales that has been introduced in our previous work [21,22]. Here, we consider only black oil production. First, we identify play **regions** in which reservoir quality is similar, see Figure 1. In each region, we identify well **classes** by different completion technologies. Finally, a well class in a region constitutes a well **sample**. We ensure that oil production from all wells in each sample is statistically uniform, that is, has a unimodal distribution. For each well sample, we then identify well **cohorts** with **at least** $1, 2, \ldots$ years on production. In general, well cohorts contain different sets of wells that satisfy the minimum time on production required for each cohort. It turns out that each cohort of wells is superbly characterized by its unique Generalized Extreme Value (GEV) distribution (see Appendix A) of annualized well rates or cumulative well production. Different cohorts in the same sample have different GEV distributions, each with its unique expected value, median and mode. Here we choose the somewhat better GEV fits of the production rate distributions. Each GEV distribution is statistically superior to the corresponding log-normal distribution at the 95% confidence level. When we plot the expected values of the GEV distributions of all wells cohorts in a sample versus elapsed time of production, we obtain this sample's average P_{50} statistical well prototype that is purely field data-driven.

Now we fit each statistical well prototype with a physical scaling curve that extends this prototype to 30 years on production. The physical scaling curves are based on an analytical solution of the pressure diffusion equation in the hydrofractured horizontal shale well geometry. In previous papers, we comprehensively detailed the physical scaling solutions for shale gas wells [23,24] and shale oil wells [22,25]. Late-time flow from outer reservoir encompassing the stimulated reservoir volume (SRV) [26] was also quantified. We have verified that our physical scaling is equivalent to detailed numerical reservoir simulations [27]. This scaling is much simpler to set up and runs almost infinitely faster than the corresponding reservoir simulations.

At this point, for each well sample in every play region, we have obtained a unique hybrid mean (P_{50}) well prototype with 30 years on production. In each play region, we know how many wells were drilled and completed each month up until current time. We then multiply each well prototype by the number of wells completed per month and stack them up to represent the total historical field rate and future production decline. In this manner, we obtain a 'base' case forecast for all existing wells. This base case forecast is a 'do nothing' scenario with no new wells drilled in the future. For all other forecasts of future field production rate, we first determine the infill potential or the number of wells that can be drilled in the future without causing significant interwell pressure interference (fracture hits). We cover each region with a fish-net grid that consists of one-square-mile pixels. We then calculate the infill potential as the number of wells that can be drilled so that the total number of wells in each pixel is less than the maximum allowable number of wells without fracture hits. Next, based on the infill potentials for all regions, we create future drilling programs to obtain plausible forecasts of oil or gas production. Based on current rig count in the Bakken, we assume a constant overall drilling rate. Finally, we assign the correct well prototype to every future well that will be drilled during each month of a postulated drilling schedule, and sum them up to obtain a forecast scenario.

In this paper, we select the Bakken shale, the current second-largest oil producer in the U.S. with 1.5 million bbl of oil per day, as an illustration. Being one of the oldest shale oil plays, Bakken has been a field laboratory to test drilling and completion technologies and increase well productivity. Currently, Bakken has ~15,000 active hydrofractured horizontal wells with a few wells that have 18 years of production data. In a previous paper [22], we scaled well-by-well all ~15,000 wells in the Bakken. We accounted for well refracturing and/or changes in downhole pressure. It turns out that the 12 well prototypes obtained with our hybrid GEV–physical scaling method are as good in duplicating the total field rate as the super-precise scaling of each individual well in our previous work [22]. Given the results of our analyses that are free of bias, policy-makers should not assume that the production boom in the Bakken shale will last decades longer.

2. Results

Our approach is as follows:

- Divide all 14,678 horizontal oil wells in the Bakken shale into 12 samples in which oil production is statistically uniform;
- Fit a generalized extreme value distribution to all wells in every sample and obtain 12 stable mean P_{50} well prototypes;
- Fit the physics-based scaling curves to every statistical well prototype and extend these prototypes smoothly to 30 years on production;
- Replace oil production from all existing wells with the 12 extended well prototypes and obtain a 'base' forecast;

- Calculate the infill potential for each of the 12 regions in the Bakken; and
- Create the plausible infill drilling schedules and forecast total field oil production rate up to the year 2050.

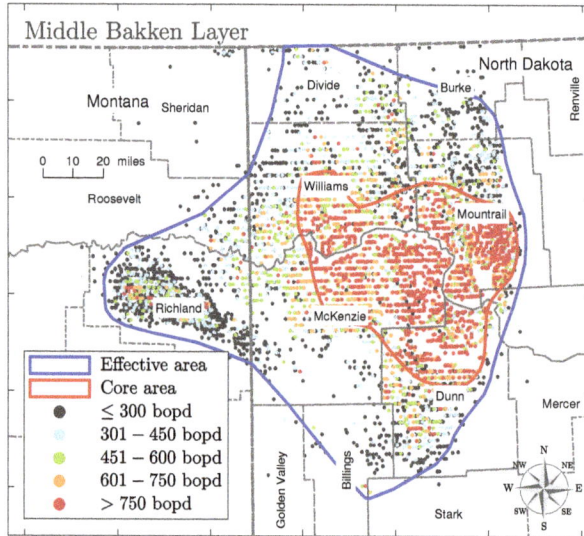

(a)

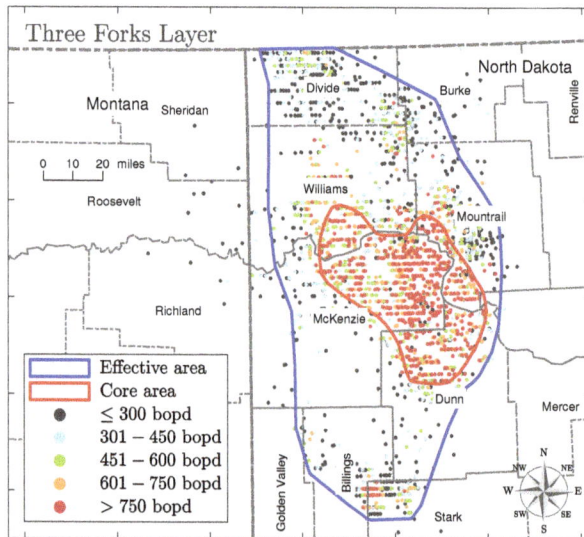

(b)

Figure 2. Maps of all 14,678 active horizontal wells completed in the Middle Bakken formation (**a**) and Three Forks formation (**b**) colored by maximum daily oil rate. The red lines define the *core* areas for each reservoir and delineate best producing wells with more than 750 barrels of oil per day. The blue lines define the *effective* areas for drilling neglecting a few poor producing wells outside. We define the *noncore* areas as the difference between the effective and core areas. In the Middle Bakken, there are currently 5732 and 4128 wells located in the core and noncore areas, respectively. The other 2672 and 2146 wells are located in the core and noncore areas of Three Forks, respectively.

2.1. Design of Well Regions

In the previous work [21], we divided all 13,141 horizontal wells in the Barnett shale into 6 *static* regions that corresponded to the geographical borders of 6 major gas-producing counties. We justified this approach as follows: (i) Productive Barnett shale is a single gas reservoir; (ii) the Barnett play area is relatively small, and the division into counties captures different reservoir qualities; and (iii) completion technologies have not been revolutionized over time.

However, in the Bakken shale, we introduce a more complicated set of 12 static well samples classified by reservoir, geography, and completion dates. In Bakken, there are two different producing reservoirs, the Middle Bakken and Three Forks. Accordingly, we first split all Bakken wells into the Middle Bakken and Three Forks groups. To account for different reservoir quality, we then divide each of these two well groups into two macro regions: the core and noncore areas, see Figure 2. Finally, we further refine the four Bakken macro regions by splitting them into three classes by completion date intervals, 2000–2012, 2013–2016 and 2017–2019. These classes reflect advancements in well completion technology, such as longer wells, more fracture stages, fewer clusters and perforation shots, bigger hydrofractures, more proppant, and so forth. In the end, we divide the existing 14,678 horizontal oil wells in the Bakken into 12 unique samples listed in Table 1.

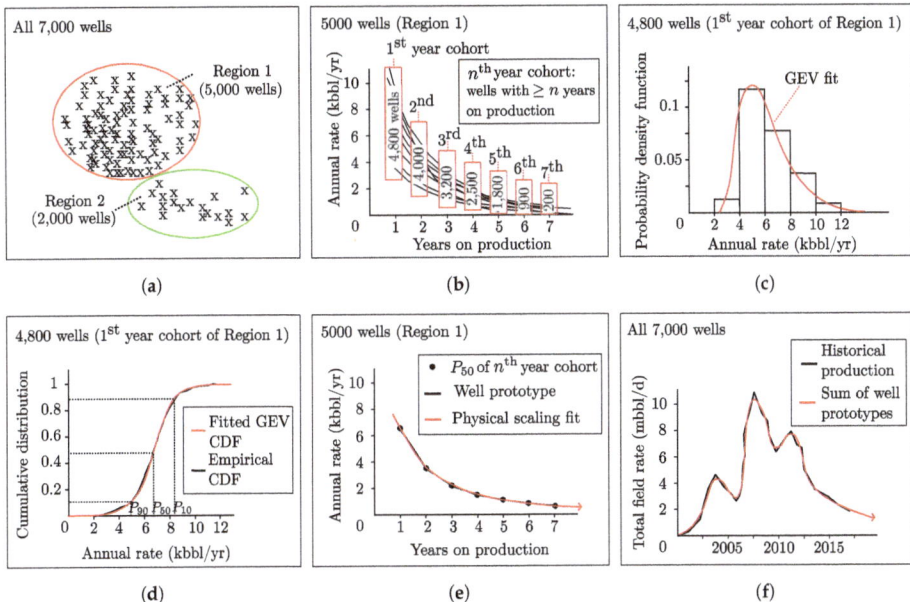

Figure 3. The procedure of arriving at the Generalized Extreme Value (GEV) well prototypes in a given shale play. (**a**) Define *static* regions in which oil production is statistically uniform. (**b**) For each region, gather the dynamic cohorts of wells with $\geq i$ years on production, $i = 1, 2, 3, \ldots$. (**c**) Fit GEV probability density function (PDF) to each well cohort. (**d**) From the corresponding cumulative distribution function (CDF) pick the P_{10}, P_{50}, and P_{90} values for each cohort. (**e**) Construct time-lapse P_{50} well prototype for each region, by connecting all P_{50} values of all cohorts. (**f**) Time-shift and superpose the GEV well prototypes to match past production. Reproduced from our previous work [21].

2.2. Statistical Well Prototypes

For each of the 12 Bakken well samples defined in Section 2.1, we construct a statistical well prototype based on the Generalized Extreme Value (GEV, see Appendix A) mean or the P_{50} values of annual rates from all well cohorts that exist in that particular region. Figure 3 illustrates the procedure of arriving at the GEV well prototypes. For every GEV distribution fit we acquire values of the location parameter (μ), scale parameter (σ), shape parameter (ξ), confidence interval (CI), mean (P_{50}), median, mode, P_{10} and P_{90}. In this paper, we show only two GEV fit examples (Figures 4 and 5). Please see Supporting Online Materials-1 to find all GEV fits for all 12 regions of the Bakken shale. The resulting statistical well prototypes are plotted versus the square root of time on production in Figure 6. In the initial year on production, for some well cohorts, time was shifted by subtracting the first 1–3 months on production (see the second column of Table 2) to discount all possible initial disturbances in well performance. After these time shifts, each mean or P_{50} well prototype starts from a straight line versus the square root of time on production, as expected in linear flow regime. Later, cumulative oil production bends down due to the inter-hydrofracture pressure interference and exponential decline of production rate [23]. For our GEV distribution fits, the mean or P_{50} prototype is always higher than the median, and the median is higher than the mode. The P_{10} and P_{90} prototypes diverge from the mean with time, indicating that with time the best wells grow better and worst wells worse. A similar trend was also observed in the Barnett shale [21].

Table 1. Static regions for the Bakken shale play.

Well Sample	Reservoir	Area	Completion Date	Sample Size
1	Middle Bakken	Noncore	2000–2012	2550
2	Middle Bakken	Noncore	2013–2016	1355
3	Middle Bakken	Noncore	2017–2019	223
4	Three Forks	Noncore	2000–2012	735
5	Three Forks	Noncore	2013–2016	1204
6	Three Forks	Noncore	2017–2019	207
7	Middle Bakken	Core	2000–2012	2086
8	Middle Bakken	Core	2013–2016	2534
9	Middle Bakken	Core	2017–2019	1112
10	Three Forks	Core	2000–2012	428
11	Three Forks	Core	2013–2016	1502
12	Three Forks	Core	2017–2019	742
TOTAL				14,678

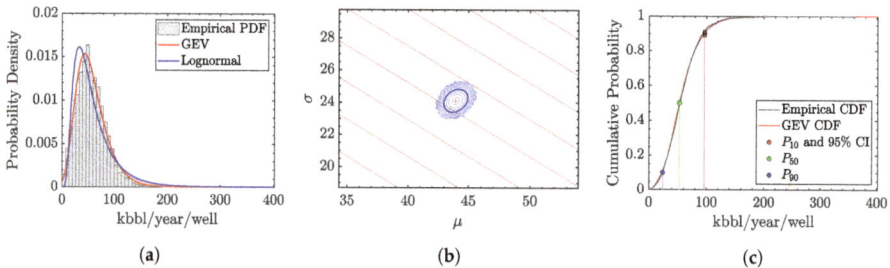

(a) (b) (c)

Figure 4. Distribution of oil production rates for 2540 horizontal wells with one year on production, completed between the year 2000 and 2012 in the Middle Bakken noncore area. (a) GEV PDF: $\xi = -0.0274$, $\mu = 43.9421$, $\sigma = 24.0635$. (b) Maximum Likelihood Estimate, 95% confidence interval (CI) for μ and σ. (c) GEV CDF with the 95% CI on the residual for the P_{10} well.

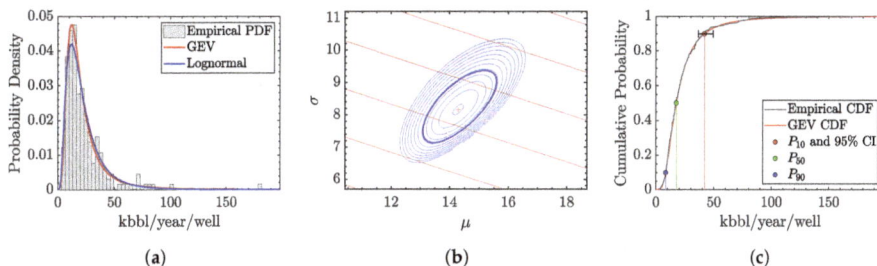

Figure 5. Distribution of oil production rates for 197 horizontal wells with seven years on production, completed between the year 2000 and 2012 in the Three Forks core area. (**a**) GEV PDF: $\xi = 0.3368$, $\mu = 14.2825$, $\sigma = 8.1499$. (**b**) Maximum Likelihood Estimate, 95% confidence interval (CI) for μ and σ. (**c**) GEV CDF with the 95% CI on the residual for the P_{10} well.

2.3. Physical Scaling Fits

For each region, we fit the statistical P_{50} well prototype with a physical scaling curve, see Appendix B, to extend this prototype to 30 years on production. The fitting procedure is detailed in the Materials and Methods section. Briefly, by least squares fit, we find optimum values of the scaling parameters τ and $\mathcal{M}$ so that the physics-based well prototypes match the statistical well prototypes. The results are shown in Figure 6. The red solid lines show the most pessimistic versions of each physics-based prototype, which assume that oil is produced only from the interior of the stimulated reservoir volume (SRV). The green solid lines show a more realistic forecast with an assumption that at late times there will be additional exterior flow of $a\sqrt{t}$ towards the SRV. Finally, one obtains the Estimated Ultimate Recovery (EUR) from each region by identifying the endpoint of cumulative production for the physics-scaling prototype. In Table 2, the scaling parameters, τ, $\mathcal{M}$ and a, and EUR values are listed for each region with and without exterior flow. Notice that in general the newest wells have the shortest pressure interference times, τ, and fastest decline rates.

Table 2. Scaling parameters for an average well in the Bakken regions.

Well Sample	Physical Scaling				Physical Scaling with Exterior Flow			
	Shift Years	τ Years	$\mathcal{M}$ kbbl/Well	EUR kbbl/Well	τ Years	$\mathcal{M}$	a	EUR kbbl/Well
Middle Bakken noncore [2000–2012]	0.25	24.5	454.3	271.1	19.9	286.7	0.050	282.2
Middle Bakken noncore [2013–2016]	0.00	11.8	341.3	211.7	10.9	267.2	0.026	236.4
Middle Bakken noncore [2017–2019]	0.00	5.0	346.9	215.5	5.0	306.6	0.015	259.5
Three Forks noncore [2000–2012]	0.15	13.3	288.4	207.5	10.7	181.2	0.058	238.3
Three Forks noncore [2013–2016]	0.00	10.8	271.9	196.1	9.9	212.4	0.030	221.1
Three Forks noncore [2017–2019]	0.00	5.0	270.8	195.4	5.0	238.4	0.018	236.5
Middle Bakken core [2000–2012]	0.20	25.0	810.8	482.5	22.6	536.3	0.050	512.2
Middle Bakken core [2013–2016]	0.00	9.1	543.8	337.7	8.2	420.7	0.026	389.7
Middle Bakken core [2017–2019]	0.00	5.0	581.9	361.5	5.0	513.6	0.015	434.6
Three Forks core [2000–2012]	0.27	25.0	681.1	470.9	25.0	473.4	0.058	511.6
Three Forks core [2013–2016]	0.00	10.3	450.1	324.6	9.6	353.1	0.030	369.7
Three Forks core [2017–2019]	0.00	5.0	462.8	334.0	5.0	406.9	0.018	403.5

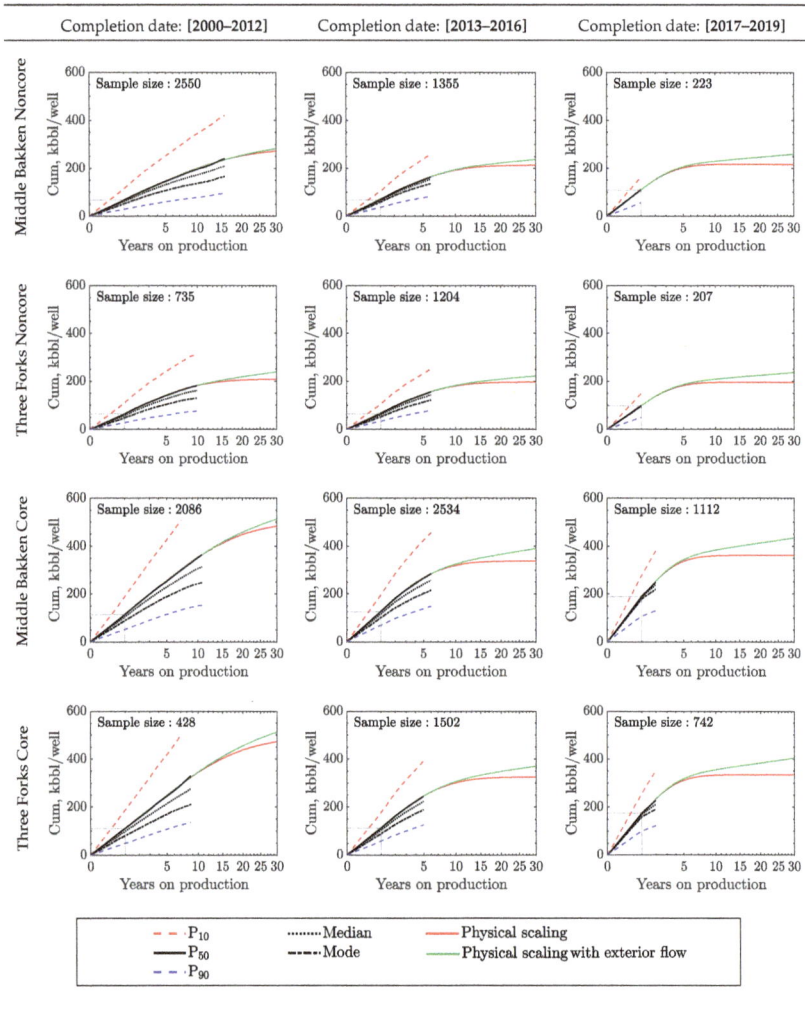

Figure 6. Average wells in 12 regions in the Middle Bakken and Three Forks. These wells are located in both the core and noncore areas and have three different completion periods, (2000–2012), (2013–2016) and (2017–2019). Each year in the past, every average well traces the expected values of the Generalized Extreme Value (GEV) distributions of all active horizontal wells in each well cohort, which have at least 1, 2, ..., 15 years on production. The dashed lines labeled P_{10} and P_{90} denote wells whose cumulative production is exceeded by 10% and 90% of wells in each region. The red and green lines are the physics-based scaling curves that match each average well, respectively, with and without exterior flow during late time production. In general, ultimate oil recovery from the core areas is higher than that from the noncore ones. The Middle Bakken wells are slightly more productive than the Three Forks wells. The newly completed wells have much higher initial oil production but they decline faster, resulting in more or less the same ultimate recovery as older wells.

2.4. Base or 'Do Nothing' Forecasts

We have replaced actual field production rate from all existing wells in each region with this region's well prototype multiplied by the historic numbers of completed wells, and we time-shifted the products. The superposed prototypes match the historical field rate. Figure 7 shows the sum of the well prototypes with and without exterior flows (green and red lines) versus the total field historical oil rate and cumulative oil (black lines). The results are rather satisfactory. Our GEV prototypes are robust. They capture all peaks of total historical oil rate and flawlessly match the cumulative production. As stated before, the red line forecast is pessimistic. It assumes that there will be no production contribution from the reservoir exterior to the SRVs of individual wells. This pessimistic case may hold if a shale play, here Bakken, is already overdrilled. For the Bakken shale, we can infer that the 14,678 existing wells will ultimately produce 4.5–5.3 billion of oil by 2050. This is a 'base' or 'do nothing' case with no future drilling in the Bakken shale.

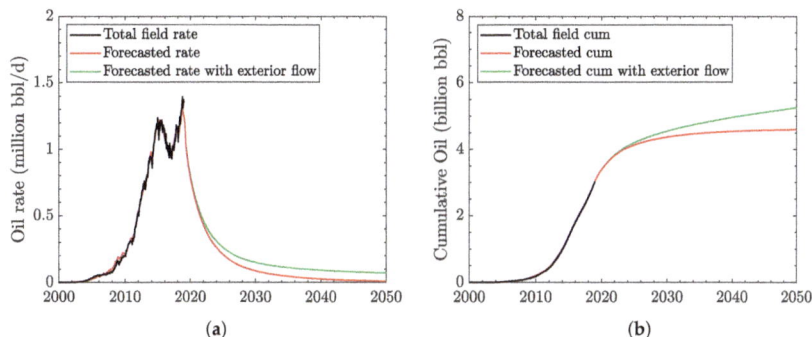

Figure 7. The actual and forecasted total field rate (**a**) and cumulative oil (**b**) in the Bakken shale. Total field cumulative production curves were obtained by stacking calendar-shifted average wells. Total production rates were obtained by differencing cumulative production. The red and green curves are the physical scaling forecasts with and without exterior flow, while the black line shows the historical production from the existing 14,678 wells. The red physical scaling gives the lower bound of EUR estimate with about 4.5 billion bbl by 2050. Assuming reasonable exterior flow, EUR prediction becomes slightly more than 5 billion bbl.

2.5. Infill Potentials

We have calculated the number of wells that can be drilled in the future in every one-square-mile pixel of the grid that covers the entire Bakken play. In order to calculate infill potentials, one should first determine well density. However, the publicly available data rarely provide information about the bottomhole locations of the wells. Instead, only surface locations are reported as latitude-longitude coordinates. Therefore, we have developed an algorithm that allows us to predict the bottomhole well density from surface well locations. See Figure 8 and the Materials and Methods section for more details. As mentioned before in Section 2.1, the Bakken shale has two reservoirs, the Middle Bakken and Three Forks and two significantly different reservoir qualities, the core and noncore areas. Thus the infill potentials should be calculated for these four parts of the Bakken. Supporting Online Materials-2 shows the gridding, well locations, wellhead densities, well densities and infill potentials for the four Bakken parts. Table 3 lists the numbers of wells that can potentially be drilled in the future. We note in passing that infilling the less productive noncore areas to the same well density as the core areas is likely too optimistic.

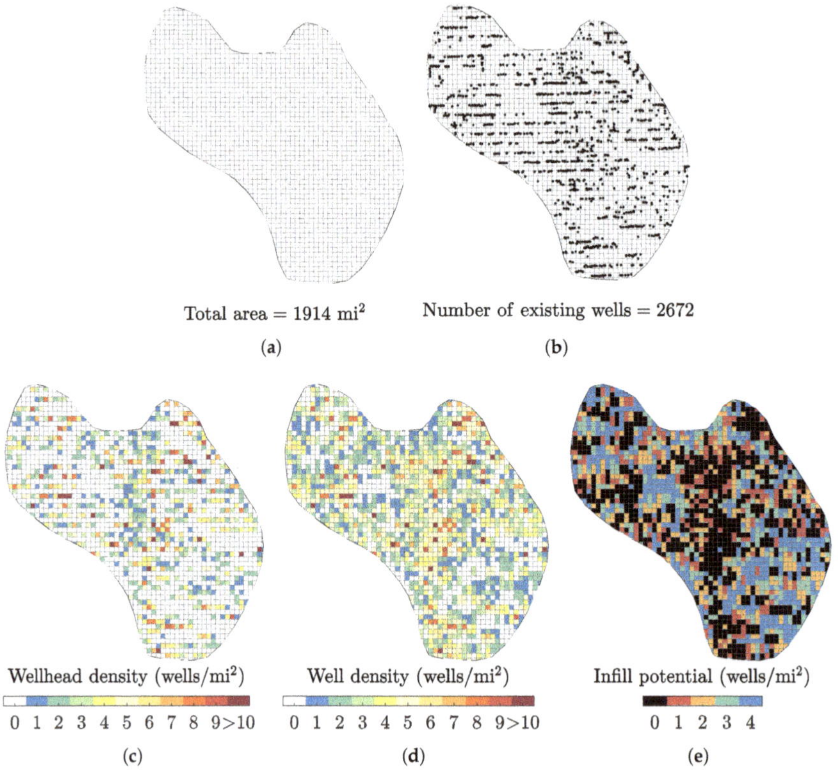

Total area = 1914 mi^2

(a)

Number of existing wells = 2672

(b)

Wellhead density (wells/mi^2)

0 1 2 3 4 5 6 7 8 9>10

(c)

Well density (wells/mi^2)

0 1 2 3 4 5 6 7 8 9>10

(d)

Infill potential (wells/mi^2)

0 1 2 3 4

(e)

Figure 8. A procedure to calculate infill well potential for each part of the Bakken play. This figure shows only the Three Forks core area. For other areas and reservoirs, please see the Supporting Online Materials. The procedure is as follows: (**a**) Create a one-square-mile fishnet inside the boundary of each area. In this case, there are 1914 grid squares that translate into the total area of 1914 mi^2. (**b**) Search all existing wells located inside the boundary. The black dots show the surface locations of 2672 existing wells in Three Forks core area. (**c**) Calculate wellhead density by counting the number of wells on each of the one-square-mile squares. This map is not the real well density map, because it only shows the wellhead density. (**d**) Calculate an approximate well density map from wellhead density map. The algorithm is as follows: (1) For each well, record its lateral length and calculate the number of squares, n, intercepted by the lateral. For example, a 5000 ft lateral will occupy one square and a 10,000 ft lateral will occupy two squares because one mile is 5280 ft. (2) Search for the least occupied n squares in all possible directions (i.e., north, northeast, east, southeast, south, southwest, west and northwest). (3) Increase the value of well density by 1 well/mi^2 for every least dense square found. (4) Repeat the process until all wells in the area of interest are exhausted. (**e**) Calculate an infill potential map by subtracting the calculated well density from the maximum number of wells, N_{max} (e.g., $N_{max} = 4$ wells/mi^2 to avoid frac hits). The summation of all values in the map is the infill potential for the one-square-mile grid. In this case, we obtain 3650 wells. As most wells have 10,000 ft laterals and occupy two one-square-mile squares, we divide 3650 by 2 to obtain 1825 infill potential wells in the Three Forks core area.

Table 3. Infill potential for Bakken region.

Reservoir	Region Type	Total Area sq Miles	Existing Wells	Infill Potential
Middle Bakken	Noncore	8534.5	4128	14,171 *
Three Forks	Noncore	6905.9	2146	12,341 *
Middle Bakken	Core	3382.5	5732	2577
Three Forks	Core	1914.1	2672	1825
TOTAL		20,737	14678	30,914

* Given the low reservoir quality in noncore areas, our infill potentials there are probably too high.

2.6. Future Drilling Scenarios and Infill Forecasts

We have created several future drilling scenarios based on infill potentials listed in Table 3. We assume that the rig count in the Bakken will not significantly change from the current value and 120 wells will be drilled each month between now and 2041 (Figure 9a). By looking at the current trend, many operators have narrowed their drilling choices to only the core areas to avoid spending money on the less productive wells with high watercuts typical of the noncore areas [22]. Thus, it is reasonable to create a drilling schedule that exhausts all potential wells in the core areas first, before moving out to the less productive noncore areas. Figure 9b shows this scenario predicting that the Bakken core area will be completely drilled out by 2022. Later, the drilling in noncore areas can last up to 2041, according to our calculation. See Supporting Online Materials-2 for the more detailed schedules.

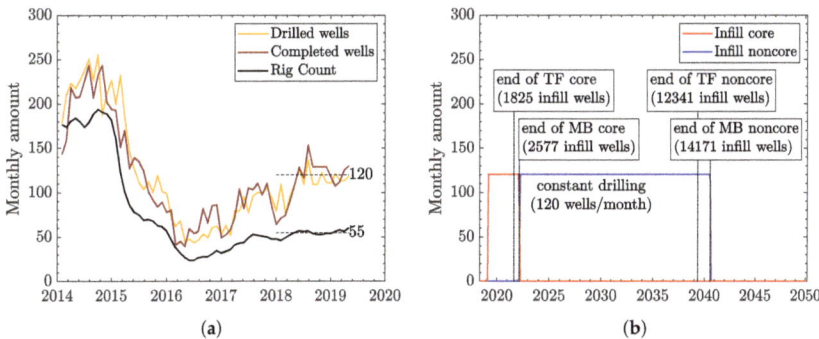

Figure 9. (**a**) The number of drilled wells, completed wells and rig count per month for the Bakken play from the U.S. Energy Information Administration (EIA). The drilled and completed wells are almost the same for each month and are strongly correlated with the number of rigs available. These data reveal an increase in drilling efficiency. In 2015, one rig could drill about 1.2 wells per month, while in 2019 so far, one rig could drill two wells per month. The current drilling rate is constant at about 120 wells/month. (**b**) A future drilling scenario for the Bakken region up to 2041. The plan is to continue at the current drilling rate of 120 wells/month for both core and noncore areas. The numbers of wells to be infilled in each region were previously calculated using the procedure detailed in Figure 8. The results show that the core areas will be fully drilled by mid 2022, leaving the less productive noncore areas for infilling until 2041.

Next, we assign each new well scheduled to be drilled in the future to its corresponding region and class prototype. The results are shown in Figure 10. The black solid lines are the historical total field oil rate and cumulative oil. The purple lines denote the 'base' forecast from all 14,678 existing wells displayed in Section 2.4. The red lines show the result of adding 4402 new wells in the core area. Because all wells in the core area are high-grade wells, the total oil production rate climbs up its highest production peak at 1.6 million bbl/d in the year 2022. After 2022, there is no further drilling in the core area and the production rate declines by a factor of two within one year. Ultimately, the total of 19,080 wells (existing + new core) will produce 7 billion bbl of oil by the year 2050. The blue lines show the most optimistic case with not only the core areas but also the noncore areas will be drilled out by the operators. The drilling of less productive wells in the noncore areas will maintain the plateau oil rate of 1 million bbl/d up to the year 2041. However, this plateau will require the drilling of additional 26,500 wells (almost twice the current number of wells) which will have high water cuts. Ultimately, the total of 45,580 wells (existing + core + noncore) will produce 13 billion bbl of oil by 2050.

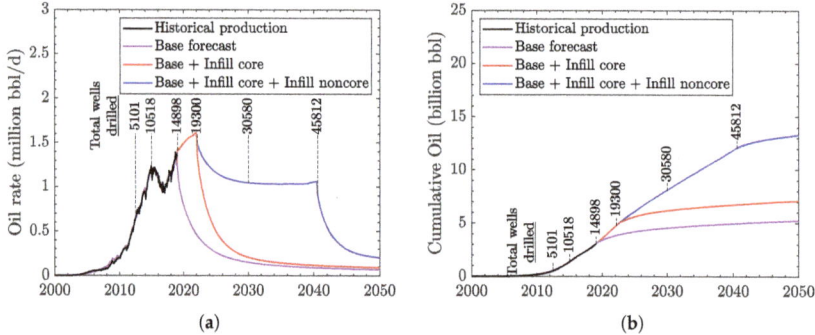

Figure 10. The forecasted total field rate (**a**) and cumulative oil (**b**) for Bakken based on the drilling scenario in Figure 9b. Since we plan to infill the core areas first, the field oil rate will reach the all-time production peak of about 1.6 million bbl/d by mid 2022, leaving no 'sweet spots' in the core areas. The continuous infill of the less productive wells in the noncore areas will decline the production to a plateau of 1 million bbl/d. After 2041, no more drilling locations will be left in the Bakken region and the oil rate will steeply decline to half a million bbl/d by 2042 and 0.2 million bbl/d by 2050. Ultimately, the 14,898 existing wells will give an EUR of 5 billion bbl. By adding 4402 wells in the core areas, EUR will increase to 7 billion bbl. By adding another 26,512 wells in the noncore areas, EUR will increase to 13 billion bbl.

3. Discussion

We have presented an alternative to the current industry-standard empirical forecasts of oil production from hydrofractured horizontal wells in shales. With our hybrid modeling approach, we have matched current oil production in the Bakken rather accurately. We have also delivered an optimal prediction of possible futures of the Bakken shale play for up to three decades.

Our Bakken oil forecasting method extends the previous work on predicting fieldwide gas production in the Barnett shale [21] and merges it with our new scaling of oil production in the Bakken [22]. Our field data-driven statistical well prototypes are conditioned by well attrition, hydrofracture deterioration, pressure interference, well interference, progress in technology, and so forth. With *no* physical scaling, these prototypes follow the exact physics of linear transient oil flow with pressure interference. Therefore these statistical well prototypes serve as templates to calibrate the parameters of our physical scaling model (τ and $\mathcal{M}$) [23,24] and obtain a *smooth* time-extrapolation of oil production that is mechanistic and not merely an empirical decline curve. At late times, we add to our extended prototypes some radial inflow from the outside of well SRVs [26].

The extended P_{50} well prototypes in Figure 6 can be used to compare ultimate recovery in each of the static regions we have identified in the Bakken. The lower bounds are the extended P_{50} prototypes without exterior flow (red lines). In most cases, wells completed in the upper Three Forks reservoir are somewhat less productive than those in the Middle Bakken reservoir. The reasons for this difference are: (1) higher water saturation and water cut in Three Forks, (2) faster decline rate (lower τ) and (3) lower initial oil in place (lower $\mathcal{M}$) [22]. This difference is consistent with the stratigraphic column of the Bakken total petroleum system, where Middle Bakken is sandwiched between two world-class source rocks with 10% TOC, the Upper Bakken Shale and the Lower Bakken Shale. On the other hand, the Three Forks formation is below the Lower Bakken Shale member and is exposed to water-bearing formations beneath [3,28–30]. For the same reasons, the *core* and *effective* drilling areas in the Three Forks are smaller than those in the Middle Bakken.

In both reservoirs, production from the *core* areas is superior to that from the *noncore* areas. The core area located in the center of Williston Basin has been known as the most oil-prolific location in the Bakken region in North Dakota [3,28–30]. Since the 1950s, oil has been produced there in the thickest, naturally fractured Middle Bakken formation in the Nesson anticline. One inexpensive vertical well drilled in the core area in the 1950s has had ultimate recovery of 200 kbbl, the same as a $10 million hydrofractured horizontal well drilled in the Middle Bakken noncore area nowadays. The noncore areas are less productive because the three Bakken formations: Upper, Middle and Lower Bakken are pinching out upward (are thinner and less mature) near the edges of the Williston basin. Consequently, the noncore areas are producing more water than oil, with watercut exceeding 50% on average.

The newly completed wells have much higher initial oil rates than the older ones, because they have: (1) longer lateral lengths, (2) bigger hydraulic fractures and (3) more fracture stages [22,31–34]. However, the newly completed wells decline faster and have essentially the same ultimate recovery as the older wells. The reasons for this behavior have been described elsewhere [35]. Interestingly, in most cases, older wells completed in 2000–2012 have higher ultimate recovery than newer ones, even though their initial production rates are lower. These older wells might have been drilled in the best locations ever in the Bakken region. In addition, shorter lateral lengths and fewer fracture stages may help in maintaining a stable pressure drawdown and prevent reservoir degassing that is unfavorable for future production. For comparison, the average lateral length in 2005 was 5000 ft while the average lateral length in 2019 doubled to 10,000 ft. Historically, the number of hydrofracture stages in the Bakken has increased over time from 8 stages in 2007 [36] to 18 stages in 2009 [31], 35 stages in 2016 [32] and to as many as 60 stages in 2019 [33].

According to our records, more than 90% of the wells completed after 2017 are located in the core areas only. Operators have learned to drill only the best parts of the Williston Basin and avoid the less mature noncore areas. However, after calculating the infill potentials of all areas, we predict that by 2021 there will be no well locations left for future drilling in the core areas. Assuming a constant current drilling rate of 120 wells per month, the total field oil rate in the Bakken will reach record level of about 1.6 million bbl/d in 2021. Without further drilling, production will decline by one-half within a year. Later, operators will be forced to drill in the less productive, high watercut noncore areas along the edges of the Williston Basin. Our findings suggest that policy-makers should not assume that the shale oil boom in the Bakken will last for several decades longer. We recommend that operators not focus only on increasing the initial oil rate. Maintenance of reservoir pressure above the bubble point by preventing over-drilling is key to increasing ultimate oil recovery.

4. Materials and Methods

We mined public well data in the Bakken region from DrillingInfo database, the Montana Department of Natural Resources and Conservation–Board of Oil and Gas Conservation website and the North Dakota Department of Mineral Resources–Oil and Gas Division website. From DrillingInfo, we selected only 14,678 horizontal oil wells that were completed in the Middle Bakken and Three Forks formations after January 2000. We designed an integrated MATLAB® software package to perform data cleanup, consolidation and unit conversions. The clean production data for each well consist of a vector of elapsed times on production, vector of oil and water production rates and records of wellhead location (latitude and longitude), lateral length, completion date, reservoir description and maximum oil rate. The average reservoir and well lateral properties are listed in Appendix C. Our approach can be divided into six main steps:

I.   Define play regions in which oil production is statistically uniform. In each region, follow the steps below:

(a)   Divide all wells in a play into $i = 1, 2, \ldots, n$ well groups, where n is the number of reservoirs. In the Bakken play, for example, there are two main reservoirs. Thus, at this step, we identify two groups of wells,

Well Sample	Reservoir	Sample Size
1	Middle Bakken	9860
2	Three Forks	4818

and $n = 2$ in the Bakken shale.

(b)   Further subdivide each of these well groups into j_i, $i = 1, 2, \ldots, n$, areas with different reservoir qualities. For example, in the Bakken play, the center of the basin has the thickest oil-prolific layer and hence the highest oil production [3,22,28–30]. Thus, we have delineated an area at the center of the basin (with maximum oil rate >750 bopd) as *core* area and the rest as *noncore* area, see Figure 2 for more details. At this step, we have four well groups that fall into four distinct, static play regions,

	Reservoir	Region	Sample Size
1	Middle Bakken	core	5732
2	Middle Bakken	noncore	4128
3	Three Forks	core	2672
4	Three Forks	noncore	2146

In Bakken, $j_1 = 2$ and $j_2 = 2$, that is, there are two reservoir qualities (core and noncore) for each of the reservoirs (Middle Bakken and Three Forks).

(c)   Subdivide wells in each of the four *regions* (two reservoirs × two reservoir qualities each) by time interval *classes* that encompass significant changes in well completion technology. In the Bakken play, for instance, the newly completed wells have longer lateral lengths, bigger hydraulic fractures and more fracture stages [22,31–34]. Thus, we classify the wells in each of the four regions by three completion date intervals: [2000–2012], [2013–2016] and [2017–2019]. In the end, we have divided all 14,678 horizontal wells in the Bakken into 12 well groups (4 regions × 3 completion date classes) listed in Table 1. In Bakken, each of the 4 play regions has 3 completion classes, finally yielding 12 static well *samples*.

II.   For each well sample, obtain a P_{50} well prototype by fitting a Generalized Extreme Value (GEV) distribution to all qualifying sample wells as follows:

(a)   From a given static well sample-k (a region further subdivided by completion dates), consider a dynamic cohort l_k that contains all wells that have *at least* l_k years on production ($l_k = 1, 2, \ldots, t_{\max_k}$ and $k = 1, 2, \ldots, N_{\text{sample}}$). For example: (i) There are 2550 wells in *Middle Bakken noncore [2000–2012]* group. However, there are only 2540 wells with at least one year on production (the other 10 wells have production records with less than 12 months). Thus, we retain these 2540 wells as cohort-1 of this particular group, see Figure 4. (ii) There are 428 wells in *Three Forks core [2000–2012]* group. But, only 197 wells have production records of at least seven years. As such, these 197 wells are qualified as cohort-7 of this particular well group, see Figure 5. For detailed GEV fits for all dynamic well cohorts and static well groups in the Bakken play, see Supporting Online Materials-1. For Bakken, $N_{\text{sample}} = 12$.

(b)   Define a set X_{l_k} that takes values of oil production rate (kbbl/year) from all wells with at least l_k years on production in sample k.

(c)   For every X_{l_k}, fit a Generalized Extreme Value (GEV) distribution using Equation (A1) and obtain the location parameter, μ, scale parameter, σ and shape parameter, ξ. (All of these parameters are further subscripted by l_k but we will skip this complication in the notation.)

(d) Calculate $\bar{x}_{l_k}$ as the GEV mean of X_{l_k} for each year-l_k, $l_k = 1, 2, \ldots, t_{max_k}$, using Equation (A3).

(e) For each well sample k, obtain the P_{50_k} well prototype. We can use the same procedure to obtain other statistic parameters for each well cohort. The P_{10_k} and P_{90_k} values can be calculated as the ninetieth and tenth percentiles of X_{l_k}, respectively. The median and mode can be calculated using Equations (A5) and (A6). by connecting GEV mean values of all years $l_k = 1, 2, \ldots, t_{max_k}$

$$P_{50_k} = (\bar{x}_1, \bar{x}_2, \ldots, \bar{x}_{t_{max_k}}), k = 1, 2, \ldots, N_{\text{sample}}, \quad \text{kbbl/year}. \tag{1}$$

III. Extend P_{50_k} (we will subsequently skip the implied subscript k) well prototypes by fitting to them physical scaling curves as follows:

(a) For a each well sample k, calculate the *observed* cumulative mass produced, **m** (ktons) from Equation (A13). Fix $q_{o,ST} = P_{50_k}$ in kbbl/year and $\Delta t_i = 1$ year.

(b) Adjust τ in Equation (A22) or Equation (A7) (with or without exterior flow) and $\mathcal{M}$ in Equation (A11), so that the physical scaling curve matches the observed **m**. Use $c_t/S_{oi}(p_i - p_f)$ and a values from the average Bakken reservoir properties listed in Tables A1 and A2. The matching values of τ and $\mathcal{M}$ for two both scaling curves and for all well samples are detailed in Table 2.

(c) Get the extended cumulative mass produced, **m̂** (ktons) by multiplying the fitted $\mathcal{M}$ and $RF(t/\tau)$ from the master curve where t is now calculated each month The benefit of matching P_{50} with a physics-based scaling curve is that we can interpolate and extrapolate production data precisely. Thus, we can change time intervals from years to months and forecast production decades into the future. We recommend to use monthly intervals for precise forecasts that is, $t = \frac{1}{12}, \frac{2}{12}, \ldots, 50$ years.

(d) Obtain the extended P_{50} well prototypes for well sample k by differencing **m̂** converted to volume

$$\hat{P}_{50} = \frac{\Delta \hat{\mathbf{m}}}{\Delta t (\rho_{o,ST} + R_s \rho_{g,ST})}, \quad \text{kbbl/month} \tag{2}$$

IV. Obtain base forecast of total field oil production for existing wells as follows:

(a) Create a calendar date series with monthly intervals for example, (1/1/2000, 2/1/2000, ..., 12/1/2050) and assign N_{dates} as the length of this series.

(b) Create an empty vector **s** with the length of N_{dates}: ($\mathbf{s} = [s_1 \ s_2 \ \cdots \ s_{N_{\text{dates}}}] = [0 \ \cdots \ 0]$).

(c) For each well, find index ii in the calendar date series that brackets well completion date.

(d) For each well, add to vector **s** its corresponding $\hat{P}_{50}$ right-shifted by ii.

$$\mathbf{s} = \mathbf{s} + [\underbrace{0 \quad \cdots \quad 0}_{ii \text{ zeros}} \ \hat{P}_{50}(1) \quad \cdots \quad \hat{P}_{50}(N_{\text{dates}} - ii + 1)] \tag{3}$$

(e) Calculate total field oil rate, **q** and total field cumulative oil, **Q** for existing wells as follows:

$$\mathbf{q} = \mathbf{s}/(365.25/12) \times 10^{-3} \quad \text{million bbl of oil per day} \tag{4}$$

$$\mathbf{Q} = \begin{bmatrix} s_1 & s_1 + s_2 & \cdots & \sum_{i=1}^{N_{\text{dates}}} s_i \end{bmatrix} \times 10^{-6} \quad \text{billion bbl} \tag{5}$$

V. Calculate infill potential as follows:

(a) Create a one-square-mile fishnet inside the boundary of each area defined in step I(b).

(b) Search all existing wells located inside the boundary.

(c)   Calculate wellhead density by counting the number of wells on each of the one-square-mile squares.

(d)   Calculate approximate well density following algorithm below.

        i.   For each well, calculate the number of squares, n, intercepted by the lateral For example, a 5000 ft lateral occupies one square, a 9000 ft lateral occupies two squares and a 14,000 ft lateral occupies three squares because one mile is 5280 ft.

        ii.   Search for the least-occupied n squares in all possible directions.

        iii.   Increase the value of well density by 1 well/mi² for every least dense n square found.

        iv.   Repeat points i–iii until all wells in the area of interest are exhausted.

(e)   Calculate infill potential by subtracting the calculated well density from the maximum allowable number of wells that still avoid frac hits. For example, in the Bakken play, the tip-to-tip hydraulic fracture length is roughly 1200 ft. 5280 ft/1200 ft $\approx$ 4. Therefore, the maximum allowable number of wells to avoid frac hits is 4 wells per square mile. Suppose that at some location, the well density already is 3 wells per square mile. Then, the infill potential is $4 - 3 = 1$ well per square mile.

VI.   Obtain an infill forecast of total field oil production after adding future wells.

(a)   Create monthly drilling schedules for every infill potential and assume a constant drilling rate based on current rig availability, see Figure 9b and SOM-2.

(b)   Use the same calendar date series as in step IV(a) with N_{dates} as the length of date series.

(c)   Create an empty vector $\mathbf{s_f}$ with the length of N_{dates} ($\mathbf{s_f} = [s_{f,1}\ s_{f,2} \cdots s_{f,N}] = [0 \cdots 0]$).

(d)   For every drilling schedule, find index ii in the calendar date series that brackets infill schedule date. Let N_f be the number of wells to be drilled on that date.

(e)   For every drilling schedule, add to vector $\mathbf{s_f}$ its corresponding $\hat{P}_{50}$ after right-shifting it by ii and multiplying by N_f.

$$\mathbf{s_f} = \mathbf{s_f} + N_f \times [\underbrace{0 \quad \cdots \quad 0}_{ii\ \text{zeros}} \quad \hat{P}_{50}(1) \quad \cdots \quad \hat{P}_{50}(N_{dates} - ii + 1)] \tag{6}$$

(f)   Calculate total field oil rate, $\mathbf{q_f}$ and total field cumulative oil, $\mathbf{Q_f}$ after infill drilling as follows

$$\mathbf{q_f} = \mathbf{q} + \mathbf{s_f}/(365.25/12) \times 10^{-3} \text{ million bbl of oil per day} \tag{7}$$

$$\mathbf{Q_f} = \mathbf{Q} + \left[s_{f,1} \quad s_{f,1} + s_{f,2} \quad \cdots \quad \sum_{i=1}^{N_{dates}} s_{f,i} \right] \times 10^{-6} \text{ in billion bbl} \tag{8}$$

where $\mathbf{q}$ and $\mathbf{Q}$ are total field rate and field cumulative from existing wells calculated before with Equations (4) and (5).

5. Conclusions

- We have provided a transparent hybrid method of forecasting oil production at shale basin scale.
- Our statistical approach generates the non-parametric well prototype templates that are used to calibrate our physics-based flow scaling with late-time radial inflow.
- In particular, our average P_{50} well prototypes follow the physics of linear transient flow and are used to calibrate the physics-based scaling extensions to 30 years on production.
- A combination of GEV statistics with physical scaling matches historical production data almost perfectly and gives a smooth, physics-based estimate of future production.

- Our prediction of the Bakken future is optimal in the least square sense [37–40]; in other words, our prediction is as good as it gets given all data at hand and first order physics of oil recovery in the Bakken shale.
- Regulators may want to consider our approach as a prerequisite to booking reserves in oil shales.
- Newly completed wells have almost the same ultimate recovery as the older ones, despite their much higher initial oil rates.
- Ultimately, we predict that the 14,678 existing wells in the Bakken will produce 5 billion bbl of oil by 2050 (∼340 kbbl/well).
- After drilling additional 4400 new wells at the rate of 120 wells/month, the core area of the Bakken will be drilled out by 2021 and ultimate recovery will be 7 billion barrels of oil.
- With 26,500 more wells to be drilled in the noncore area until 2041, ultimate recovery in the Bakken might be 13 billion barrels of oil but drilling of such scale is unlikely to happen.
- Policy-makers should not of assume that oil boom in the Bakken shale will last decades longer.

Supplementary Materials: The following are available online at http://www.mdpi.com/1996-1073/12/19/3641/s1. There are two files. The first one shows full details of infill potential maps for every sub-region in the Bakken shale and a table listing drilling schedule program until year 2050. The second file shows 225 unique plots showing all the GEV distributions used to match all well cohorts in the Bakken shale.

Author Contributions: Conceptualization, T.P. and W.S.; methodology, T.P. and W.S.; software, W.S. and T.P.; validation, T.P. and W.S.; formal analysis, W.S. and W.K.; investigation, W.K. and W.S.; resources, T.P.; data curation, W.K.; writing–original draft preparation, W.S.; rewriting manuscript–review and editing, T.P. and W.S.; visualization, W.S.; supervision, T.P. and W.K.; project administration, T.P.; funding acquisition, T.P.

Funding: Wardana Saputra (PhD student) was supported by baseline research funding from KAUST to Tad Patzek. Wissem Kirati (Research Engineer) was supported by the Division of Computer, Electrical and Mathematical Science at KAUST.

Acknowledgments: The authors thank the Ali I. Al-Naimi Petroleum Engineering Research Center (ANPERC) at KAUST for supporting this research. We thank the reviewers for their thorough, informative and timely reviews.

Conflicts of Interest: The authors declare no conflict of interest.

Abbreviations

The following abbreviations are used in this manuscript:

CDF	Cumulative Distribution Function
EUR	Estimated Ultimate Recovery
GEV	Generalized Extreme Value
PDF	Probability Density Function
SRV	Stimulated Reservoir Volume

Appendix A. The Generalized Extreme Value (GEV) Distributions

The Generalized Extreme Value (GEV) distribution [41] combines three "Extreme Value" distributions, Weibull [42], Gumbel [43] and Frechet [44] into a single functional form. The data decide which of the three distributions is appropriate. In our case, Fréchet distribution almost always wins.

The probability density function (PDF) of GEV distribution contains three parameters: location, μ; scale, σ; and shape, ξ

$$f(x) = \frac{1}{\sigma} t(x)^{\xi+1} e^{-t(x)}, \text{ where } t(x) = \begin{cases} \left(1 + \xi\left(\frac{x-\mu}{\sigma}\right)\right)^{-1/\xi} & : \text{if } \xi \neq 0 \\ e^{-(x-\mu)/\sigma} & : \text{if } \xi = 0 \end{cases} \quad \text{(A1)}$$

If ξ is zero, Gumbel distribution results, followed by Weibull distribution, if ξ is negative and by Fréchet distribution if ξ is positive (Figure A1).

Integrating the GEV PDF, one obtains the cumulative distribution function (CDF)

$$F(x) = \exp(-t(x)) \tag{A2}$$

The expected value (mean) of GEV distribution is defined as

$$E(X) = \begin{cases} \mu + \sigma(g_1 - 1)/\xi & : \text{if } \xi \neq 0 \text{ and } \xi < 1, \\ \mu + \sigma \gamma & : \text{if } \xi = 0, \\ \infty & : \text{if } \xi \geq 1 \end{cases} \tag{A3}$$

and the variance is

$$V(X) = \begin{cases} \sigma^2 (g_2 - g_1^2)/\xi^2 & : \text{if } \xi \neq 0 \text{ and } \xi < \frac{1}{2}, \\ \sigma^2 \frac{\pi^2}{6} & : \text{if } \xi = 0, \\ \infty & : \text{if } \xi \geq \frac{1}{2} \end{cases} \tag{A4}$$

where $g_k = \Gamma(1 - k\xi)$ and γ is Euler's constant. Notice that for $\xi \geq 1$, the expected value becomes infinite and for $\xi \geq \frac{1}{2}$ the variance goes to infinity.

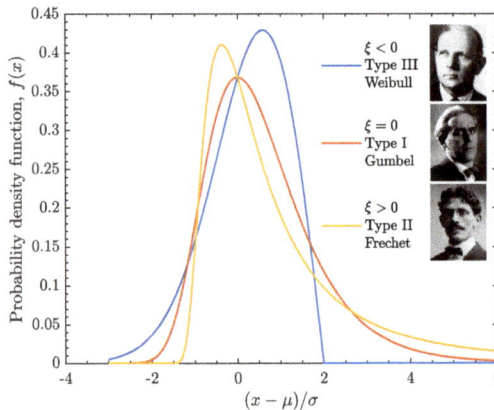

Figure A1. Examples of three GEV distributions with the location parameter, $\mu = 0$, the scale parameter, $\sigma = 1$. and three values of the shape parameter i.e., $\xi < 0$ (Weibull), $\xi = 0$ (Gumbel) and $\xi > 0$ (Fréchet).

The median of GEV distribution is defined as

$$\text{Median}(X) = \begin{cases} \mu + \sigma \frac{(\ln 2)^{-\xi} - 1}{\xi} & : \text{if } \xi \neq 0, \\ \mu - \sigma \ln(\ln 2) & : \text{if } \xi = 0 \end{cases} \tag{A5}$$

and the mode is

$$\text{Mode}(X) = \begin{cases} \mu + \sigma \frac{(1+\xi)^{-\xi} - 1}{\xi} & : \text{if } \xi \neq 0, \\ \mu & : \text{if } \xi = 0 \end{cases} \tag{A6}$$

Appendix B. Physical Scaling Approach to Forecasting Oil Production in Hydrofractured Shales

The physical scaling approach used in this paper was first derived by Patzek et al. to predict gas production from thousands of hydrofractured horizontal gas wells in the Barnett shale [23,24]. Later, Patzek et al. derived the scaling curve solution for oil production in the Eagle Ford shale [25].

Eftekhari et al. extended the model to include gas inflow from outside of SRVs for the Barnett shale wells [26].

Appendix B.1. Physical Scaling without Exterior Flow

Patzek et al. [25] derived the solution of oil production inside SRV based on a simple model illustrated in Figure A2a. This model assumes bi-linear flow towards hydraulic fracture planes inside SRV with the volume $H \times 2d \times 2L$. Briefly, by solving a one-dimensional nonlinear pressure diffusion equation analytically, we obtain a master curve equation for the *interior* oil production problem

$$\mathrm{RF}_\mathrm{I}(\tilde{t}) = \frac{c_t}{S_{oi}}(p_i - p_f)\left(1 - \frac{8}{\pi^2}\sum_{n=0}^{\infty}\frac{1}{(2n+1)^2}\,\mathrm{e}^{(-(2n+1)^2\pi^2\tilde{t}/4)}\right) \tag{A7}$$

where $\tilde{t}$ is the dimensionless time defined as the ratio between elapsed time on production, t and the characteristic time of pressure diffusion, τ.

$$\tilde{t} = \frac{t}{\tau} \tag{A8}$$

Here τ is determined by the onset of pressure interference between two hydrofractures $2d$ apart

$$\tau = \frac{d^2}{\alpha_i} \tag{A9}$$

and α_i is the constant hydraulic diffusivity at initial reservoir conditions

$$\alpha_i = \frac{k}{\phi c_t \mu_{oi}}\frac{\rho_{oi}}{\rho_{\text{fluid},i}} \tag{A10}$$

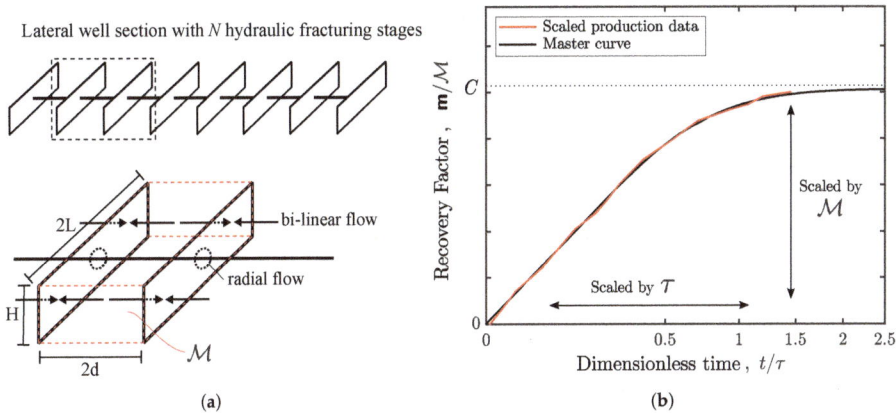

Lateral well section with N hydraulic fracturing stages

2L — bi-linear flow

radial flow

H

2d — $\mathcal{M}$

(a)

Recovery Factor, $\mathbf{m}/\mathcal{M}$

Scaled production data
Master curve

$\mathcal{C}$

Scaled by $\mathcal{M}$

Scaled by $\mathcal{T}$

Dimensionless time, t/τ

(b)

Figure A2. (a) Schematic of bi-linear flow towards hydraulic fractures in a shale well inside the SRV with volume $H \times 2d \times 2L$ (*interior* problem). The early radial flow is neglected because the hydrofracture permeability is much higher than that of the matrix. Reproduced from Patzek et al. [25] (b) Illustration of physical scaling approach of the interior oil production problem. Reproduced from Saputra et al. [27]

The fractional oil recovery factor, RF, is the ratio between the cumulative oil mass production, $\mathbf{m}$ and the stimulated mass contained in the SRV, $\mathcal{M}$

$$\mathrm{RF}(\tilde{t}) = \frac{\mathbf{m}}{\mathcal{M}} \tag{A11}$$

where

$$\mathcal{M} = \rho_{oi} V = \rho_{oi} A_f (2d) \phi S_{oi} \tag{A12}$$

and

$$\mathbf{m} = \left(\rho_{o,ST} + R_s \rho_{g,ST} \right) \left(\sum_{k=1}^{i} q_{o,ST}(t_k) \Delta t_k \right)_{i=1}^{i=t_{max}} \tag{A13}$$

The physical scaling approach is as follows. By adjusting τ and $\mathcal{M}$, minimize the square of the difference of $\mathcal{M}RF(\tilde{t})$ minus the cumulative mass produced with Equation (A13). A least square fit procedure was used. Figure A2b shows the master curve and scaled cumulative mass produced as the black and red lines.

Appendix B.2. Physical Scaling with Exterior Flow

Eftekhari et al. [26] solved numerically the *exterior* gas production problem in the Barnett shale. We later extended the same concept to oil shales and proposed an analytical solution of this problem. Figure A3a illustrates interior flow towards the hydraulic fractures inside SRV and exterior flow from the outer reservoir towards SRV. We write a new master curve equation for both interior and exterior flow as follows

$$RF_T(\tilde{t}) = RF_I(\tilde{t}) + RF_E(\tilde{t}) \tag{A14}$$

where $RF_I(\tilde{t})$ is the master curve equation for interior oil production problem (Equation (A7)). Eftekhari et al. [26] defined the recovery factor of the exterior problem RF_E as a function of the new scaled time, $\tilde{t}' = t/\tau'$ where $\mathbf{m}_E(\tilde{t}')$ is the cumulative exterior production and $\mathcal{M}$ is the stimulated mass inside SRV defined in Equation (A12)

$$RF_E(\tilde{t}') = \frac{\mathbf{m}_E(\tilde{t}')}{\mathcal{M}} \tag{A15}$$

Here τ' is the characteristic time scale for exterior production, defined as the time it takes low pressure to diffuse from the wellbore over a distance equal to fracture half-length L

$$\tau' = \frac{L^2}{\alpha_i} = \frac{d^2}{\alpha_i} \left(\frac{L}{d} \right)^2 = \tau \left(\frac{L}{d} \right)^2 \tag{A16}$$

The new scaled time for exterior production, $\tilde{t}'$ can be written as a function of the interior scaled time, $\tilde{t}$ in Equation (A8):

$$\tilde{t}' = \frac{t}{\tau'} = \frac{t}{\tau} \left(\frac{d}{L} \right)^2 = \tilde{t} \left(\frac{d}{L} \right)^2 \tag{A17}$$

Because oil flow from the exterior reservoir is another one-dimensional pressure diffusion problem, one can assume the square root of time solution [26] with a constant slope c_1

$$RF_E(\tilde{t}') \approx c_1 \sqrt{\tilde{t}'} \tag{A18}$$

Because both the interior and exterior reservoir likely contain similar shale matrix with similar reservoir properties, the constant c_1 can be assumed to be the slope of the interior master curve versus square-root of time (Equation (A7) and Figure A2b), which is constant at early times, say, $0 < \tilde{t}^* < 0.2$.

$$c_1 = \frac{4c_t \sqrt{\tilde{t}^*}}{S_{oi}} (p_i - p_f) \sum_{n=0}^{\infty} e^{(-(2n+1)^2 \pi^2 \tilde{t}^*/4)} = 1.1284 \frac{c_t}{S_{oi}} (p_i - p_f) \tag{A19}$$

RF_E can be written as a function of the scaled time for the interior production by substituting Equations (A17) and (A18):

$$RF_E(\tilde{t}) = c_1 \frac{d}{L} \sqrt{\tilde{t}} = a\sqrt{\tilde{t}} \tag{A20}$$

where a is defined as

$$a = 1.1284 \left(\frac{d}{L}\right) \frac{c_t}{S_{oi}} (p_i - p_f) \tag{A21}$$

Recalling Equation (A14), we obtain a new master curve equation that embodies interior and exterior oil flow as follows

$$\text{RFT}(\tilde{t}) = \frac{c_t}{S_{oi}}(p_i - p_f)\left(1 - \frac{8}{\pi^2}\sum_{n=0}^{\infty}\frac{1}{(2n+1)^2}e^{(-(2n+1)^2\pi^2\tilde{t}/4)}\right) + a\sqrt{\tilde{t}} \tag{A22}$$

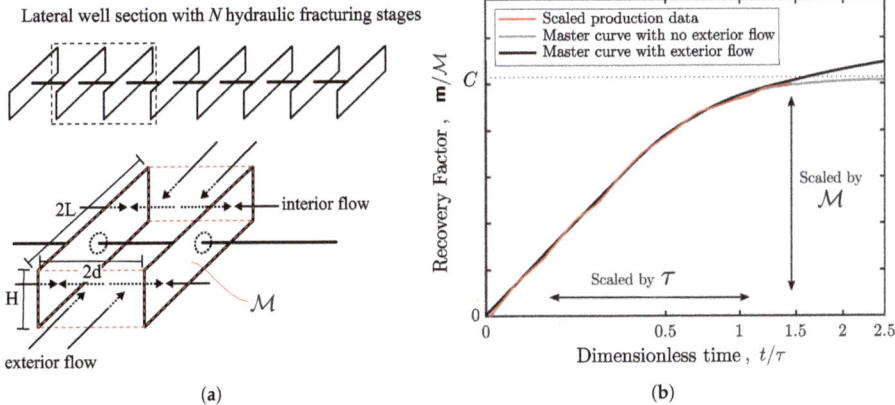

(a)   (b)

Figure A3. (a) Illustration of *interior* flow towards the hydraulic fractures of a shale well inside SRV and *exterior* flow from the reservoir beyond SRV. Adapted from Patzek et al. [25] and Eftekhari et al. [26] (b) Illustration of physical scaling approach of the interior and exterior oil production problem. Adapted from Saputra et al. [27]

The physical scaling approach for this problem is as follows. By adjusting τ and $\mathcal{M}$, minimize the square of the difference of $\mathcal{M}\text{RFT}(\tilde{t})$ minus the cumulative mass produced with Equation (A13). A least square fit procedure was used. Figure A3b shows the scaled cumulative mass as the red line and the new master curve as the black line that deviates from the interior master curve in Equation (A7) (shown as the gray line) at late times on production.

Appendix C. Reservoir Properties of the Bakken Shale

Table A1. Reservoir properties used in scaling oil production in the Middle Bakken and Three Forks (part-1).

Parameter	Middle Bakken		Three Forks		Data Source
	SI Units	Field Units	SI Units	Field Units	
Initial pressure, p_i	36.8 Mpa	5340 psia	37.1 Mpa	5380 psia	[22]
Fracture pressure, p_f	3.4 Mpa	500 psia	3.4 Mpa	500 psia	[22]
Connate water saturation, S_{wc}	0.57	0.57	0.65	0.65	[22]
Initial oil saturation, S_{oi}	0.43	0.43	0.35	0.35	[22]
Rock porosity, ϕ	0.046	0.046	0.058	0.058	[22]
Rock permeability, k	4.4×10^{-17} m^2	0.045 md	4.6×10^{-17} m^2	0.047 md	[22]
Rock compressibility, c_ϕ	4.3×10^{-10} Pa^{-1}	3.0×10^{-6} psi^{-1}	4.3×10^{-10} Pa^{-1}	3.0×10^{-6} psi^{-1}	[45]
Water compressibility, c_w	4.3×10^{-10} Pa^{-1}	3.0×10^{-6} psi^{-1}	4.3×10^{-10} Pa^{-1}	3.0×10^{-6} psi^{-1}	[45]
Oil compressibility, c_o	1.4×10^{-9} Pa^{-1}	1.0×10^{-5} psi^{-1}	1.4×10^{-9} Pa^{-1}	1.0×10^{-5} psi^{-1}	[45]
Total compressibility, c_t	6.3×10^{-10} Pa^{-1}	9.0×10^{-6} psi^{-1}	5.9×10^{-10} Pa^{-1}	8.5×10^{-6} psi^{-1}	
Oil viscosity, $\mu_{o,i}$	3.9×10^{-5} Pa s	0.392 cp	2.8×10^{-5} Pa s	0.276 cp	[46]
API gravity	42.2°API	42.2°API	38.7°API	38.7°API	[46]
$(c_t/S_{oi})(p_i - p_f)$	0.1014	0.1014	0.1178	0.1178	

Table A2. Reservoir properties used in scaling oil production in the Middle Bakken and Three Forks (part-2).

Completion Date	Number of Frac Stages, N	2d (ft)	$\frac{d}{L}$	a	
				Middle Bakken	Three Forks
2000–2012	18 [31]	528	0.44	0.050	0.058
2013–2016	35 [32]	271	0.23	0.026	0.030
2017–2019	60 [33]	158	0.13	0.015	0.018

References

1. Arps, J.J. Analysis of decline curves. *Trans. AIME* **1945**, *160*, 228–247. [CrossRef]
2. Zuo, L.; Yu, W.; Wu, K. A fractional decline curve analysis model for shale gas reservoirs. *Int. J. Coal Geol.* **2016**, *163*, 140–148. [CrossRef]
3. Gaswirth, S.B.; Marra, K.R.; Cook, T.A.; Charpentier, R.R.; Gautier, D.L.; Higley, D.K.; Klett, T.R.; Lewan, M.D.; Lillis, P.G.; Schenk, C.J. Assessment of undiscovered oil resources in the Bakken and Three Forks formations, Williston Basin Province, Montana, North Dakota, and South Dakota, 2013. *US Geol. Surv. Fact Sheet* **2013**, *3013*, 1–4.
4. Cook, T.A. *Procedure for Calculating Estimated Ultimate Recoveries of Bakken and Three Forks Formations Horizontal Wells in the Williston Basin*; US Geol. Surv. Open-File Report; U.S. Geological Survey: Reston, VA, USA, 2013; pp. 1–14.
5. EIA. Assumptions to the Annual Energy Outlook 2019: Oil and Gas Supply Module. *US Energy Inf. Adm. Annu. Energy Outlook* **2019**, *7*, 4–7.
6. Hughes, J. A reality check on the shale revolution. *Nature* **2013**, *494*, 307–308. [CrossRef] [PubMed]
7. Hughes, J. *2016 Tight Oil Reality Check*; Post Carbon Institude: Santa Rosa, CA, USA, 2016.
8. Hughes, J. *Shale Reality Check*; Post Carbon Institude: Santa Rosa, CA, USA, 2018.
9. Hughes, J. *How Long Will the Shale Revolution Last? Technology Versus Geology and the Lifecycle of Shale Plays*; Post Carbon Institude: Santa Rosa, CA, USA, 2019.
10. Ilk, D.; Anderson, D.M.; Stotts, G.W.; Mattar, L.; Blasingame, T. Production Data Analysis–Challenges, Pitfalls, Diagnostics. *SPE Reserv. Eval. Eng.* **2010**, *13*, 538–552. [CrossRef]
11. Valko, P.; Lee, W. A Better Way To Forecast Production From Unconventional Gas Wells. In Proceedings of the SPE Annual Technical Conference and Exhibition, Florence, Italy, 19–22 September 2010.
12. Clark, A.J.; Lake, L.W.; Patzek, T.W. Production forecasting with logistic growth models. In Proceedings of the SPE Annual Technical Conference and Exhibition, Denver, CO, USA, 30 October–2 November 2011.
13. Zhang, H.; Rietz, D.; Cagle, A.; Cocco, M.; Lee, J. Extended exponential decline curve analysis. *J. Nat. Gas Sci. Eng.* **2016**, *36*, 402–413. [CrossRef]
14. Du, D.; Zhao, Y.; Guo, Q.; Guan, X.; Hu, P.; Fan, D. Extended Hyperbolic Decline Curve Analysis in Shale Gas Reservoirs. In Proceedings of the International Field Exploration and Development, Chengdu, China, 21–22 September 2017; pp. 1491–1504.
15. Gong, X.; Gonzalez, R.; McVay, D.A.; Hart, J.D. Bayesian probabilistic decline-curve analysis reliably quantifies uncertainty in shale-well-production forecasts. *SPE J.* **2014**, *19*, 1–47. [CrossRef]
16. Yu, W.; Tan, X.; Zuo, L.; Liang, J.; Liang, L. A new probabilistic approach for uncertainty quantification in well performance of shale gas reservoirs. *SPE J.* **2016**, *21*, 2038–2048. [CrossRef]
17. Paryani, M.; Awoleke, O.O.; Ahmadi, M.; Hanks, C.; Barry, R. Approximate Bayesian Computation for Probabilistic Decline-Curve Analysis in Unconventional Reservoirs. *SPE Reserv. Eval. Eng.* **2017**, *20*, 478–485. [CrossRef]
18. Fulford, D.S.; Bowie, B.; Berry, M.E.; Bowen, B.; Turk, D.W. Machine learning as a reliable technology for evaluating time/rate performance of unconventional wells. *SPE Econ. Manag.* **2016**, *8*, 23–39. [CrossRef]
19. de Holanda, R.W.; Gildin, E.; Valkó, P.P. Mapping the Barnett Shale Gas with Probabilistic Physics-Based Decline Curve Models and the Development of a Localized Prior Distribution. In Proceedings of the Unconventional Resources Technology Conference, Houston, TX, USA, 23–25 July 2018; pp. 217–232.

20. Holanda, R.W.d.; Gildin, E.; Valko, P.P. Combining physics, statistics, and heuristics in the decline-curve analysis of large data sets in unconventional reservoirs. *SPE Reserv. Eval. Eng.* **2018**, *21*, 683–702. [CrossRef]

21. Patzek, T.W.; Saputra, W.; Kirati, W.; Marder, M. Generalized Extreme Value Statistics, Physical Scaling and Forecasts of Gas Production in the Barnett Shale. *Energy Fuels* **2019**, in review. Available online: https://doi.org/10.26434/chemrxiv.8326898.v1 (accessed on 27 June 2019).

22. Saputra, W.; Kirati, W.; Patzek, T.W. Prediction of ultimate recovery in the existing Bakken Shale wells using a physical scaling approach. *J. Petr. Sci. Eng.* **2019**, In review.

23. Patzek, T.W.; Male, F.; Marder, M. Gas production in the Barnett Shale obeys a simple scaling theory. *Proc. Natl. Acad. Sci. USA* **2013**, *110*, 19731–19736. [CrossRef]

24. Patzek, T.; Male, F.; Marder, M. A simple model of gas production from hydrofractured horizontal wells in shales. *AAPG Bull.* **2014**, *98*, 2507 – 2529. [CrossRef]

25. Patzek, T.; Saputra, W.; Kirati, W. A Simple Physics-Based Model Predicts Oil Production from Thousands of Horizontal Wells in Shales. In Proceedings of the SPE Annual Technical Conference and Exhibition, San Antonio, TX, USA, 9–11 October 2017.

26. Eftekhari, B.; Marder, M.; Patzek, T.W. Field data provide estimates of effective permeability, fracture spacing, well drainage area and incremental production in gas shales. *J. Nat. Gas Sci. Eng.* **2018**, *56*, 141–151. [CrossRef]

27. Saputra, W.; Albinali, A.A. Validation of the Generalized Scaling Curve Method for EUR Prediction in Fractured Shale Oil Wells. In Proceedings of the SPE Kingdom of Saudi Arabia Annual Technical Symposium and Exhibition, Dammam, Saudi Arabia, 3–26 April 2018.

28. Sonnenberg, S.A.; Pramudito, A. Petroleum geology of the giant Elm Coulee field, Williston Basin. *AAPG Bull.* **2009**, *93*, 1127–1153. [CrossRef]

29. Mason, J. Oil production potential of the North Dakota Bakken. *Oil Gas J.* **2012**, *110*, 1–12.

30. Meissner, F.F. Petroleum geology of the Bakken Formation Williston Basin, North Dakota and Montana. In *1991 Guidebook to Geology and Horizontal Drilling of the Bakken Formation*; Montana Geological Society: Butte, MT, USA, 1991.

31. Buffington, N.; Kellner, J.; King, J.G.; David, B.L.; Demarchos, A.S.; Shepard, L.R. New technology in the Bakken Play increases the number of stages in packer/sleeve completions. In Proceedings of the SPE Western Regional Meeting, Anaheim, CA, USA, 27–29 May 2010.

32. Weddle, P.; Griffin, L.; Pearson, C.M. Mining the Bakken: Driving Cluster Efficiency Higher Using Particulate Diverters. In Proceedings of the SPE Hydraulic Fracturing Technology Conference and Exhibition, The Woodlands, TX, USA, 24–26 January 2017.

33. Martin, E. United States Tight Oil Production 2018. Available online: www.allaboutshale.com/united-states-tight-oil-production-2018-emanuel-omar-martin (accessed on 25 March 2019).

34. Weijermars, R.; van Harmelen, A. Shale Reservoir Drainage Visualized for a Wolfcamp Well (Midland Basin, West Texas, USA). *Energies* **2018**, *11*, 1665. [CrossRef]

35. Haider, S.; Patzek, T.W. The key physical factors that yield a good horizontal hydrofractured gas well in mudrock. *J. Nat. Gas Sci. Eng.* **2019**, Submitted. [CrossRef]

36. Paneitz, J. Evolution of the Bakken Completions in Sanish Field, Williston Basin, North Dakota. In Proceedings of the SPE Applied Technology Workshop (ATW), Keystone, CO, USA, 6 August 2010.

37. Chorin, A.J.; Kast, A.; Kupferman, R. Optimal prediction of underresolved dynamics. *Proc. Natl. Acad. Sci. USA* **1999**, *95*, 4094–4098. [CrossRef] [PubMed]

38. Chorin, A.J.; Kast, A.; Kupferman, R. Unresolved computation and optimal prediction. *Commun. Pure Appl. Math.* **1999**, *52*, 1231–1254. [CrossRef]

39. Chorin, A.J.; Kast, A.; Kupferman, R. On the prediction of large-scale dynamics using unresolved computations. *Contemp. Math.* **1999**, *238*, 53–75.

40. Chorin, A.J.; Hald, O.; Kupferman, R. Optimal prediction and the Mori-Zwanzig representation of irreversible processes. *Proc. Natl. Acad. Sci. USA* **2000**, *97*, 2968–2973. [CrossRef] [PubMed]

41. Gumbel, E.J. *Statistics of Extremes*; Columbia University Press: New York, NY, USA, 1958.

42. Weibull, W. A statistical distribution function of wide applicability. *J. Appl. Mech.* **1951**, *18*, 293–297.

43. Gumbel, E.J. Statistical theory of extreme values and some practical applications. *NBS Appl. Math. Ser.* **1954**, *33*, 1–60.

44. Fréchet, M. Sur la loi de probabilité de l'écart maximum. *Ann. Soc. Math. Polon.* **1927**, *6*, 93–116.

Energies **2019**, *12*, 3641

45. Tran, T.; Sinurat, P.D.; Wattenbarger, B.A. Production characteristics of the Bakken shale oil. In Proceedings of the SPE Annual Technical Conference and Exhibition, Denver, CO, USA, 30 October–2 November 2011.

46. Kurtoglu, B. *Integrated Reservoir Characterization and Modeling in Support of Enhanced Oil Recovery for Bakken*; Colorado School of Mines: Golden, CO, USA, 2013.

energies

MDPI

Article

An Efficient Hybrid Model for Nonlinear Two-Phase Flow in Fractured Low-Permeability Reservoir

Daigang Wang [1,*], Jingjing Sun [2], Yong Li [2] and Hui Peng [2]

[1]   Beijing International Center for Gas Hydrate, Peking University, Beijing 100871, China
[2]   Research Institute of Petroleum Exploration & Development, PetroChina, Beijing 100083, China
*   Correspondence: dgwang@pku.edu.cn; Tel.: +86-18310865935

Received: 13 May 2019; Accepted: 23 July 2019; Published: 24 July 2019

Abstract: The staged fracturing horizontal well has proven to be an attractive alternative for improving the development effect of a low permeability waterflood reservoir. Due to the coexistence of matrix, fracture, and horizontal wellbore, it remains a great challenge to accurately simulate the nonlinear flow behaviors in fractured porous media. Using a discrete fracture model to reduce the dimension of the fracture network, a two-parameter model is used to describe the nonlinear two-phase flow behavior, and the equivalent pipe flow equation is selected to estimate the horizontal wellbore pressure drop in the fractured low-permeability reservoir. A hybrid mathematical model for the nonlinear two-phase flow, including the effect of horizontal wellbore pressure drop in fractured porous media, is developed. A numerical scheme of the hybrid model is derived using the mimetic finite difference method and finite volume method. With a staggered five-spot flood system, the accuracy of the proposed model and the effect of fracture properties on nonlinear two-phase flow behaviors are further investigated. The results also show that with an increase of fracture length near injectors, the breakthrough time of injected water into the horizontal wellbore will be shorter, indicating a faster rise of the water cut, and a worse development effect. The impact of shortening fracture spacing is consistent with that of enlarging fracture length. Successful practice in modeling the complex waterflood behaviors for a 3-D heterogeneous reservoir provides powerful evidence for the practicability and reliability of our model.

Keywords: low-permeability reservoir; staged fracturing horizontal well; mimetic finite difference method; discrete fracture model; fracture properties

1. Introduction

Due to the strong heterogeneity and poor distribution of petrophysical properties, the natural oil productivity of low-permeability reservoirs is extremely low. Currently, the staged fracturing horizontal well has proven to be a viable alternative for improving the development effect of a low-permeability reservoir [1]. In general, the matrix, fracture network and horizontal wellbore are simultaneously distributed in this type of reservoir. However, the fluid flow does not obey the traditional Darcy's law, which further increases the difficulty of modeling the underlying flow dynamics in fractured porous media. To avoid too-early water breakthrough as much as possible, it is of great importance to establish a hybrid mathematical model for nonlinear two-phase flow, considering the effect of horizontal wellbore pressure drop in the fractured low-permeability waterflood reservoir.

Many attempts had been made to investigate the nonlinear flow characteristics in a low permeability waterflood reservoir, multiphase fluid flow in fractured porous media, optimal design of numerical discretization, and the analysis of variable mass transfer in horizontal wellbore. The commonly used models to describe the nonlinear flow behavior in low-permeability porous media consist of the quasi-threshold gradient model [2], the piecewise nonlinear model [3], and the continuous nonlinear models [4–7]. Since that the smooth functional relationship is efficient enough to calculate the threshold pressure gradient of nonlinear fluid flow in the subsurface, the continuous nonlinear models, especially the two-parameter model, are growing in popularity among scholars.

Typical models for the representation of fractured porous media generally rely on the dual-porosity model and its extended form (dual permeability model). Those models consider matrix block and fracture network as two parallel continuum systems coupled by crossflow function. Previous studies [8–10] have demonstrated that the dual porosity model is not suitable for multiphase flow in disconnected fractured media and mix-wet fracture media. Great trouble will also be encountered when estimating the crossflow function of different systems. To overcome the drawbacks, the discrete fracture model was first introduced for single phase flow by Noorishad and Mehran [11]. The fracture cells are geometrically simplified by using (n − 1)-dimensional cells in an n-dimensional domain, which considerably improved the computational efficiency. The discrete fracture model has been widely employed to investigate the flow mechanism of multiphase fluid in fractured media ever since [12–18]. On the other hand, strategies for numerical discretization are also key to simulating the complicated flow dynamics in fractured porous media, which mainly include the finite difference (FD) method [19], the finite element (FE) method [8,11], the finite volume (FV) method [20,21], and the mimetic finite difference (MFD) method [22–24]. The MFD method, requiring only the node and surface information of single grid cell, is theoretically feasible for any geometry, even the concave grid systems. However, there are limited studies [25–28] on the nonlinear fluid flow behaviors in fractured low-permeability porous media coupling the MFD method with a discrete fracture model.

The analysis of variable mass transfer in a horizontal wellbore supports insights that predict the production performance, design well trajectory, and optimize well completion parameters. Taking the pressure drop caused by the pipe friction into consideration, Dikken [29] coupled the flux variation in a horizontal wellbore and fluid flow in porous media and then presented a semi-analytical mathematical model, which was then used to calculate the variable mass single-phase pressure drop. Based on extensive theoretical or experimental studies, several analytical models for frictional pressure drop [30–34] have been developed over the last decades. Nevertheless, the accuracy of analysis on the variable mass pressure drop using the analytical models cannot be guaranteed due to the idealized hypothesis. To resolve this issue, the equivalent percolation model of pipe flow was introduced by Wu et al [35] in order to develop a coupled correlation between the variable mass flow in a horizontal wellbore and fluid flow in reservoir. Thereafter, Birchenko et al. [36] and Wang et al. [37] addressed further studies on the equivalent representation models of variable mass flow.

In this study, we aim to propose a novel hybrid mathematical model for two-phase nonlinear flow in fractured low-permeability waterflood reservoir with considering the effect of horizontal wellbore pressure drop. Firstly, using the discrete fracture model to reduce the dimension of fracture network explicitly, the two-parameter model is selected to reflect the nonlinear flow behavior in low-permeability reservoir, and the equivalent percolation model of pipe flow is used to calculate the wellbore pressure drop. Then, a novel mathematical hybrid model for two-phase flow in fractured pore media is established by coupling the governing equations satisfied by matrix, fracture and horizontal well, respectively. Combing the MFD method and FV method, we derive and validate the numerical scheme of the proposed model with a synthetic staggered five-spot flood system. The effect of fracture properties on the nonlinear flow behavior in a fractured low- permeability waterflood reservoir is extensively investigated. Ultimately, the proposed model is applied to a 3-D heterogeneous low-permeability waterflood reservoir to validate its practicability and feasibility.

2. Mimetic Finite Difference Method

As illustrated in Figure 1, the reservoir area $\Omega \in R^d$ is subdivided by a group of non-overlapping polygon meshes (d = 2) or polyhedron meshes (d = 3) $\Omega_h = \{\Omega_i\}$. For arbitrary grid cell Ω_i, grid Ω_j is the adjacent grid cell, $A_k = \Omega_i \cap \Omega_j$ is the interface, n_k is the area-weighted normal vector to face number k. The grid cell pressure p_i^e and boundary surface pressure p_k^f are defined at the central point of grid cell x_i and the boundary surface central point of grid cell x_k, respectively, which take the form of.

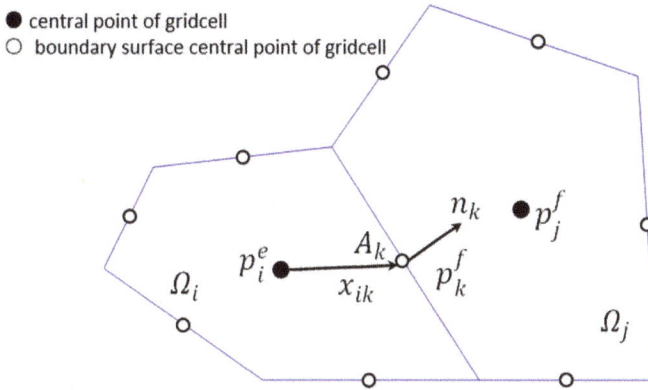

Figure 1. Schematic of grid analysis for mimetic finite difference method.

$$p_i^e = \frac{1}{|\Omega_i|} \int_{\Omega_i} p\,d\Omega, \ p_k^f = \frac{1}{|A_k|} \int_{A_k} p\,dA \tag{1}$$

The total flux normal to the faces can be described as:

$$q = T_i(e_i p_i^e - p_k^f) \tag{2}$$

where T_i is the transmissibility matrix; $q = [q_1, \cdots, q_m]^T$; m is the number of borders of grid cell Ω_i; $e_i = [1, \cdots, 1]^T$. The key to the MFD method is to construct the matrix T_i. A linear pressure field can be obtained in the form $p_i^e = a \cdot x + b$ for a constant vector a and scalar b. Then the flux and pressure drop are given by

$$q_k = -\mu^{-1}|A_k|n_k \cdot K \cdot \nabla p = -\mu^{-1}|A_k|n_k \cdot K \cdot a \tag{3}$$

In addition, $p_i^e - p_k^f = a \cdot (x_i - x_k)$. By substituting Equation (2) into Equation (3), we can get the following equation.

$$q = T_i \begin{bmatrix} x_1 - x_i \\ \vdots \\ x_k - x_i \\ \vdots \\ x_m - x_i \end{bmatrix} \cdot a = \mu^{-1} \begin{bmatrix} |A_1|\vec{n}_1 \\ \vdots \\ |A_k|\vec{n}_k \\ \vdots \\ |A_m|\vec{n}_m \end{bmatrix} \cdot K \cdot a \Rightarrow T_i X = \mu^{-1} NK \tag{4}$$

where $N^T X = |\Omega_i|E_d$, E_d is the unit matrix of d-th order, and $|\Omega_i|$ is the area of Ω_i. Then, the matrix T_i can be represented as follows.

$$T_i = \frac{1}{\mu|\Omega_i|} NKN^T + T_2 \tag{5}$$

where $T_2 X = 0$. To guarantee the existence of inverse of T_i, the Brezzi-Lipnikov-Simoncini theorem [38] is used to construct the matrix T_2:

$$T_i = \frac{1}{\mu|\Omega_i|} [NKN^T + \frac{6}{d} trace(K) A(E_m - QQ^T)A] \tag{6}$$

where $Q = orth(AX)$, and A is the diagonal matrix with A_i (i.e., the face area of the *i*th face)

Based on the divergence theorem and the integral procedure on arbitrary grid cell Ω_i, the following equation can be obtained.

$$\sum_{k=1}^{m} q_k = \int_{\Omega_i} q_i d\Omega \tag{7}$$

Considering the flux continuous condition of boundary, numerical scheme of the MFD method is established ultimately by coupling Equations (2) and (7).

$$\begin{bmatrix} B & -C & D \\ C^T & 0 & 0 \\ D^T & 0 & 0 \end{bmatrix} \begin{bmatrix} q \\ p \\ \pi \end{bmatrix} = \begin{bmatrix} 0 \\ f \\ 0 \end{bmatrix} \tag{8}$$

where $q = [q_k]$; $p = [p_i^e]$; $\pi = [p_k^f]$; $f = [f_i]$, and $f_i = \int_{\Omega_i} q_i d\Omega$. Noting that, the first row of Equation (8) denotes the Darcy's law, the second row denotes the mass conservation, and the third row denotes the normal flux continuous condition of borders. Coefficient matrices of Equation (8) can be written as:

$$B = \begin{pmatrix} T_1^{-1} & & \\ & \ddots & \\ & & T_{Ne}^{-1} \end{pmatrix}, C = \begin{pmatrix} e_1 & & \\ & \ddots & \\ & & e_{Ne} \end{pmatrix}, D = \begin{pmatrix} I_1 & & \\ & \ddots & \\ & & I_{Ne} \end{pmatrix} \tag{9}$$

where N_e is the total number of grid cells; $I_i = E_m$. As can be seen from Equation (9), all the coefficient matrices of Equation (8) are subjected to the petrophysical properties and geometric information of grid cells, but insensitive to the geometric shape.

3. Hybrid Model for Two-Phase Flow in Fractured Media

Using the discrete fracture model to reduce the dimension of the fracture network explicitly, the two-parameter model is used to reflect the nonlinear flow behavior of two-phase fluid, and the equivalent percolation model of pipe flow is selected to calculate the wellbore pressure drop of the horizontal well. Ultimately, a hybrid mathematical model for two-phase flow in a fractured low-permeability waterflood reservoir is established by combing the governing equations satisfied by the matrix, fracture and horizontal well, respectively.

3.1. Discrete Fracture Model

As previously suggested [11], fluid flow through fractures can be modelled as a laminar flow between parallel plates. The parallel-plate solution for the Naiver-Stokes equations satisfies the commonly used law that flow rate is proportional to the cube of the fracture aperture. All the variables remain constant along the direction of fracture aperture. Therefore, the dimension-reduced processing of a fracture network by using (n − 1)-dimensional grid cells in an n-dimensional domain greatly improves the computational efficiency. In 2D space, fracture networks are simplified as line elements (see Figure 2). In 3D space, the fractures are represented by the matrix grid cell interfaces, which are 2D. In this study, the Delaunay triangulation method is adopted to establish the grid system.

As shown in Figure 2, the fractured media is composed of matrix and fracture simultaneously, and the whole domain can be expressed as $\Omega = \Omega_m + \sum_i a_i \times (\Omega_f)_i$, where the subscripts m and f denote the matrix and fracture, respectively; a_i is the aperture of the *i*th fracture. Only if the representative elementary volumes (REV) of matrix and fracture exist, the constituents of flow equations F are feasible to the whole domain. For the discrete fracture model, the integral form of F is described as.

$$\int_{\Omega} Fd\Omega = \int_{\Omega_m} Fd\Omega_m + \sum_i a_i \times \int_{(\Omega_f)_i} Fd(\Omega_f)_i \tag{10}$$

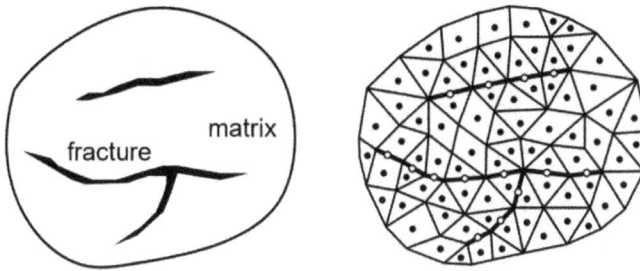

Figure 2. Schematic of a typical fractured media (**left**); the unstructured grid (**right**).

3.2. Flow Governing Equations

The basic equations of incompressible two-phase flow include the mass conservation equation, the generalized Darcy's law, the saturation equation and the capillary pressure function, which are described as:

$$\phi \frac{\partial S_w}{\partial t} + \nabla \cdot v_w = q_w \tag{11}$$

$$v_w = f_w[v + K\lambda_o \cdot \nabla p_c + K\lambda_o \cdot (\rho_w - \rho_o)G] \tag{12}$$

$$v = -K\lambda \cdot \nabla p + K \cdot (\lambda_w \rho_w + \lambda_o \rho_o)G, \ \nabla \cdot v = q \tag{13}$$

$$p = p_o - \int_1^{S_w} f_w(\xi) \frac{\partial p_c}{\partial S_w}(\xi) d\xi \tag{14}$$

$$S_w + S_o = 1 \tag{15}$$

$$p_c = p_o - p_w \geq 0 \tag{16}$$

where ϕ is the porosity; v_o and v_w are the oil and water velocity, respectively, and total velocity $v = v_o + v_w$, m/s; q_o and q_w are the source/sink term of oil and water, respectively, and total source/sink term $q = q_w + q_o$, 1/s; $\lambda_l = k_{rl}/\mu_l$ ($l = w, o$) is the flow coefficient, and total flow coefficient $\lambda = \lambda_w + \lambda_o$; $f_w = \lambda_w/\lambda$ is the water fractional flow function; ρ_o and ρ_w are the oil and water density, respectively, kg/m^3; $G = -g\nabla z$ is the gravitational force term, and g is the gravity acceleration, m/s^2; z is the vertical coordinate with positive direction upward, m; p_c is the capillary pressure, Pa; p_o and p_w are the pore pressure for oil and water, respectively, Pa; S_o and S_w are the oil and water saturation, respectively.

Defining the flow potential function $\Phi_l = p_l + \rho_l gz$, the initial and boundary conditions are written as follows.

(i) Initial conditions

$$\Phi_l(x,t)_{t=0} = \Phi_l(x), \quad S_l(x,t)_{t=0} = S_l(x), \quad l = w, o \tag{17}$$

(ii) Dirichlet boundary conditions

$$\Phi_l(x,t) = \Phi_l, \ S_l(x,t) = S_l \tag{18}$$

(iii) Neumann boundary conditions. The boundary conditions used in this paper are assumed to be impervious.

$$v_l \cdot \vec{n} = (-K\lambda_l \nabla \Phi_l) \cdot \vec{n} = 0, \ \nabla S_l \cdot \vec{n} = 0 \tag{19}$$

where $\vec{n}$ denotes the outer normal unit vector.

3.3. Description of Nonlinear Fluid Flow

The two-parameter model is used to describe the nonlinear flow behavior of multiphase fluid in porous media and takes the form of.

$$v = \frac{K}{\mu}\nabla P - \frac{K}{\mu}\frac{\nabla P}{a + b|\nabla P|} \tag{20}$$

where b is the reciprocal of quasi-threshold pressure gradient, $10^{-6}(\text{Pa/m})^{-1}$; a is a dimensionless parameter to determine the shape of nonlinear concave curve segment, $a > 0$. Both parameters can be obtained from displacement experiments. When the dimensionless parameter a is equal to zero, the model is equivalent to the quasi-threshold pressure gradient model, among which, the nonlinear seepage segment satisfy the linear law and its intersection with x-axis is located at $1/b$; when a is larger than zero and less than 1.0, the intersection with x-axis is located at $(1 - a)/b$, which is virtually the minimum threshold pressure gradient. When b tends to be infinite, the interaction between rock and reservoir fluid is so weak that it equals to zero approximately. In this case, Equation (20) will be transformed as the Darcy's law.

3.4. Calculation of Wellbore Pressure Drop

Generally, the flow regime in horizontal wellbore consists of the spindle flow in horizontal wellbore and the radial flow from reservoir to horizontal well. Due to the radial flow, the flux of lateral segment from toe end to root end varies gradually, in other words, it is a variable mass flow, which is shown as Figure 3, where v_R denotes the mass flux of the radial flow from reservoir to horizontal well and $v(x)$ denotes the variable mass flux of the spindle flow in horizontal wellbore.

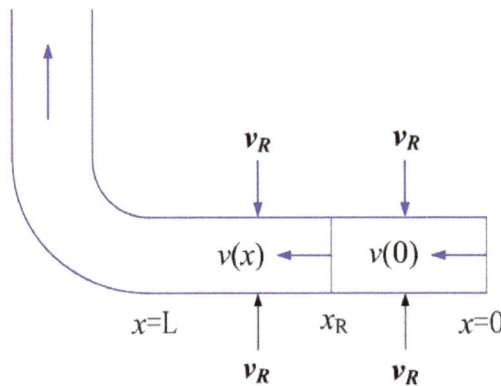

Figure 3. Sketch map of horizontal wellbore flow.

In this study, the equivalent percolation model of pipe flow is employed to calculate the pressure drop caused by variable mass flow in horizontal wellbore. Namely, the pipe flow can be interpreted as an equivalent seepage problem with constant permeability. Equation (21) is thus proposed to describe the relationship.

$$v = -\frac{K_{we}}{\mu}\frac{\Delta p}{\Delta x} \tag{21}$$

where K_{we} is the equivalent permeability of horizontal wellbore.

For laminar flow

$$K_{we} = \frac{r^2}{8} \tag{22}$$

For turbulent flow:

$$K_{we} = \mu \left(\frac{2D}{f\rho} \frac{1}{\Delta p / \Delta x} \right)^{1/2} \tag{23}$$

where the fraction factor f can be calculated by Colebrook-White equation and satisfies the following equation.

$$\frac{1}{\sqrt{f}} = -2\log_{10}[\frac{\varepsilon/D}{3.7} - \frac{5.02}{Re}\log_{10}(\frac{\varepsilon/D}{3.7} + \frac{13}{Re})] \tag{24}$$

For transitional flow:

$$K_{we} = \lambda K_{wel} + (1 - \lambda)K_{wet} \tag{25}$$

where λ is the weighting coefficient, [0.1, 0.3]; K_{wel} is the equivalent permeability in procedure of lamilar flow; K_{wet} is the equivalent permeability in procedure of turbulent flow.

3.5. Hybrid Mathematical Model

The novel hybrid model for nonlinear two-phase flow in fractured porous media is established by combining the governing equations satisfied by matrix, fracture and horizontal wellbore, respectively.

For the matrix system, considering that nonlinear flow behavior in fractured low-permeability reservoir, the following equation is given by:

$$v = -K\lambda \cdot \nabla p \left(1 - \frac{1}{a + b|\nabla p|} \right) + K \cdot (\lambda_w \rho_w + \lambda_o \rho_o)G, \quad \nabla \cdot v = q \tag{26}$$

For the fracture system, the dimension-reduced procedure of the fracture network by using (n − 1)-dimension grid cells in an n-dimensional domain is carried out. Assume that flow in fracture observes the standard Darcy's law, the system of Equations (11)–(16) is still suitable.

For the horizontal wellbore, the equation describing the correlation between velocity v and pressure p is given by:

$$v = -K_{we}\lambda \cdot \nabla p + K_{we} \cdot (\lambda_w \rho_w + \lambda_o \rho_o)G, \quad \nabla \cdot v = q \tag{27}$$

When considering the pressure drop in horizontal wellbore, the flux continuous condition of horizontal wellbore is derived as follows.

$$\begin{cases} q_{m,E} + q_{m,E'} + q_{f,F} + q_{f,F'} = M_h \\ \Sigma q_h = M_h + q_h \end{cases} \tag{28}$$

where $q_{m,E}$ is flux from the upward matrix grid cell E to the ith lateral segment; $q_{m,E'}$ is flux from the downward matrix grid cell E' to the ith lateral segment; $q_{f,F}$ is flux from the upward fracture grid cell F to the ith lateral segment; $q_{f,F'}$ is flux from the downward fracture grid cell F' to the ith lateral segment; M_h is total flux penetrated into the ith lateral segment; Σq_h is the total flux; q_h is the source or sink term.

4. Numerical Discretization and Solution

Using the MFD method, numerical discretization of the hybrid mathematical model is performed in order to obtain the pressure and saturation distribution as a function of time in fractured low-permeability waterflood reservoir according to the implicit pressure and explicit saturation (IMPES) procedure.

4.1. Numerical Discretization

Based on the mimetic finite difference method, the linear algebraic systems satisfied by matrix block and fracture network are finally established, which are written as Equations (29) and (30), respectively. As to 2D problems, due to the fracture networks being simplified as a series of line elements, the governing equations can be described as:

$$
\begin{bmatrix}
B_m & -C_m & D_m \\
C_m^T & 0 & 0 \\
D_m^T & 0 & 0
\end{bmatrix}
\begin{bmatrix}
q_m \\
p_m \\
\pi_m
\end{bmatrix}
=
\begin{bmatrix}
0 \\
f_m \\
0
\end{bmatrix}
\tag{29}
$$

$$
\begin{bmatrix}
B_f & -C_f & D_f \\
C_f^T & 0 & 0 \\
D_f^T & 0 & 0
\end{bmatrix}
\begin{bmatrix}
q_f \\
p_f \\
\pi_f
\end{bmatrix}
=
\begin{bmatrix}
0 \\
f_f \\
0
\end{bmatrix}
\tag{30}
$$

where the subscripts m and f are matrix and fracture system, respectively.

The system given in linear algebraic equations indicates that p_f is a part of π_m. Based on the flux continuity principle, the volumetric flux of fracture system is composed of matrix infiltration and fracture source or sink term. By coupling the flow equations between the matrix and fracture systems, numerical discretization is achieved, which is given by:

$$
\begin{bmatrix}
B_m & -C_m & D_m & 0 & 0 \\
C_m^T & 0 & 0 & 0 & 0 \\
D_m^T & 0 & 0 & -C_f^T & 0 \\
0 & 0 & -C_f & B_f & D_f \\
0 & 0 & 0 & D_f^T & 0
\end{bmatrix}
\begin{bmatrix}
q_m \\
p_m \\
\pi_m \\
q_f \\
p_f
\end{bmatrix}
=
\begin{bmatrix}
0 \\
f_m \\
-f_f \\
0 \\
0
\end{bmatrix}
\tag{31}
$$

When matrix, fracture, and horizontal wells exist simultaneously, the equivalent percolation model of pipe flow is used to calculate the wellbore pressure drop, and the total volumetric flux in horizontal wellbore is composed of matrix infiltration, fracture infiltration and source or sink term. In accordance with Equation (31), numerical discretization scheme of the hybrid mathematical model can be developed combining the flow equations satisfied by the matrix, fracture, and horizontal wellbore, respectively, and takes the form of:

$$
\begin{bmatrix}
B_m & -C_m & D_m & 0 & 0 & 0 & 0 \\
C_m^T & 0 & 0 & 0 & 0 & 0 & 0 \\
D_m^T & 0 & 0 & -C_f^T & 0 & C_h^T & 0 \\
0 & 0 & -C_f & B_f & D_f & 0 & 0 \\
0 & 0 & 0 & D_f^T & 0 & 0 & 0 \\
0 & 0 & C_h & 0 & 0 & B_h & D_h \\
0 & 0 & 0 & 0 & 0 & D_h^T & 0
\end{bmatrix}
\begin{bmatrix}
q_m \\
p_m \\
p_{im} \\
q_f \\
p_{if} \\
q_h \\
p_{ih}
\end{bmatrix}
=
\begin{bmatrix}
0 \\
f_m \\
-f_f - f_h \\
0 \\
0 \\
0 \\
0
\end{bmatrix}
\tag{32}
$$

where subscript h denotes the horizontal well; q_m denotes the boundary flux of matrix grid cell; p_m denotes the pressure at the central point of matrix grid cell; p_{im} denotes the boundary pressure of matrix grid cell; q_f denotes the boundary flux of fracture grid cell; p_{if} denotes the boundary pressure of fracture grid cell; q_h denotes the boundary flux of horizontal well grid cell; p_{ih} denotes the boundary pressure of horizontal well grid cell.

4.2. Solution of Pressure and Saturation

The implicit pressure explicit saturation (IMPES) procedure is used for calculation, which mainly include the sequential solution of the decoupled pressure and saturation equations. The pressure equations are solved implicitly with the preconditioned conjugate gradient and the saturation maps are determined explicitly using the FV method.

According to the FV method and θ-principle, discretization of mass conservation equation given by Equation (11) is performed, which is described as:

$$\frac{\varphi_i}{\Delta t}\left(S_i^{k+1} - S_i^k\right) + \frac{1}{|\Omega_i|}\sum_{r_{ij}}\left(\theta F_{ij}\left(S^{k+1}\right) + (1-\theta)F_{ij}\left(S^k\right)\right) = q_w\left(S_i^k\right) \tag{33}$$

where $F_{ij}(S)$ is the numerical approximation of flux at the edge r_{ij}.

$$F_{ij}(S) = \int_{\gamma_{ij}} f_w(S)_{ij}(\boldsymbol{v}\cdot\vec{\boldsymbol{n}}_{ij} + K\lambda_n\cdot\nabla p_c\cdot\vec{\boldsymbol{n}}_{ij} + K\lambda_n\cdot(\rho_w - \rho_n)\boldsymbol{G}\cdot\vec{\boldsymbol{n}}_{ij})d\Gamma$$

Moreover, the upstream weighted method is used to calculate the fractional flow function at boundary surface Γ.

$$f_w(S)_{ij} = \begin{cases} f_w(S_i) & \text{if } \boldsymbol{v}\cdot\vec{\boldsymbol{n}}_{ij} \geq 0 \\ f_w(S_j) & \text{if } \boldsymbol{v}\cdot\vec{\boldsymbol{n}}_{ij} < 0 \end{cases} \tag{34}$$

The time step is determined by a CFL condition, which is described as follows:

$$\Delta t \leq \frac{\varphi_i|\Omega_i|}{v_i^{in}\max\{f_w'(S)\}_{0\leq S\leq 1}}, \tag{35}$$

and

$$v_i^{in} = \max(q_i, 0) - \sum_{r_{ij}}\min(v_{ij}, 0), \quad \frac{\partial f_w}{\partial S} = \frac{\partial f_w}{\partial S^*}\frac{\partial S^*}{\partial S} = \frac{1}{1 - S_{wc} - S_{or}}\frac{\partial f_w}{\partial S^*}$$

where S^* is the standardized water saturation; S_{wc} is the irreducible water saturation; S_{or} is the residual oil saturation.

5. Results and Analysis

We use a synthetic two-dimensional staggered five-spot flood system to validate the accuracy of the proposed model in this paper, as shown in Figure 4. The Delaunay triangulation grid system shown as Figure 5 is firstly constructed, with a total of 994 grid cells. The rock and fluid properties are listed in Table 1. The well pattern is one staged fracturing horizontal producer located in the center face and four vertical injectors located in corner faces. Both the injection and production are performed at a constant surface liquid rate, which are the same as the controlling conditions in actual oilfield development. There exist eight hydraulic fractures distributed evenly along the direction of the horizontal wellbore. The effects of capillary pressure and gravitational force will be neglected.

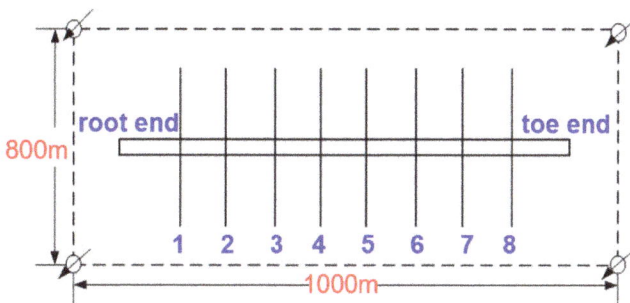

Figure 4. Scheme of staggered five spot water-injection model.

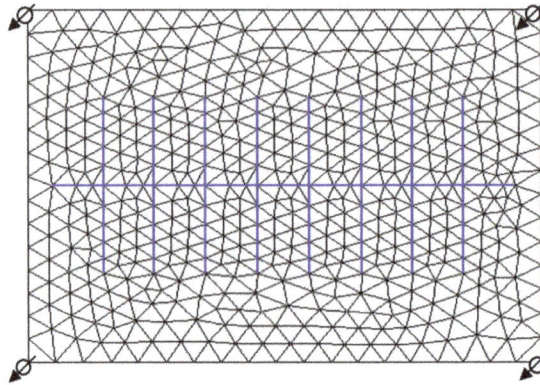

Figure 5. Delaunay triangulation of the water-injection model.

Table 1. Rock and fluid properties for the staggered five spot water-injection model.

Parameter	Value	Parameter	Value
reservoir dimensions	1000 m × 800 m	water viscosity	1 mPa·s
matrix porosity	0.1	oil viscosity	5 mPa·s
matrix permeability	1×10^{-3} μm^2	water density	1000 kg/m^3
fracture length	400 m	oil density	800 kg/m^3
fracture spacing	100 m	inject rate	0.05 PV/day
fracture aperture	1 mm	production rate	0.05 PV/day
fracture permeability	8.33×10^4 μm^2	irreducible water saturation	0.3
length of lateral segment	900 m	residual oil saturation	0.1

The oil-water relative permeability functions shared by the matrix and fracture systems are described as follows.

$$K_{rw} = S_e; \quad K_{ro} = 1 - S_e \tag{36}$$

where S_e is the normalized water saturation defined by the irreducible water saturation S_{wc} and the residual water saturation S_{or}, which takes the form of.

$$S_e = \frac{S_w - S_{wc}}{1 - S_{wc} - S_{or}} \tag{37}$$

Figure 6 displays the flux of different lateral segments in horizontal wellbore. It can be referred that the mass flux in horizontal wellbore gradually increases from the toe end to the root end, and there is a distinct difference in the mass flux of different lateral segments due to the variable mass flow. The effect of fracture distribution on the mass flux change in the horizontal wellbore is relatively large.

Figure 7 shows the distribution of friction pressure drop along direction of horizontal wellbore. It indicates that, the change of friction pressure drop from the toe end to root end is consistent with the mass flux of different lateral segments in horizontal wellbore. The main reason for this phenomenon is explained as follows: the variation of the friction pressure drop is proportional to that of the variable mass flux, and the frictional pressure drop caused by the spindle flow and the radial flow is aggravated as the variable mass flux increases. In addition, due to the great infiltration at the fractures' face, the degree of velocity variation is further improved, which usually results in a greater frictional pressure drop.

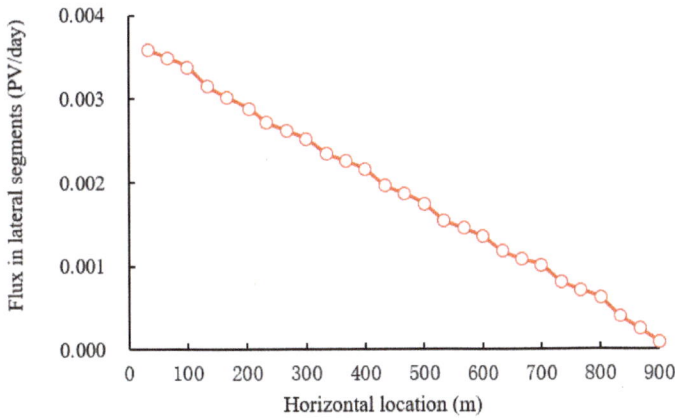

Figure 6. Mass flux distribution of different lateral segments in horizontal well.

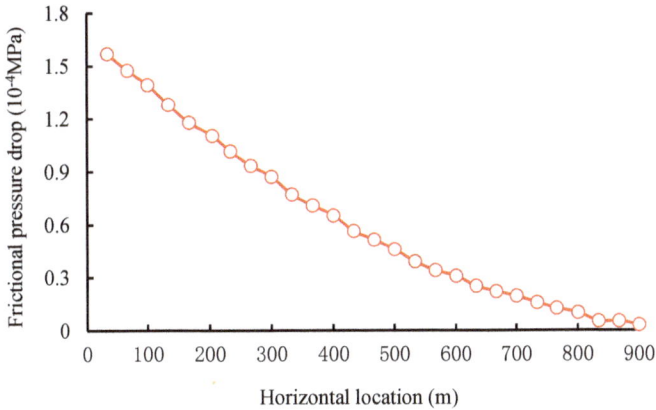

Figure 7. Evolution of frictional pressure drop across the horizontal wellbore.

As shown in Figure 8, we present the water saturation profiles when the pore volume of water injected into the reservoir is fixed at 0.05, 0.45, 0.80 and 1.0 PV, respectively. The results demonstrate that the waterflood front advances evenly at the earlier stage, while the injected water will transport quickly through the fractures and cause too-early water breakthrough. When the injected water crosses into all the fractures, large quantities of residual oil remain unexploited between the fractures. The fractures nearby the injectors are the major flow pathways of injected water to the staggered fractured horizontal wellbore, which govern the local velocity of waterfront directly. Therefore, the middle regions' fractures along the direction of the horizontal wellbore are extremely essential to achieve a higher oil productivity in the fractured low-permeability waterflood reservoir.

A comparison of oil production rate (OPR), cumulative oil production (OPT) and water cut (*fw*) as a function of PV injected is displayed in Figures 9 and 10. The accuracy of our proposed model has been validated by comparing with Eclipse. Results also demonstrate that oil production rate of the staged fracturing horizontal well at the earlier stage is relatively high, and the rising rate of cumulative oil production is rapid. However, the oil production rate gradually decreases as the waterfront advances in low-permeability porous media. When a total of 0.5 PV water is injected into the reservoir, the oil production rate tends to be constant, and the increase of cumulative oil production becomes slight. It is due to the short distance between end regions' fractures and injectors that water can channel into the horizontal wellbore along the fractures quickly, thus resulting in a rapid rise of water cut. Therefore, the fracture properties are governing factors for the development effect of a staged fracturing horizontal well in a low-permeability waterflood reservoir.

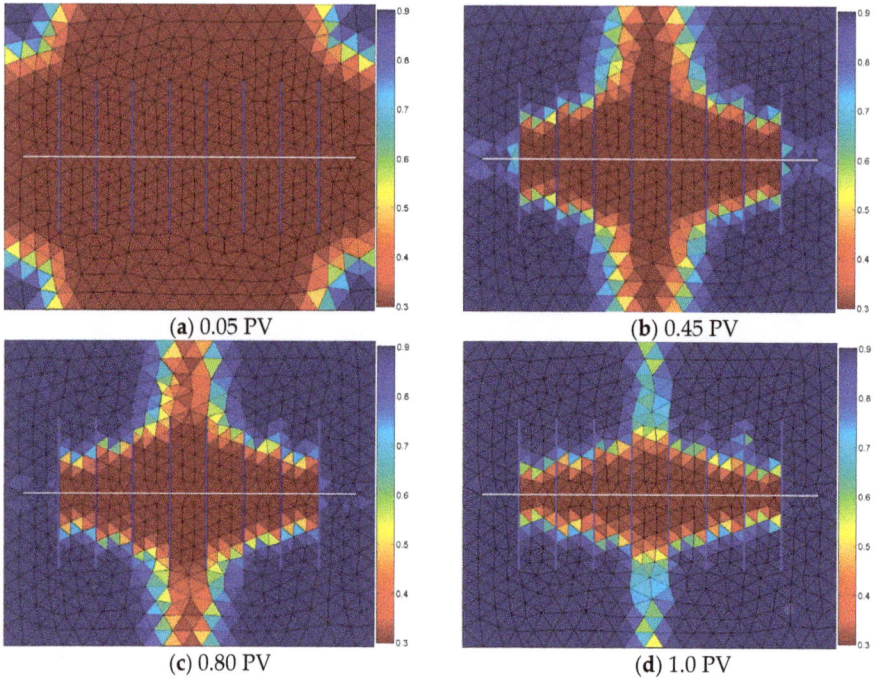

Figure 8. Water saturation profiles at different injection pore volume multiples.

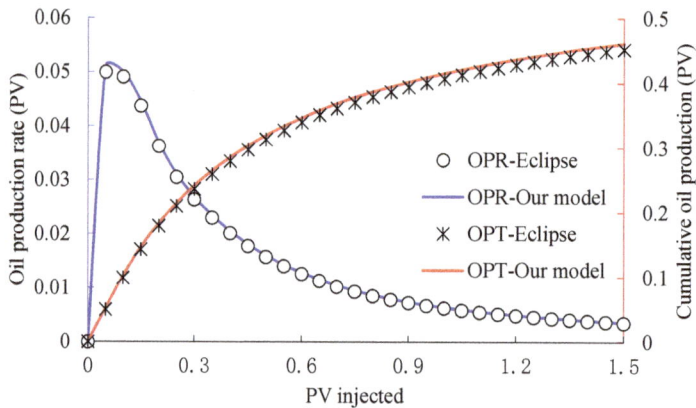

Figure 9. Oil production performance of staged fractured horizontal well.

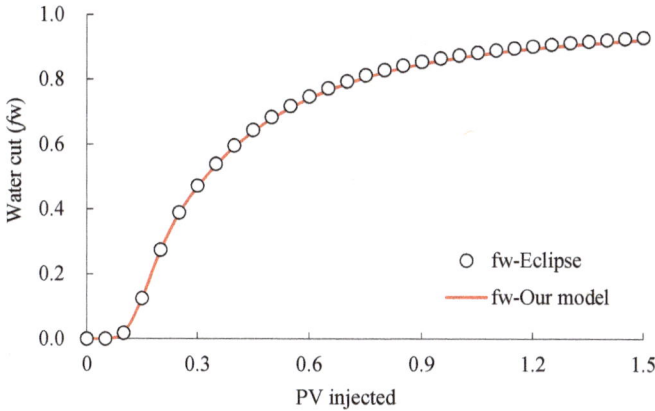

Figure 10. Water cut variation of staged fractured horizontal well.

5.1. Effect of Fracture Length

To further understand the influence of fracture length on the production performance of staged fracturing horizontal well, three patterns with different fracture length are considered, as shown in Figure 11. The fracture length of pattern 1 and pattern 3 equals 400 m and 200 m, respectively. The ratio of the middle regions' and end regions' fracture length is interpreted as the fracture length ratio for short. For pattern 2, the middle regions' fracture length is limited to 400 m, and the fracture length ratio is 2.0. The other properties are identical to those of the basic scheme. Based on the proposed model, the production performance of the staged fracturing horizontal well under different patterns of fracture length are extensively investigated. The results are shown in Figures 12 and 13.

From Figures 12 and 13, it is referred that, the time needed for waterflood-front channeling into the horizontal wellbore becomes shorter while increasing the length of fractures nearby injectors, and a higher rise of water cut will be observed. Therefore, when conducting the optimal design of fracture properties, the length of fractures nearby the injectors should be shortened appropriately to avoid a too-early breakthrough of injected water.

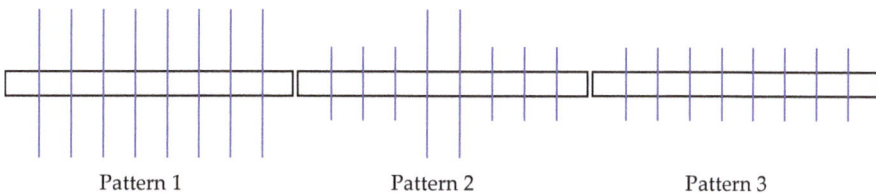

Figure 11. Schematic of different fracture length patterns.

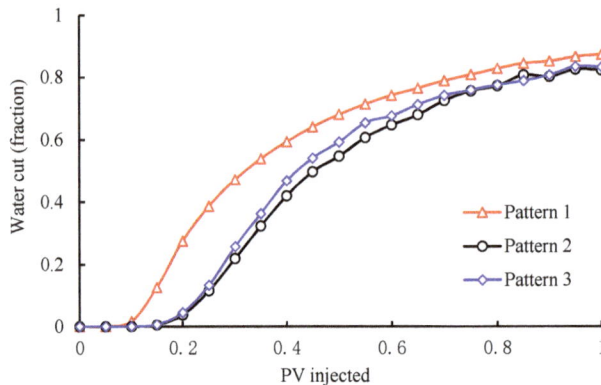

Figure 12. Water cut variation under different fracture length patterns.

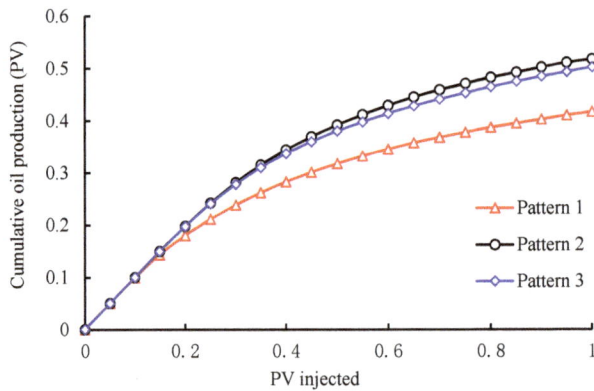

Figure 13. Cumulative oil production variation under different fracture lengths.

5.2. Effect of Fracture Spacing

To understand the effect of fracture spacing on development effect of staged fracturing horizontal well, three patterns with different fracture spacing are considered, as shown in Figure 14. The fracture spacing between the first fracture and the 8th fracture is limited to 700 m, and the ratio of middle regions' and end regions' fracture spacing along the direction of the horizontal wellbore is defined as the fracture spacing ratio for short. In this study, the fracture spacing ratio of three different patterns is selected to be 1.0, 2.0 and 0.5, respectively. The other properties are identical to those of the basic scheme. Based on the proposed model, the production performance of the staged fracturing horizontal well under different patterns of fracture spacing are thoroughly investigated. The results are shown in Figures 15 and 16.

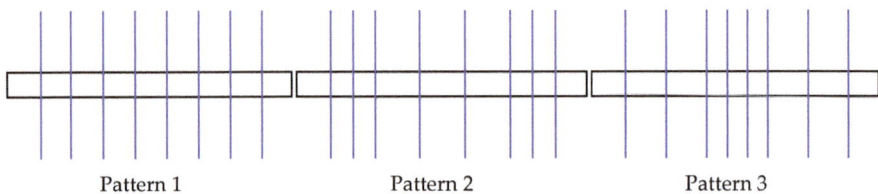

Figure 14. Schematic of different fracture spacing patterns.

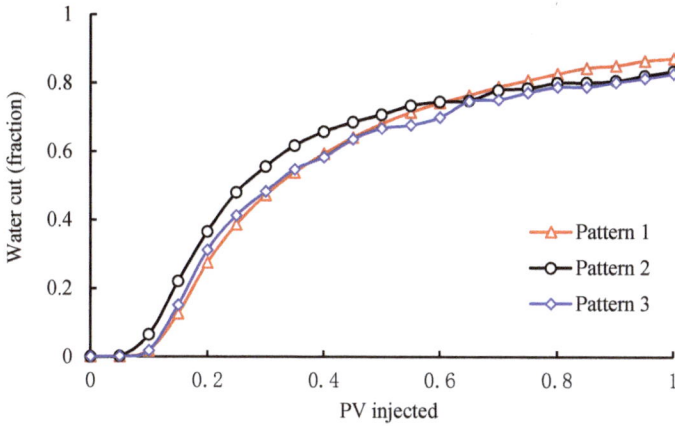

Figure 15. Water cut curve under different fracture spacing patterns.

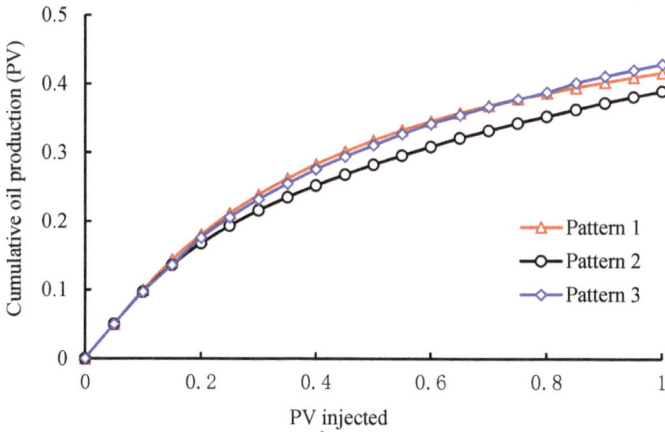

Figure 16. Cumulative oil production variation under different fracture spacings.

As seen from Figures 15 and 16, the shortest breakthrough time of injected water is obtained when fracture spacing in the midst is two times the fracture spacing in the end, and the minimum cumulative oil production is achieved. It can be explained as follows: in pattern 2, the fracture spacing nearby injectors is relatively small so that the injected water channels into the end regions' fractures more quickly, which results in a larger rising rate of water cut and worse development effect. Therefore, it is of great importance to enlarge the end regions' fracture spacing to delay the breakthrough time of injected water.

6. Case Study

To further validate the practicability of the proposed model, a 3-D heterogeneous low permeability five-spot waterflood reservoir is established, as shown in Figure 17. The waterflood reservoir size is 2000 m × 1000 m × 24 m, which is vertically divided into 2 layers with a total of 30,149 grid cells. The traditional Kriging interpolation algorithm is then used to generate a heterogenous distribution of matrix permeability, as depicted in Figure 18. The rock and fluid properties for the staggered five-spot model are displayed in Table 2. The well pattern is one staged fracturing horizontal producer located in the center and four vertical injectors located in the corner faces. Water injection is achieved at a constant surface liquid rate of 60 m^3/d, and oil production is triggered with a constant bottom hole pressure of 5.0 MPa. Based on our proposed model, the production performance of the 3-D heterogenous low-permeability waterflood reservoir is well documented.

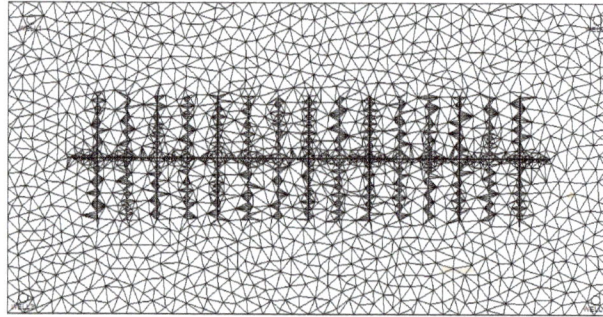

Figure 17. Delaunay triangulation of the heterogeneous waterflood model.

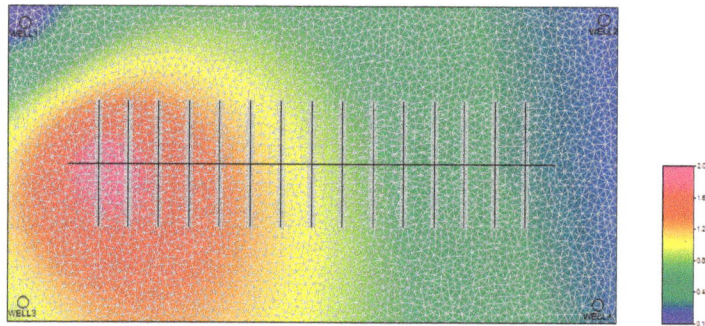

Figure 18. Heterogeneous distribution of matrix permeability in five-spot system.

Table 2. Rock and fluid properties for the 3-D heterogeneous waterflood reservoir.

Parameter	Value	Parameter	Value
matrix porosity	0.1	water viscosity	1.0 mPa·s
matrix permeability	0.1~2.0 mD	oil viscosity	5.0 mPa·s
fracture conductivity	5.0 D·cm	water density	1000 kg/m^3
fracture length	400 m	oil density	750 kg/m^3
fracture spacing	100 m	inject rate	60 m^3/d
length of lateral segment	1600 m	BHP pressure	5.0 MPa
initial reservoir pressure	20 MPa	irreducible water saturation	0.35
bubble point pressure	28 MPa	residual oil saturation	0.22

Figure 19 illustrates the 2-D and 3-D water saturation profile and overall pressure field of low-permeability reservoir when a total of 0.085 PV water is injected. It shows that, due to the weak transport capacity of injected water in low-permeability reservoir, a relatively uniform waterfront advancing behavior is achieved before water breakthrough despite the heterogeneous permeability distribution, and majorities of remaining oil are still unexploited underground. This finding also agrees well with the actual production performance of low-permeability reservoirs commonly found in the Ordos basin, northwestern China, which provides powerful evidence for the applicability and reliability of our proposed model.

(**a**) 2-D water saturation profile

(**b**) 2-D reservoir pressure field

(**c**) 3-D water saturation profile

(**d**) 3-D reservoir pressure field

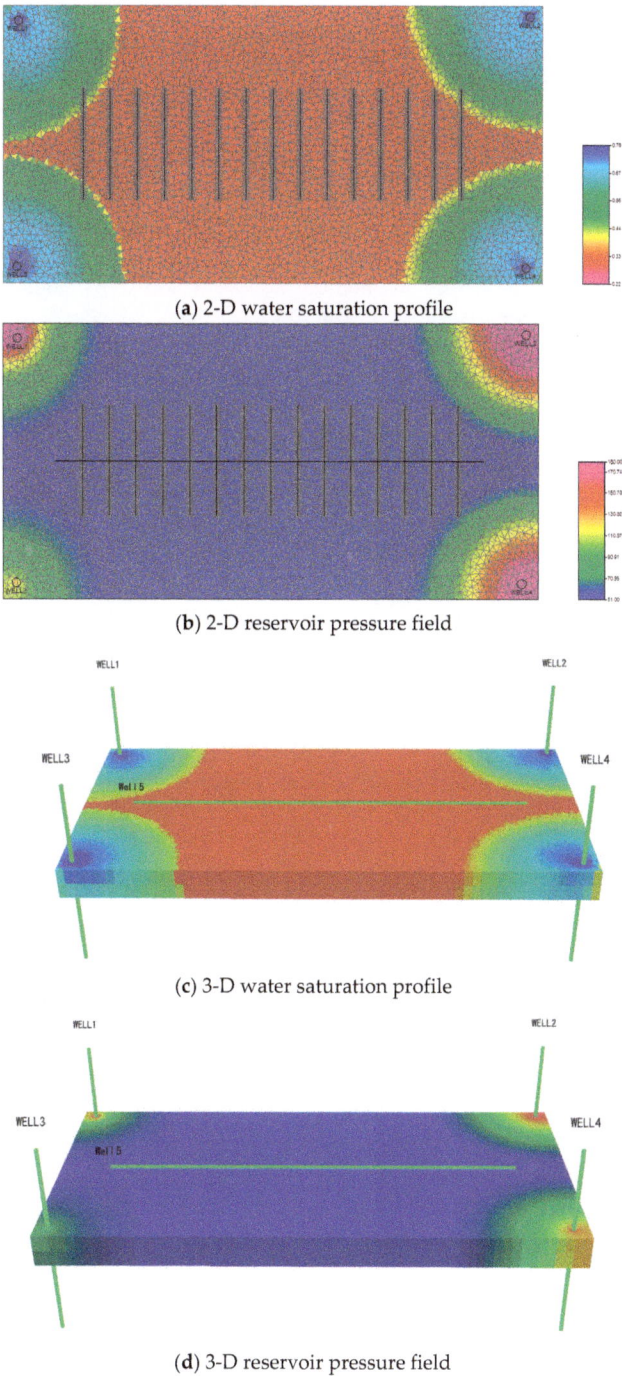

Figure 19. Water saturation and reservoir pressure profile at 0.085PV water injected.

7. Conclusions

(1) Using a discrete fracture model to reduce the dimension of the fracture network explicitly, the two-parameter model is used to represent the nonlinear flow behavior of multiphase fluid in porous media, and the equivalent percolation model of pipe flow is selected to calculate the wellbore pressure drop in a horizontal wellbore. A novel hybrid mathematical model for nonlinear two-phase flow in a fractured low-permeability waterflood reservoir is developed by combining the governing equations satisfied by the matrix, fracture and horizontal wellbore, respectively. By combing the MFD method and FV method, the numerical discretization of the hybrid model is derived and validated using a synthetic staggered five-spot flood system. The effect of fracture properties on nonlinear flow behaviors in fractured low-permeability reservoir are ultimately investigated.

(2) The results show that with an increase of fracture length near injectors, injected water will cross into the horizontal wellbore more easily, resulting in a faster increase of water cut, and a worse development effect. The effect of shortening fracture spacing is consistent with that of increasing fracture length. When performing the optimization design of fracture parameters, it is necessary to shorten the length of fractures nearby injectors and enlarge the fracture spacing of end regions to avoid too early a breakthrough of injected water. Successful practice in modeling the complex waterflood behaviors for a 3-D heterogeneous reservoir provides powerful evidence for the practicability and reliability of our model.

Author Contributions: Conceptualization, D.W. and J.S.; methodology, D.W.; software, H.P.; validation, J.S., H.P. and Y.L.; formal analysis, D.W.; investigation, D.W.; resources, Y.L.; data curation, J.S.; writing—original draft preparation, D.W.; writing—review and editing, Y.L.; visualization, H.P.; supervision, Y.L.; project administration, D.W.; funding acquisition, D.W.

Funding: This research was funded by the National Natural Science Foundation of China (Grant No. 41806070 and 51874346), the PetroChina Innovation Foundation (Grant No. 2018D-5007-0201), the funding provided by the China Postdoctoral Science Foundation (Grant No. 2018M641069 and No. and 2019T120022), the Opening Fund of Key Laboratory of Unconventional Oil & Gas Development (China University of Petroleum (East China)), Ministry of Education, and the Fundamental Research Funds for the Central Universities.

Acknowledgments: The authors sincerely acknowledge the Research Institute of Petroleum Exploration and Development, PetroChina for permission to publish this paper. We are also grateful to all the anonymous reviewers for their constructive comments.

Conflicts of Interest: The authors declare no conflict of interest.

References

1. Li, G.; Sheng, M.; Tian, S.; Huang, Z.; Li, Y.; Yuan, X. Multistage hydraulic jet acid fracturing technique for horizontal wells. *Pet. Explor. Dev.* **2012**, *39*, 100–104. [CrossRef]
2. Liu, R.; Li, B.; Jiang, Y.; Yu, L. A numerical approach for assessing effects of shear on equivalent permeability and nonlinear flow characteristics of 2-D fracture networks. *Adv. Water Resour.* **2018**, *111*, 289–300. [CrossRef]
3. Deng, Y.E.; Liu, C.Q. Mathematical model of nonlinear flow in low permeability porous media and its application. *Acta Pet. Sin.* **2001**, *22*, 72–77.
4. Li, S.Q.; Cheng, L.S.; Li, X.S.; Hao, F. Nonlinear seepage flow models of ultra-low permeability reservoirs. *Pet. Explor. Dev.* **2008**, *35*, 606–612. [CrossRef]
5. Yu, R.; Bian, Y.; Li, Y.; Zhang, X.; Yan, J.; Wang, H.; Wang, K. Non-Darcy flow numerical simulation of XPL low permeability reservoir. *J. Pet. Sci. Eng.* **2012**, *92–93*, 40–47.
6. Shu, W.B.; Xu, H.H.; Wan, J.Y. Numerical research on dynamic effect of capillary pressure in low permeability reservoirs. *Chin. J. Hydrodyn.* **2014**, *29*, 189–196.
7. Liu, W.; Yao, J.; Chen, Z.; Zhu, W. An exact analytical solution of moving boundary problem of radial fluid flow in an infinite low-permeability reservoir with threshold pressure gradient. *J. Pet. Sci. Eng.* **2019**, *175*, 9–21. [CrossRef]
8. Shahraeeni, E.; Moortgat, J.; Firoozabadi, A. High-resolution finite element methods for 3D simulation of compositionally triggered instabilities in porous media. *Comput. Geosci.* **2015**, *19*, 899–920. [CrossRef]

9. Cordero, J.A.R.; Sanchez, E.C.M.; Roehl, D.; Rueda, J.; Mejia, E. Integrated discrete fracture and dual porosity—Dual permeability models for fluid flow in deformable fractured media. *J. Pet. Sci. Eng.* **2019**, *175*, 644–653. [CrossRef]

10. Abbasi, M.; Madani, M.; Sharifi, M.; Kazemi, A. Fluid flow in fractured reservoirs: Exact analytical solution for transient dual porosity model with variable rock matrix block size. *J. Pet. Sci. Eng.* **2018**, *164*, 571–583. [CrossRef]

11. Noorishad, J.; Mehran, M. An upstream finite element method for solution of transient transport equation in fractured porous media. *Water Resour. Res.* **1982**, *18*, 588–596. [CrossRef]

12. Karimi-Fard, M.; Firoozabadi, A. Numerical simulation of water injection in fractured media using discrete-fracture model and the Galerkin method. *SPE Reserv. Eval. Eng.* **2003**, *6*, 117–126. [CrossRef]

13. Hoteit, H.; Firoozabadi, A. An efficient numerical model for incompressible two-phase flow in fractured media. *Adv. Water Resour.* **2008**, *31*, 891–905. [CrossRef]

14. Hauge, V.L.; Aarnes, J.E. Modeling of two-phase flow in fractured porous media on unstructured non-uniformly coarsened grids. *Transp. Porous Media* **2009**, *77*, 373–398. [CrossRef]

15. Zhang, N.; Yao, J.; Huang, Z.; Wang, Y. Accurate multiscale finite element method for numerical simulation of two-phase flow in fractured media using discrete-fracture model. *J. Comput. Phys.* **2013**, *242*, 420–438. [CrossRef]

16. Zhou, F.Q.; Shi, A.F.; Wang, X.H. An efficient finite difference model for multiphase flow in fractured reservoirs. *Pet. Explor. Dev.* **2014**, *41*, 262–266. [CrossRef]

17. Khoei, A.R.; Hosseini, N.; Mohammadnejad, T. Numerical modeling of two-phase fluid flow in deformable fractured porous media using the extended finite element method and an equivalent continuum model. *Adv. Water Resour.* **2016**, *94*, 510–528. [CrossRef]

18. Hosseinimehr, M.; Cusini, M.; Vuik, C.; Hajibeygi, H. Algebraic dynamic multilevel method for embedded discrete fracture model (F-ADM). *J. Comput. Phys.* **2018**, *373*, 324–345. [CrossRef]

19. Baca, R.G.; Arnett, R.C.; Langford, D.W. Modeling fluid flow in fractured-porous rock masses by finite element techniques. *Int. J. Numer. Methods Fluids* **1984**, *4*, 337–348. [CrossRef]

20. Zidane, A.; Firoozabadi, A. An efficient numerical model for multicomponent compressible flow in fractured porous media. *Adv. Water Resour.* **2014**, *74*, 127–147. [CrossRef]

21. Stefansson, I.; Berre, I.; Keilegavlen, E. Finite-volume discretizations for flow in fractured porous media. *Transp. Porous Media* **2018**, *124*, 439–462. [CrossRef]

22. Jiang, L.; Moulton, J.D.; Svyatskiy, D. Analysis of stochastic mimetic finite difference methods and their applications in single-phase stochastic flows. *Comput. Methods Appl. Mech. Eng.* **2012**, *217–220*, 58–76. [CrossRef]

23. Huang, Z.; Yan, X.; Yao, J. A two-phase flow simulation of discrete fractured media using mimetic finite difference method. *Commun. Comput. Phys.* **2014**, *16*, 799–816. [CrossRef]

24. Yan, X.; Huang, Z.; Yao, J.; Li, Y.; Fan, D.; Sun, H.; Zhang, K. An efficient numerical hybrid model for multiphase flow in deformable fractured-shale reservoirs. *SPE J.* **2018**, *23*, 1412–1437. [CrossRef]

25. Yan, X.; Huang, Z.; Yao, J.; Li, Y.; Fan, D. An efficient embedded discrete fracture model based on mimetic finite difference method. *J. Pet. Sci. Eng.* **2016**, *145*, 11–21. [CrossRef]

26. Zhang, Q.; Huang, Z.; Yao, J.; Wang, Y.; Li, Y. Multiscale mimetic method for two-phase flow in fractured media using embedded discrete fracture model. *Adv. Water Resour.* **2017**, *107*, 180–190. [CrossRef]

27. Jing, P.; Omar, A.H.; Mary, F.W. Data assimilation method for fractured reservoirs using mimetic finite differences and ensemble Kalman filter. *Comput. Geosci.* **2017**, *21*, 781–794.

28. Gyrya, V.; Lipnikov, K. The arbitrary order mimetic finite difference method for a diffusion equation with a non-symmetric diffusion tensor. *J Comput. Phys.* **2017**, *348*, 549–566. [CrossRef]

29. Dikken, B.J. Pressure drop in horizontal wells and its effect on their production performance. *J. Pet. Technol.* **1990**, *42*, 1426–1433. [CrossRef]

30. Jansen, J.D. A semi-analytical model for calculating pressure drop along horizontal wells with stinger completions. *SPE J.* **2003**, *8*, 138–146. [CrossRef]

31. Yao, S.; Zeng, F.; Liu, H.; Zhao, G. A semi-analytical model for multi-stage fractured horizontal wells. *J. Hydrol.* **2013**, *507*, 201–212. [CrossRef]

32. Chen, Z.; Liao, X.; Zhao, X.; Zhu, L.; Liu, H. Performance of multiple fractured horizontal wells with consideration of pressure drop within wellbore. *J. Pet. Sci. Eng.* **2016**, *146*, 677–693. [CrossRef]

Energies **2019**, *12*, 2850

33. Li, H.; Tan, Y.; Jiang, B.; Wang, Y.; Zhang, N. A semi-analytical model for predicting inflow profile of horizontal wells in bottom-water gas reservoir. *J. Pet. Sci. Eng.* **2018**, *160*, 351–362. [CrossRef]

34. Vodorezov, D.D. Estimation of horizontal-well productivity loss caused by formation damage on the basis of numerical modeling and laboratory-test data. *SPE J.* **2019**, *24*, 44–59. [CrossRef]

35. Wu, S.H.; Liu, X.E.; Guo, S.P. A simplified model of flow in horizontal wellbore. *Pet. Explor. Dev.* **1999**, *26*, 64–66.

36. Birchenko, V.M.; Usnich, A.V.; Davies, D.R. Impact of frictional pressure losses along the completion on well performance. *J. Pet. Sci. Eng.* **2010**, *73*, 204–213. [CrossRef]

37. Wang, Z.; Zhang, Q.; Zeng, Q.; Wei, J. A unified model of oil/water two-phase flow in the horizontal wellbore. *SPE J.* **2017**, *22*, 353–364. [CrossRef]

38. Lipnikov, K.; Manzini, G.; Moulton, J.D.; Shashkov, M. The mimetic finite difference method for elliptic and parabolic problems with a staggered discretization of diffusion coefficient. *J. Comput. Phys.* **2016**, *305*, 111–126. [CrossRef]

energies

MDPI

Article

History Matching and Forecast of Shale Gas Production Considering Hydraulic Fracture Closure

Juhyun Kim [1], Youngjin Seo [1], Jihoon Wang [2,*] and Youngsoo Lee [1,*]

[1] Department of Mineral Resources and Energy Engineering, Chonbuk National University, 567 Baekje-daero, Deokjin-gu, Jeonju-si, Jeollabuk-do 561-756, Korea; jump1604@jbnu.ac.kr (J.K.); yjseo@jbnu.ac.kr (Y.S.)

[2] Department of Petroleum and Natural Gas Engineering, New Mexico Tech, 801 Leroy Place, Socorro, NM 87801, USA

* Correspondence: jihoon.wang@nmt.edu (J.W.); youngsoo.lee@jbnu.ac.kr (Y.L.);
Tel.: +01-(575)-835-5289 (J.W.); +82-(63)-270-2392 (Y.L.); Fax: +82-(63)-270-2366 (Y.L.)

Received: 7 March 2019; Accepted: 27 April 2019; Published: 29 April 2019

Abstract: Most shale gas reservoirs have extremely low permeability. Predicting their fluid transport characteristics is extremely difficult due to complex flow mechanisms between hydraulic fractures and the adjacent rock matrix. Recently, studies adopting the dynamic modeling approach have been proposed to investigate the shape of the flow regime between induced and natural fractures. In this study, a production history matching was performed on a shale gas reservoir in Canada's Horn River basin. Hypocenters and densities of the microseismic signals were used to identify the hydraulic fracture distributions and the stimulated reservoir volume. In addition, the fracture width decreased because of fluid pressure reduction during production, which was integrated with the dynamic permeability change of the hydraulic fractures. We also incorporated the geometric change of hydraulic fractures to the 3D reservoir simulation model and established a new shale gas modeling procedure. Results demonstrate that the accuracy of the predictions for shale gas flow improved. We believe that this technique will enrich the community's understanding of fluid flows in shale gas reservoirs.

Keywords: shale gas; stimulated reservoir volume; microseismic; hydraulic fracture closure; production history matching

1. Introduction

Global energy consumption is steadily increasing, and as of 2017, natural gas has become a vital resource, supplying 28% of the world's energy [1]. Natural gas offers an additional significant advantage in that it generates only half of the greenhouse gases of other fossil fuel sources [2]. In 2012, carbon dioxide emissions in the U.S. decreased to their lowest levels in 20 years, which can be attributed to the replacement of coal-fired power plants with natural-gas-fired power plants [3]. Consequently, natural gas has garnered more interest as an alternative and environmentally friendly energy source. Shale gas, in particular, has since emerged as an unconventional resource. Although shale gas production accounted for only 1% of natural gas production in 2000 in the U.S., this value increased to >20% in 2010. According to the Energy Information Administration (EIA) 2018 annual energy report, most of the U.S.'s natural gas supply is expected to be produced from shale and tight reservoirs (Figure 1) [1,4].

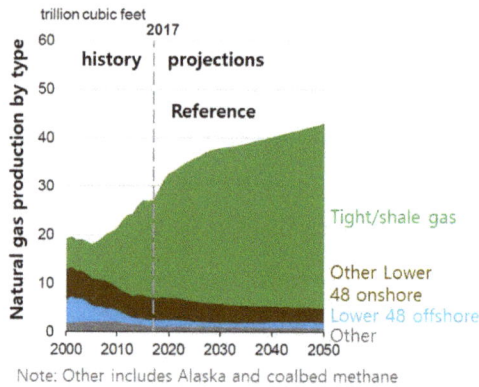

Figure 1. Natural gas production by type, 2000–2050 (trillion cubic feet) [1,4]. Reproduced from [1,4], EIA: 2018, Stevens: 2012.

Horizontal drilling and hydraulic fracturing techniques have become the standard technologies for shale gas development. Generally, shale formations have extremely low permeability, in the order of 1×10^2 nano-Darcy for liquid-rich shale and 10 nano-Darcy for dry gas shale [5]. Moreover, production forecasting of shale gas reservoirs is still very challenging because the fluid flow phenomena are very complex and the induced hydraulic fracture networks are difficult to model [6]. To overcome these obstacles, various studies focused on flow simulations that use microseismic monitoring data, which improved researchers' understanding of the shape of the hydraulic fractures as well as the flow regime during production [7–17]. These studies also revealed that numerical simulations can be used to construct hydraulic fracture geometries for reliable history matching and production forecasting [18–20]. In general, dual porosity and dual permeability models are used to describe the fluid flow through matrices and natural fractures. The dual porosity model assumes that there is no fluid flow between the matrix grids and that the rock matrix simply supplies gas to adjacent fractures (Figure 2) [21]. In contrast, the dual permeability model considers both the fluid flow within fractures and between matrix grids (Figure 3) [22]. According to Ho [23], the dual permeability model yields more reliable outcomes for shale reservoir analysis. An approach that employs seismic data for fracture network characteristics at subsurface reservoirs has been proposed; this approach can be successfully applied to production forecast simulations using the 3D discrete fracture network model [24,25].

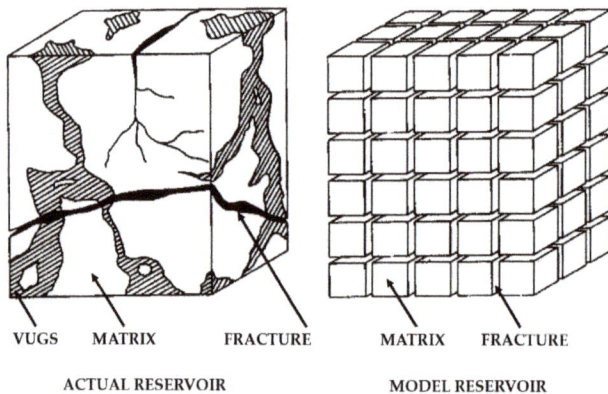

Figure 2. Schematic illustration of dual porosity model [21]. Reproduced from [21], Warren: 1963.

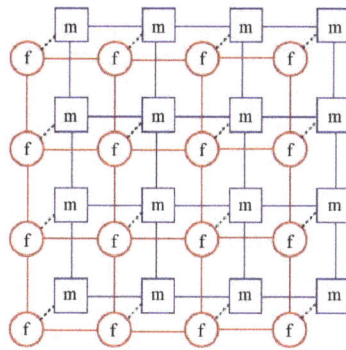

Figure 3. Schematic illustration of dual permeability model [22]. Reproduced from [22], Zeng: 2015.

Cipolla et al. [26] proposed a workflow that combined microseismic data with dynamic simulations. By constructing 3D hydraulic fracture networks as a series of very fine grids and implementing the networks with a reservoir simulation model, the authors demonstrated that fluid flow analysis enhances accuracy.

Methods for constructing hydraulic fracture grids can be classified into three types according to the complexity of the grids for the hydraulic fractures: planar, wire-mesh, and un-structured fracture model (UFM) (Figure 4). The planar model is most commonly used to represent hydraulic fractures, because it can simply describe the fractures with a set of planes. However, it cannot be well applied to the stimulated reservoir volume (SRV) because it solely focuses on fluid flows in the hydraulic fractures. In contrast, the wire-mesh model suggests more complicated fracture geometries with the assumption that the orthogonally generated planes are more reliable. To construct a more realistic model, UFM yields the most complex geometries with irregular grids to describe the actual fracture shape [26]. However, the model requires considerable computation time and much more sensitive flow analyses. Hence, the planar and wire-mesh models are widely accepted for numerical flow simulations in shale reservoirs. This study incorporates fluid flows in the fractures and the SRV; therefore, the wire-mesh model was adopted to construct the hydraulic fracture network. To reliably represent the induced fracture geometry, microseismic hypocenters and densities were used to construct the model.

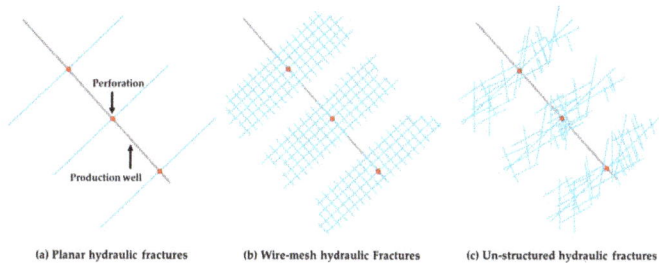

(a) Planar hydraulic fractures   (b) Wire-mesh hydraulic Fractures   (c) Un-structured hydraulic fractures

Figure 4. Illustration of various fracture geometry models to approximate a horizontal well. (**a**) Simple planar fractures (conventional approach), (**b**) wire-mesh hydraulic fractures, and (**c**) un-structured hydraulic fractures (UFM) [26]. Reproduced from [26], Cipolla: 2011.

As production continues, the fluid pressure in the hydraulic fractures decreases; the increased effective stress reduces the width of hydraulic fractures filled with proppant [27]. Since this effect is directly related to fracture permeability, the fluid pressure reduction results in deteriorated gas productivity [28]. When this phenomenon is significant, it may be extended to the proppant crush or embedment [29,30]. The fluid pressure reduction in the fractures is more considerable in the early production period, because fractures are initially pressurized as high as the reservoir pore pressure (or higher if the excessive fracture fluid pressure after the fracturing process has not yet been released).

Consequently, this process yields misleading results because production forecasting for the mid or late period is based on the reservoir properties obtained from a history matching process in the early period unless alterations in the fracture permeability are considered.

To reliably extrapolate the observations from previous studies to a field scale, we investigate the effect of hydraulic fracture closure during the production period and the actual productive reservoir volume induced by the hydraulic fractures during long-term production. Although several studies have attempted to understand the effect of stress on fractures [18–20], few field-scale studies have been performed to investigate the fracture width reduction due to stress changes. Furthermore, many studies have calculated estimated ultimate recovery (EUR) based on microseismic data, which does not always represent the productive reservoir volume. In addition, predicted long-term production shows that the SRV obtained using microseismic data are inconsistent with the actual productive volume of hydraulic fractures. To avert these issues, we use the relation between pore pressure and hydraulic fracture width via history matching and directly apply it to reservoir simulation. Consequently, our results are applicable for the more precise prediction of EUR at the early production stage.

2. Material and Methods

2.1. Dynamic Modeling Workflow

A typical reservoir simulation workflow comprises the processes shown in Figure 5 (left) [31,32]. However, additional steps are required for shale reservoir simulation because of the existence of hydraulic fractures, which provide conductive flow paths. To construct the hydraulic fracture network, the SRV must be identified. Although microseismic data are most desirable for this process, their availability is frequently restricted due to cost. If microseismic data are unavailable, the SRV can be estimated via hydraulic fracture modeling based on information obtained during fracturing treatment such as the injected volume of water and proppants and surface treating pressure. Therefore, a typical workflow for the shale gas simulation process contains additional procedures, particularly for determining the SRV and hydraulic fracture geometry, as shown in Figure 5 (right).

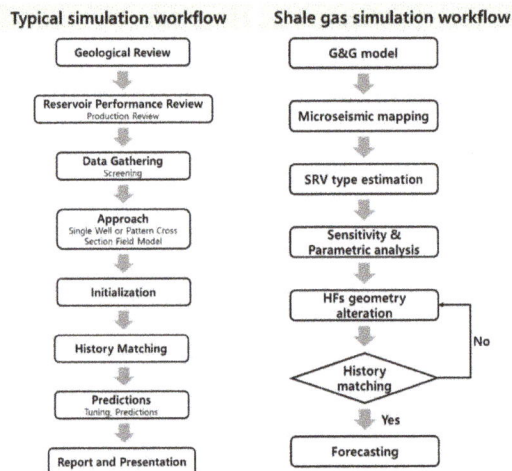

Figure 5. Workflows for typical geological simulations and shale gas simulation workflow [31,32]. Reproduced from [31,32], Ertekin: 2001, Carlson: 2003.

Reliable history matching processes are very challenging and require experience and insight from multiple disciplines. One of the greatest obstacles in the history matching of shale gas productivity is the characterization of hydraulic fractures (such as length, width, and permeability), which are the dominant parameters for productivity analysis and the most difficult to precisely compute. Regardless of the usefulness of microseismic data as an indicator of the SRV, the data still contain uncertainties because their signals do not always represent conductive fracture generation; thus, the calculated volume may be overestimated. In short, reliable flow simulations on shale reservoirs rely on the identification of hydraulic fracture properties and interpretation of the microseismic data. Accordingly, the SRV was determined by hypocenters and densities of microseismic data and the hydraulic fracture network model was constructed. Consequently, the simulation workflow for shale reservoirs was improved by considering the permeability alterations of hydraulic fractures.

2.2. Construction of Dynamic Model

The target reservoir is in northeast British Columbia, Canada. A dynamic model was constructed for the reservoir of approximately 2.5 years production. Table 1 lists the reservoir properties and model description; the reservoir model was constructed using a commercial black oil simulator (CMG, IMEX). Generally, shale gas production comprises three effects: free gas, diffusion, and desorption. In this case, we considered only free gas flow and diffusion. In the case of desorption, the total organic carbon of the target formation was low, and during the two-and-a-half-year production period, the average reservoir pressure decreased from 32,000 kPa to 16,000 kPa, which typically results in less than 10% desorption [33]. Mainly, the pressure drop occurs only in hydraulic fractures and a few adjacent matrix grids. Therefore, although large amounts of gas adsorb in whole matrix grids, it does not contribute to production. Hence, the effect of desorption is not considered in this case.

Table 1. Reservoir properties and model description.

Parameter	Value	Parameter	Value
Simulation type	Black oil	Number of grid (ea)	200,000
Top depth range (m)	1895–2177	Fluid type	Gas (CH_4 95% over)
Pressure (kPa)	32,000	Temperature (°C)	132
Initial water saturation	0.25	Initial gas saturation	0.75
Matrix porosity	0.05	Matrix permeability (md)	2.65×10^{-6}
Hydraulic fracturing spacing (m)	≈ 37	Length of the horizontal well (m)	≈ 3200

The composition of the reservoir fluid was obtained from gas analysis data, which indicate that the existing fluid is identified as dry gas containing more than 95% CH_4. Therefore, the black oil simulation scheme has been adopted for numerical simulation.

Relative permeability curves in the fractured system were first proposed by Romm (1966) [34]. Romm's model suggests that relative permeability of a fracture flow can be simplified by a linear function of saturation. However, several recent studies [35–38] emphasize that relative permeability in fractures behaves non-linearly. Chima and Geiger [39] note that relative permeability calculations using the Romm's model yield misleading results with overestimated gas production. In this study, the gas–water relative permeability curve was generated based on a non-linear mathematical model (Equation (1)) and is shown in Figure 6. For the relative permeability curve of the matrix, the end-points were selected as matching parameters in the history matching process because no experimental results are available.

$$k_{rg} = S_g^2 \left(\frac{2\mu_w S_g^2 + 3\mu_g S_w^2 + 6 S_w S_g \mu_g}{12 \mu_w} \right)$$
$$k_{rw} = S_w^2 \left(\frac{4 S_w^2 + 6 S_w S_g}{12} \right),$$

(1)

where k_{rg} and k_{rw} are the gas–water relative permeability of hydraulic fractures, S_g and S_w are the gas–water saturation, and μ_g and μ_w are the gas–water viscosity, respectively [39].

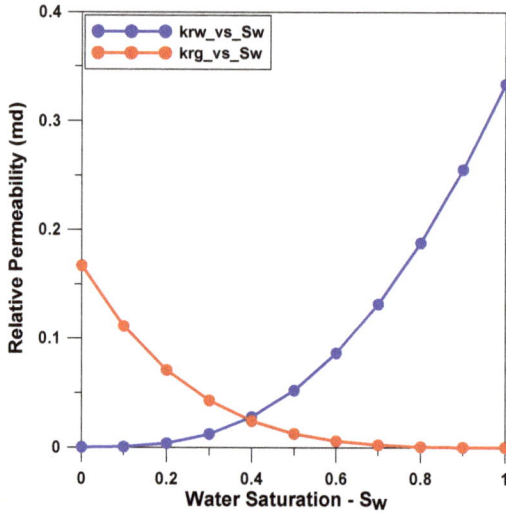

Figure 6. Relative permeability of hydraulic fractures.

2.3. Microseismic Mapping and SRV Calculation

Microseismic data are seismic signals with small magnitudes generated by rock failure during the hydraulic fracturing process. From the hypocenters, times, and magnitudes of the signals, the SRV can be estimated and the induced fracture geometry can be determined. Fracturing processes in shale reservoirs are intended to induce a fracture with a long half-length, which is directly related to the SRV. In general, fracture half-lengths determined by microseismic data yield relatively higher values than those of other diagnostic techniques (Figure 7) because microseismic signals are generated from both conductive propped fractures (proppant-filled) and non-propped fractures. The latter is more likely to close as the effective stress increases and contributes less to the reservoir productivity. Nevertheless, microseismic is a powerful tool for determining fracture geometry [40].

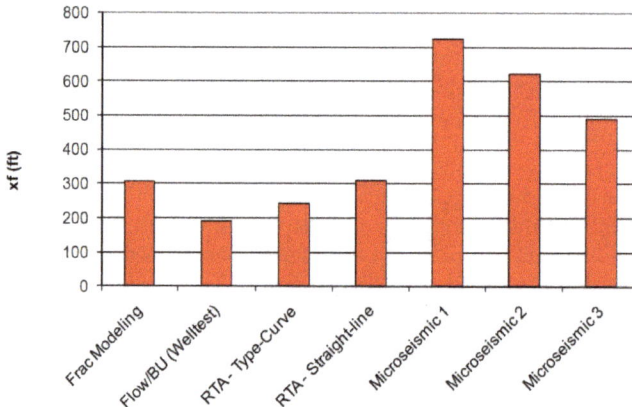

Figure 7. Comparison of fracture half-lengths (xf) derived from various sources [41]. Reproduced from [41], Clarkson: 2011.

Suliman et al. [42] propose a method for using microseismic data to define the shape and size of the SRV in a simulation model. The authors distribute the signal density into the simulation grids to quantitatively evaluate the SRV. They suggest that areas with a high density of signals are expected to be more stimulated than those with a low density. Based on the stimulation rate and connectivity of the grids, the SRV is divided into three categories as follows. First, Hydraulic SRV (HSRV) assumes that all microseismic signals are related to hydraulic fractures. Second, Conductivity SRV (CSRV) indicates that two or more microseismic signals are emitted in a single grid, and this grid will have higher permeability than a grid in HSRV. Finally, in the Flush SRV (FSRV), three or more microseismic signals are detected in a grid. Normally, the grids are located very close to the production well and have the highest permeability.

During a hydraulic fracturing process in the target reservoir, a total of 2000 signals were acquired over 31 stages. The hypocenters of the acquired signals for each stage are shown in Figure 8. Accordingly, the SRV was generated as described in Figure 9 and Table 2.

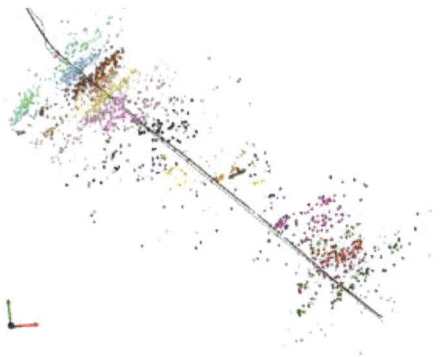

Figure 8. Microseismic signals for the each of stages in target well.

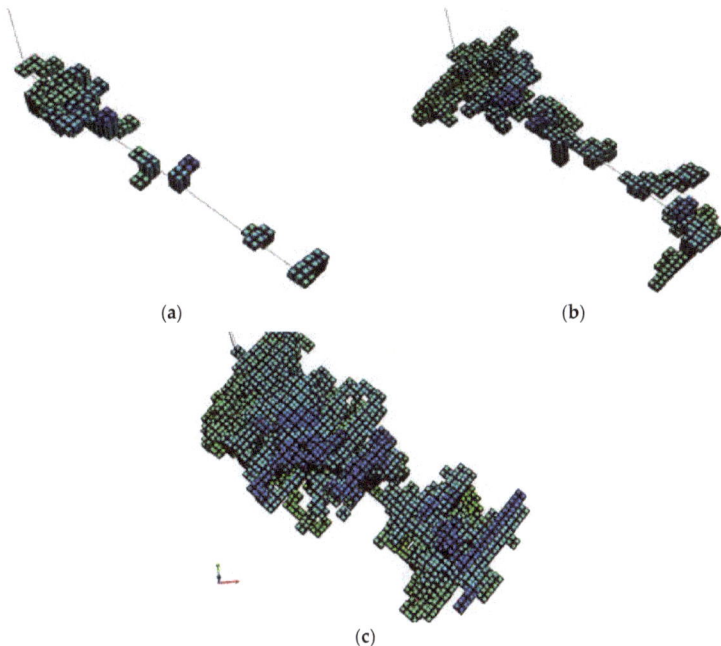

(a)

(b)

(c)

Figure 9. Constructed SRV. (**a**) Flush SRV, (**b**) Conductivity SRV, and (**c**) Hydraulic SRV for the target reservoir. (Colors of the grids indicate the top depths).

Table 2. Stimulated reservoir volume (SRV) information.

SRV type	FSRV	CSRV	HSRV
Number of blocks (ea)	200	463	1798
Volume (m³)	8,303,275	19,183,333	73,438,690

The LS-LR-DK (logarithmically spaced, locally refined, dual permeability) method was applied to describe the hydraulic fractures (Figure 10). The method generates a very fine fracture grid within a matrix grid. Since permeability values assigned to the hydraulic fracture grids are much higher than that of the matrix blocks, convergence problems occur when dimensions of the fracture grids are the same as the actual fracture size (μm scale). Therefore, the fracture grids in a simulation model usually have larger sizes (1 to 2 ft) than the actual ones. In addition, the effective permeability is calculated by Equation (2) and is used for each block, instead of the actual permeability. This method enables the fastest runtime without loss of accuracy in expressing hydraulic fractures in the reservoir simulation [43].

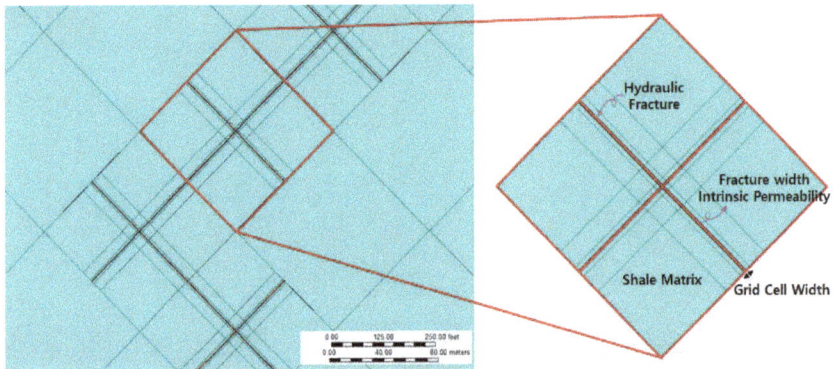

Figure 10. LS-LR-DK (logarithmically spaced, locally refined, dual permeability).

$$k_{eff} w_{grid} = k_f w_f$$
$$k_{eff} = \frac{k_f w_f}{w_{grid}},$$

(2)

In the above expressions, k_{eff} is the effective fracture permeability, w_{grid} is the grid-cell width, k_f is the intrinsic permeability, and w_f is the effective fracture width.

Although liquid flow in a porous rock can be simply described by Darcy's Law, the description is not valid for high-rate gas flow because inertial forces are not negligible. To characterize the non-Darcy flow, Darcy's equation was extended with a quadratic flow term. Equation (3) is known as the Forchheimer equation for non-Darcy flow [44]. Especially, β is the coefficient of inertial flow resistance or turbulence factor, which is a characteristic of porous rocks much like permeability and porosity. Inertial flow resistance is also related to the contrast in size between pore throats and pore bodies, which Hagoort [45] has well summarized for both non-Darcy flow and β.

$$\frac{\Delta p}{L} = \frac{\mu}{k} u + \rho \beta u^2,$$

(3)

In the above expression, Δp is the pressure difference between the inlet and outlet, L is the sample length, μ is the fluid viscosity, k is the permeability, u is the volumetric velocity ($= q_i/A$), q_i is volumetric injection rate, A is the sample cross-sectional area, and ρ is the fluid density, while β is the coefficient of inertial flow resistance or turbulence factor [45].

Although many researchers have studied β, the coefficient is difficult to apply to the reservoir simulation model precisely. To describe the non-Darcy flow in the reservoir simulation, the fracture width needs to be larger (1 to 2 ft) than the actual width (generally less than 1 mm) due to the convergence problem. Hence, the non-Darcy coefficient correction factor (κ) concept offered by CMG needs to be additionally incorporated in the reservoir simulation model [46]. This concept can help to effectively model non-Darcy flow in fine grid blocks, which are set up to describe very thin hydraulic fractures. κ was calculated using Equation (4).

$$\kappa = \left(\frac{k_f}{k_{eff}}\right)^{2-N1g} = \left(\frac{w_{grid}}{w_f}\right)^{2-N1g} \tag{4}$$

In the above expression, k_f is the intrinsic fracture permeability, k_{eff} is the effective fracture permeability, w_{grid} is the intrinsic fracture width, w_f is the effective fracture width, and $N1g$ is an exponent of the $\left(k_g \times k\right)$ term in the β factor correlation for the model in question, in which case $N1g = 1.021$.

3. Results and Analysis

3.1. Production History Matching

To identify the important parameters that affect shale gas productivity, we carried out sensitivity analyses by adjusting the ranges of several parameters. According to Novlesky et al. [43], the most sensitive variables that affect cumulative gas production are hydraulic fracture spacings, hydraulic fracture permeability, and natural fracture permeability. In an attempt to identify the most sensitive parameters using production history matching, a set of sensitivity analyses was performed (Figure 11). If it is assumed that if the SRV does not change by the parameter adjustment made during the analysis, the hydraulic fracture properties, such as hydraulic fracture width and hydraulic fracture intrinsic permeability, will most significantly impact reservoir productivity—especially in the case of hydraulic fracture width, when we construct the hydraulic fracture grid in the reservoir simulation model. We obtained the approximate value of the hydraulic fracture width using the Mangrove Stimulation Design tool from Schlumberger.

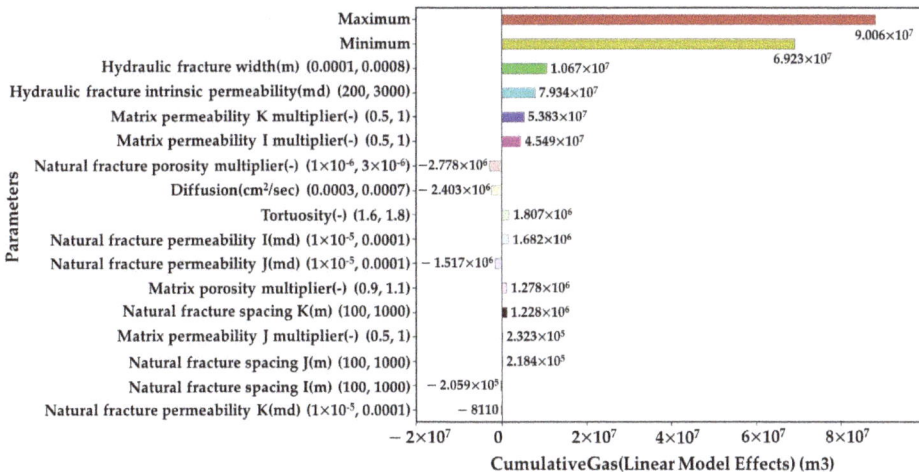

Figure 11. Tornado plot of linear effect estimates for cumulative gas production (fixed SRV) Horn River case.

At first, the final parameters were determined by production history matching, ignoring the fracture width change during production (Table 3 and Figure 12). However, when compared with the actual production data, the gas production rate with the matched parameters displayed an error of approximately 5.5%. In this case, we used averaged daily production rate to weekly production rate for the reduction of computation time that production variation was normalized. Because the goal of this study is to find the effect of hydraulic fracture closure on gas recovery, several shut-in periods (normally less than 1 week) were eliminated and overall production history was modified without major trend changes for the fast history matching. As shown in Figure 12, the production history in the early stage appears to match well with the actual data; however, the model strays from the actual data at around 100 days increasing gradually. It is expected that this phenomenon is caused by the change in hydraulic fracture geometries as the production proceeds. As a result, the simulation results overestimate the gas production rate in the late stage of the production. Thus, the model needs to be updated to consider the width and permeability change of the hydraulic fractures during the production period.

Table 3. Parameter values of matched model.

Property	Min Value	Max Value	Matched Value	Unit
Hydraulic fracture intrinsic permeability	200	3000	450	md
Hydraulic fracture width	0.0001	0.002	0.001	m
Natural fracture spacing I			550	
Natural fracture spacing J	100	1,000	460	m
Natural fracture spacing K			370	
Natural fracture permeability I			0.0001	
Natural fracture permeability J	1×10^{-5}	0.0001	1×10^{-5}	md
Natural fracture permeability K			2.8×10^{-5}	
Matrix permeability I			0.00012	
Matrix permeability J	0	0.0016	0.00015	md
Matrix permeability K			0.00008	
Natural fracture porosity	1×10^{-6}	3×10^{-6}	2.6×10^{-6}	-
Matrix porosity	5.8×10^{-7}	0.147	0.054	-
Tortuosity	1.3	1.9	1.7	-
Diffusion	0.0003	0.0007	0.00058	cm^2/s

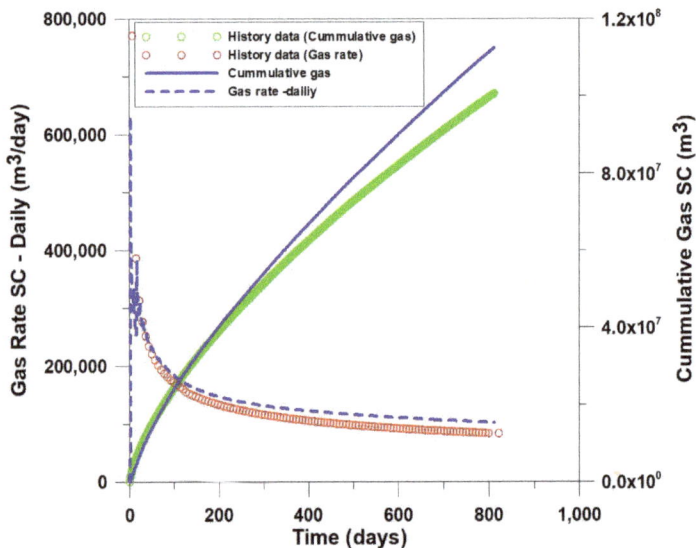

Figure 12. Production history matching result.

3.2. Fracture Width Change Due to Stress

In order to overcome the model's shortcomings as described in the previous section, the simulation model was updated to consider fracture width change.

Hydraulic fracture propagation is aligned with the direction of the maximum principal stress when the fracturing pressure exceeds the minimum principal stress, and thus hydraulic fractures are generated perpendicular to the minimum principal stress [47]. Various experiments have been conducted to investigate the closure behavior of hydraulic fractures [48–50]. Kam et al. [48] performed a set of experiments to analyze the behavior of fracture conductivity change under different confining stress levels and found that the conductivity of induced fractures decreases with the confining stress increment, while that of the natural fracture showed lower decrements (Figure 13). In addition, Palisch et al. [51] examined the fracture conductivity loss mechanisms and confirmed that the fluid pressure reduction in the fracture has a significant effect on the fracture conductivity. In that study, the authors showed that the fracture conductivity can be drastically dropped to 6–10% of the initial values.

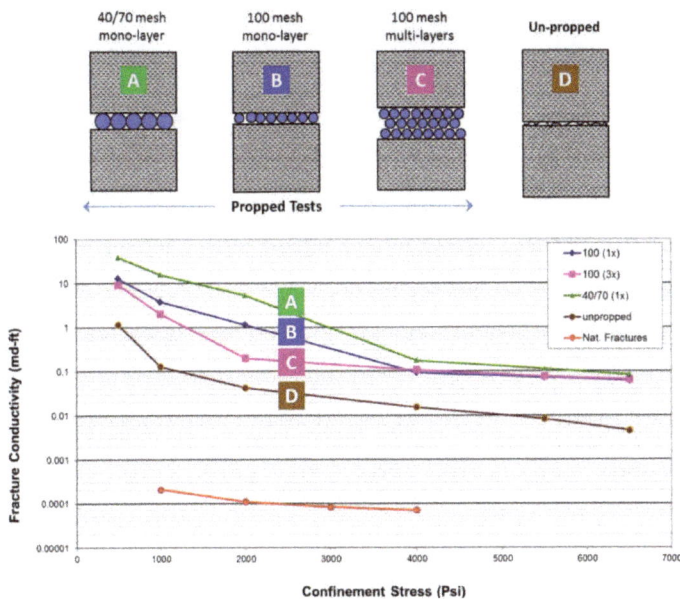

Figure 13. Fracture conductivity measurement test results that show fracture conductivity reduction by confining stress increase [48]. Reproduced from [48], Kam: 2014.

To consider these phenomena, distances affected by the pressure of the production well over time were computed, as shown in Figure 14. After 2 years of production, the fluid pressure at more than 250 m from the production well was decreased. The area affected by the production corresponds well with the FSRV. As shown in Figure 15, the 2 years of production decreased the average fluid pressure in the FSRV from 32 MPa to 17 MPa, which can be approximated with a semi-logarithmical relationship. Therefore, the fluid pressure in the fractures in the FSRV would significantly influence the production when considering the fracture permeability alterations due to the effective stress increase at each time step.

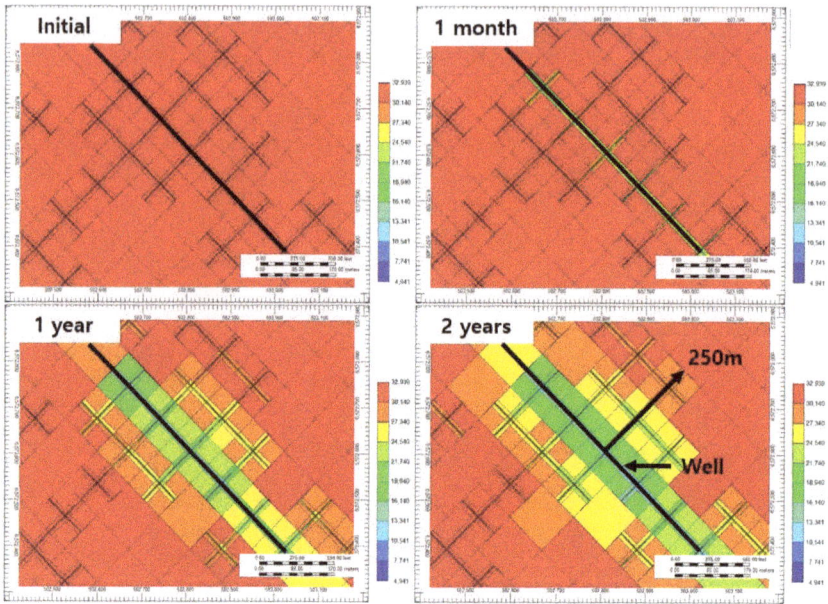

Figure 14. Pressure propagation over time. (Colors of the grids indicate the fluid pressure (kPa)).

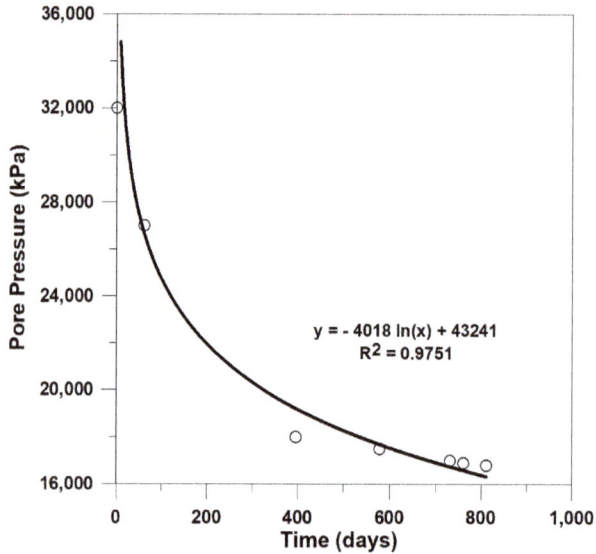

Figure 15. Average fracture fluid pressure in the flush SRV (FSRV) over time.

3.3. Improved Production History Matching and Forecast

As described in Section 3.1, the difference between the simulation results and actual production data in the late stage of production was mainly caused by ignoring the impact of the effective stress increase on the hydraulic fracture width. In order to yield more reliable results, the simulation model has been enhanced by adopting the change of the fracture width in the FSRV region when the error against the actual production data increases to 5%. Table 4 illustrates the effective permeability of hydraulic fractures during the production calculated from Equation (2), which is exponentially correlated ($R^2 = 0.9855$) with the fluid pressure in the hydraulic fractures (Figure 16). Consequently, this correlation was incorporated into the simulations in the form of a fracture closure relationship (Table 5). Using this procedure, change in the fracture width over time can be automatically incorporated into the simulation from the fluid pressure at each time step.

Table 4. Hydraulic fracture width and effective permeability change according to production time within FSRV.

Time (days)		0	43	127	239	392
	Width (m)	0.001000	0.000950	0.000930	0.000920	0.000915
FSRV	Effective Permeability (md)	3.2808	2.3376	1.9833	1.6601	1.5010

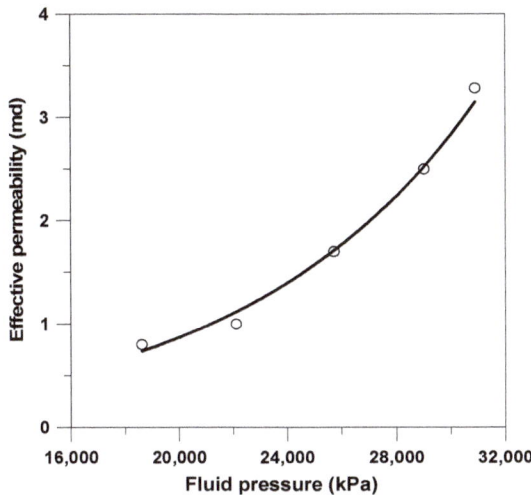

Figure 16. Effective permeability change according to fluid pressure.

Table 5. Fracture closure relationship according to fluid pressure change of hydraulic fractures.

No.	Fluid Pressure of Hydraulic Fractures	Permeability Multiplier	No.	Fluid Pressure of Hydraulic Fractures	Permeability Multiplier
1	4000	0.06	7	10,000	0.11
2	5000	0.07	8	15,000	0.18
3	6000	0.07	9	20,000	0.29
4	7000	0.08	10	25,000	0.48
5	8000	0.09	11	30,000	0.79
6	9000	0.10	12	32,302	1

With a fracture closure relationship included in the simulation, the history matching demonstrated more accurate results and the cumulative gas matching error was reduced significantly from 5.5% to under 1% (Figure 17). This clear improvement suggests that hydraulic fracture closure should be considered in shale gas simulations. In order to determine the impact on future productivity by hydraulic fracture closure, a forecasting simulation was performed over the next 20 years and the differences for both models were computed (Figure 18). According to the previous model that ignores the fracture width change, the cumulative production was 5.23×10^8 m^3, which is higher than the enhanced model with a magnitude of 0.5×10^8 m³ ($\fallingdotseq$1.8 Billions of cubic feet). These results suggest that if the fracture width is not considered, cumulative production will be significantly overestimated. Consequently, the results from our model show improved accuracy over those of the model ignoring the effective stress effect on the fracture width and permeability. Therefore, more reliable history matching and production forecasting can be achieved by considering the hydraulic fracture permeability change.

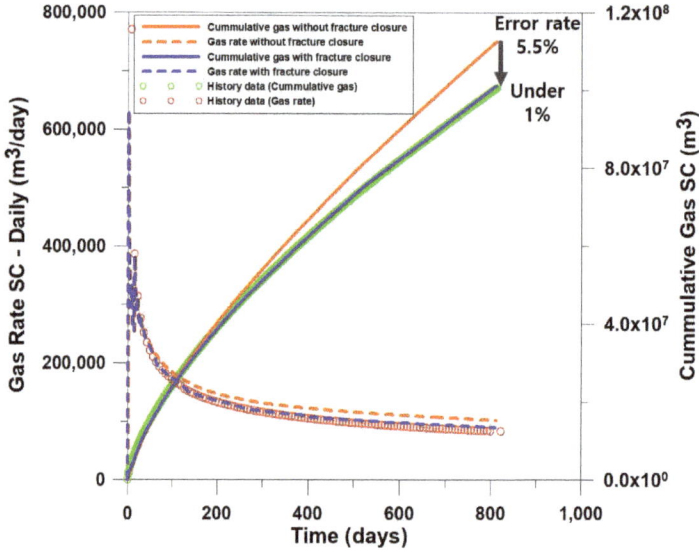

Figure 17. Comparison of production history matching results with consideration of changes in fracture width.

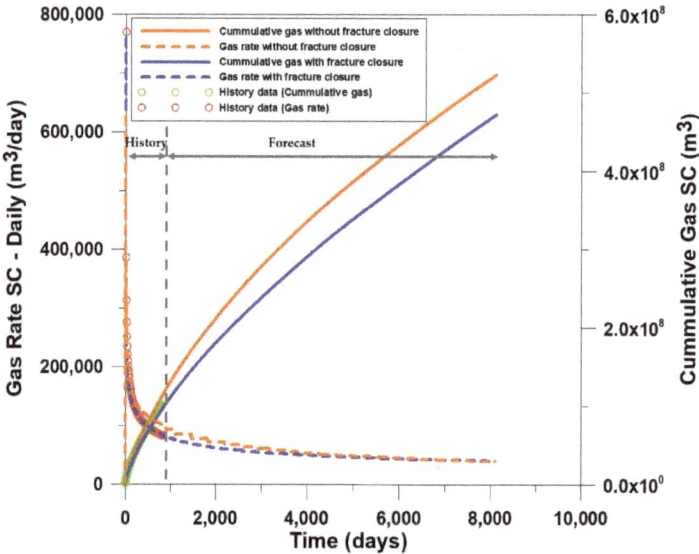

Figure 18. Comparison of production forecasting results with consideration of changes in hydraulic fracture width.

3.4. Productive Volume Analysis with Microseismic Data

The primary objective of the microseismic analysis is to determine the fracture geometry and distribution and thus, to reliably estimate the SRV. However, it is observed that the actual productive area estimated by the simulation process is frequently mismatched with the SRV derived from microseismic data.

As production progresses, the fluid pressure around the production well decreases and propagates away from the well. When fluid pressure is decreased in a grid, it indicates that the grid is contributing to reservoir productivity. If we assume grid blocks with a pressure drop of more than 10% compared to the initial pressure are involved in production, the volume contributing to the production can be observed in Figure 19. As a result, the grid block volumes after 5 and 20 years of the production are 2.57×10^7 m^3 and 7.88×10^7 m^3, respectively. In particular, the production volume (7.88×10^7 m^3) after 20 years production period is similar to the HSRV in Table 2. However, even though the volume is similar, the HSRV shape using the microseismic data and the productive volume estimated through the simulation are different from each other. The reason for this discrepancy is that HSRV is based primarily on the location at which the signal was generated, while the simulation results include the main flow path (hydraulic fractures) and surrounding matrix blocks.

To explain this concept more clearly, a comparison of the productive volume of hydraulic fractures is shown in Figure 20. In both cases, the productive volume of hydraulic fractures increased and then stabilized after a certain period, but the lower value was obtained when hydraulic fractures were closed, which indicates that as the fractures close according to the pore pressure reduction, hydraulic fractures more than a certain distance away from the production well lose their ability to flow gas.

Moreover, the stabilized productive volume of hydraulic fractures accounts for only 65% ($\fallingdotseq$210,000 m^3) of the total hydraulic fracture volume calculated from the microseismic data ($\fallingdotseq$330,000 m^3). This means that the activated hydraulic fractures involved in production are smaller than those of the microseismic-derived SRV. These results are elaborated on in Figure 21. As shown in this figure, hydraulic fractures farther than a certain distance do not contribute to the production, and the actual half-length of the hydraulic fractures contributing to the production is about 400 m. At the same time, matrix blocks that exist between fracturing stages are involved in production. As a result, the simulation techniques performed in this study can be used to calculate the optimal well spacing and fracturing intervals. In addition, the SRV obtained from the microseismic data must be distinguished from the actual productive volume because SRV is normally overestimated.

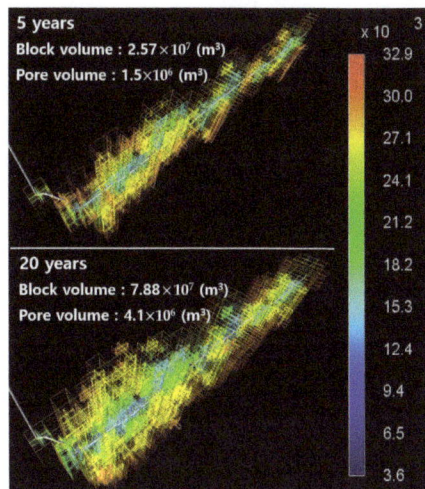

Figure 19. Productive volume map over time. (Colors of the grids indicate the kPa).

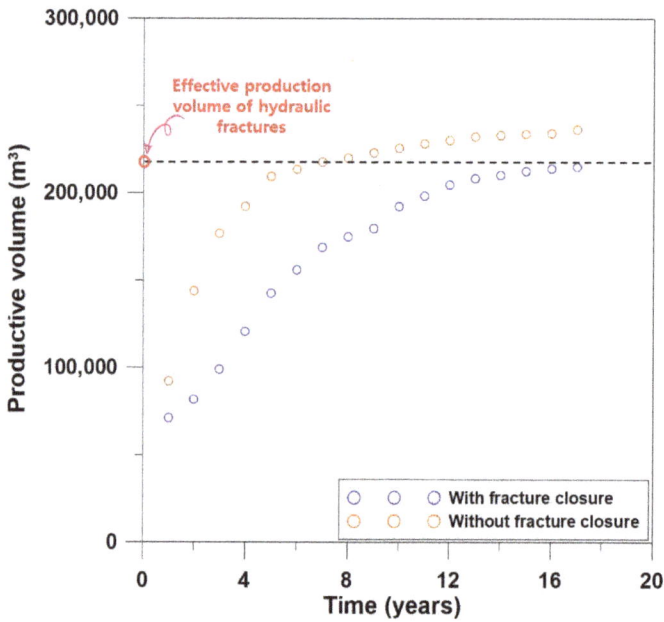

Figure 20. Simulated change in productive volume of hydraulic fractures over time, with and without fracture closure relationship.

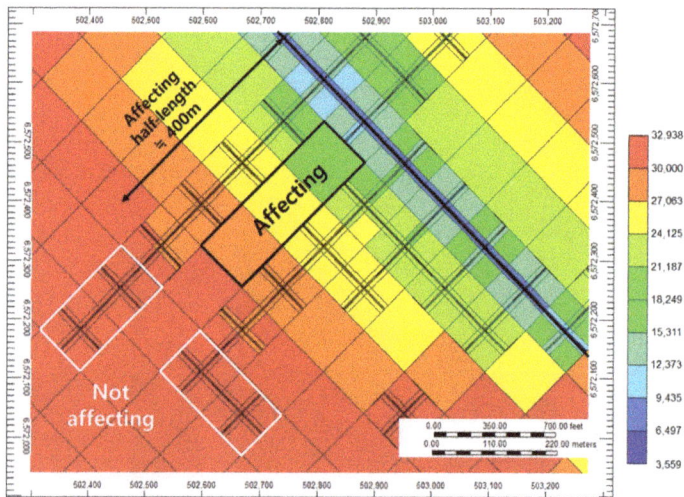

Figure 21. Map of pressure propagation after 20 years of production. (Colors of the grids indicate the kPa).

4. Discussion

Based on the observations made from the study, it is found that determination of the productive volume stimulated by the conductive induced fractures takes a major role in reliable production forecasting. Since widths of the induced fractures change with the pore pressure, and so does its conductivity, not only the fracture geometry in the initial stage but its effect on the productive volume change are crucial to the production forecasting. Although the microseismic measurement is widely accepted for determination of the induced fracture distribution, as described in the text, the determined fracture geometry does not always represent its conductivity. In addition, a direct measurement method for the fracture permeability change during the production period is not available. Therefore, production history matching can be a useful alternative route for the stimulated reservoir volume determination as well as the future production forecasting.

The pressure and rate responses during the fracturing treatment can be incorporated for more reliable analysis. The permeability alteration behaviors of the propped fracture (fracture filled with proppant) and non-propped fracture significantly differ. Therefore, if the propped portions of the induced fractures are identified by post-frac analysis, such as net pressure analysis, bottomhole pressure matching, etc., the fracture permeability change can be more precisely determined.

In addition, integrated analysis with the rate transient analysis (RTA) may enhance the reliability of the productivity forecasting. Results from reservoir simulation would be useful for determination of onset of the boundary dominating flow. Since the best way to determine end of the transient flow period is always questionable during the RTA analysis, more reliable productivity analysis is available if the reservoir simulation results are integrated.

5. Conclusions

In this study, production history matching and a production history forecast were carried out on a shale gas reservoir regarding the hydraulic fractures width change phenomenon over time. The observations made from the detailed analysis are as follows:

(a) The stimulated reservoir volume was estimated by the microseismic data and was compared with the actual productive volume obtained from numerical simulations. It was found that the deteriorated permeability of the hydraulic fractures caused by the fluid pressure reduction significantly affects the simulation results.

(b) The result suggests that if the change in fracture width is not taken into account, the cumulative production will be considerably overestimated (5.5 %). Therefore, more reliable history matching and forecasting can be achieved by adopting the fracture permeability reduction effect.

(c) As the production progresses, hydraulic fractures above a certain distance are not expected to have an influence on the production, but the matrix blocks close to the production well contribute to the productive volume. This indicates that the SRV obtained from the microseismic data is inconsistent with the actual productive volume, as the signals provide only a preliminary estimate for the hydraulically fractured area.

(d) Not only does considering alterations of the hydraulic fracture permeability enhance the accuracy of predictions on shale gas flow behavior, it can also improve the understanding of fluid flows in shale reservoirs. Moreover, the simulation procedure proposed in this study will provide great insight in estimating the productive volume, and it can be used to determine the optimal well spacing and the number of fracturing stages during shale reservoir development.

Author Contributions: J.K. wrote the paper and contributed to tuning the model and analyzing the results; Y.S. contributed to processing the raw data and making the initial model; J.W. and Y.L. suggested the main idea and supervised the work, providing continuous feedback.

Funding: This work was funded by the Energy Efficiency and Resources Core Technology Program of the Korea Institute of Energy Technology Evaluation and Planning (KETEP), granted financial resource from the Ministry of Trade, Industry and Energy, Republic of Korea (No.20132510100060).

Acknowledgments: This research was supported by the Energy Efficiency and Resources Core Technology Program of the Korea Institute of Energy Technology Evaluation and Planning (KETEP), granted financial resource from the Ministry of Trade, Industry and Energy, Republic of Korea (No.20172510102150).

Conflicts of Interest: The authors declare no conflict of interest.

References

1. EIA (US Energy Information Administration). Annual Energy Outlook 2018. Available online: https://www.eia.gov/outlooks/aeo/ (accessed on 24 December 2018).
2. Natgas. Natural Gas and the Environment. Available online: http://naturalgas.org/environment/naturalgas/ (accessed on 24 December 2018).
3. Carey, J.M. References. In *Surprise Side Effect of Shale Gas Boom: A Plunge in U.S.*; Greenhouse Gas Emissions Forbes magazine: Washington, NJ, USA, 2012.
4. Stevens, P. *The Shale Gas Revolution: Developments and Changes*; Chatham House: London, UK, 1 August 2012; pp. 2–3.
5. Moinfar, A.; Erdle, J.C.; Patel, K. Comparison of Numerical vs Analytical Models for EUR Calculation and Optimization in Unconventional Reservoirs. In Proceedings of the SPE Low Perm Symposium, Denver, CO, USA, 5–6 May 2016. SPE-180209-MS. [CrossRef]
6. Anderson, D.M.; Nobakht, M.; Moghadam, S.; Mattar, L. Analysis of production data from fractured shale gas wells. In Proceedings of the SPE Unconventional Gas Conference, Pennsylvania, PA, USA, 23–25 February 2010. SPE-131787-MS. [CrossRef]
7. Urban, E.; Yousefzadeh, A.; Virues, C.J.; Aguilera, R. Evolution and Evaluation of SRV in Shale Gas Reservoirs: An Application in the Horn River Shale of Canada. In Proceedings of the SPE Latin America and Caribbean Petroleum Engineering Conference, Buenos Aires, Argentina, 17–19 May 2017. SPE-185609-MS. [CrossRef]
8. Daniels, J.L.; Waters, G.A.; Le Calvez, J.H.; Bentley, D.; Lassek, J.T. Contacting more of the barnett shale through an integration of real-time microseismic monitoring, petrophysics, and hydraulic fracture design. In Proceedings of the SPE Annual Technical Conference and Exhibition, California, CA, USA, 11–14 November 2007. SPE-110562-MS. [CrossRef]
9. Fisher, M.K.; Wright, C.A.; Davidson, B.M.; Goodwin, A.K.; Fielder, E.O.; Buckler, W.S.; Steinsberger, N.P. Integrating fracture mapping technologies to optimize stimulations in the Barnett Shale. In Proceedings of the SPE Annual Technical Conference and Exhibition, San Antonio, TX, USA, 29 September–2 October 2002. SPE-77441-MS. [CrossRef]
10. Maxwell, S.C.; Urbancic, T.; Steinsberger, N.; Zinno, R. Microseismic imaging of hydraulic fracture complexity in the Barnett shale. In Proceedings of the SPE Annual Technical Conference and Exhibition, San Antonio, TX, USA, 29 September–2 October 2002. SPE-77440-MS. [CrossRef]
11. Rabczuk, T.; Zi, G.; Bordas, S.; Nguyen-Xuan, H. A simple and robust three-dimensional cracking-particle method without enrichment. *Comput. Methods Appl. Mech. Eng.* **2010**, *199*, 2437–2455. [CrossRef]
12. Rabczuk, T.; Belytschko, T. A three-dimensional large deformation meshfree method for arbitrary evolving cracks. *Comput. Methods Appl. Mech. Eng.* **2007**, *196*, 2777–2799. [CrossRef]
13. Rabczuk, T.; Gracie, R.; Song, J.H.; Belytschko, T. Immersed particle method for fluid–structure interaction. *Int. J. Numer. Methods Eng.* **2010**, *81*, 48–71. [CrossRef]
14. Zhou, S.; Zhuang, X.; Zhu, H.; Rabczuk, T. Phase field modelling of crack propagation, branching and coalescence in rocks. *Theor. Appl. Fract. Mech.* **2018**, *96*, 174–192. [CrossRef]
15. Zhou, S.; Zhuang, X.; Rabczuk, T. A phase-field modeling approach of fracture propagation in poroelastic media. *Eng. Geol.* **2018**, *240*, 189–203. [CrossRef]
16. Zhou, S.; Rabczuk, T.; Zhuang, X. Phase field modeling of quasi-static and dynamic crack propagation: COMSOL implementation and case studies. *Adv. Eng. Softw.* **2018**, *122*, 31–49. [CrossRef]
17. Zhou, S.; Zhuang, X.; Rabczuk, T. Phase-field modeling of fluid-driven dynamic cracking in porous media. *Comput. Methods Appl. Mech. Eng.* **2019**, *350*, 169–198. [CrossRef]
18. Van Dam, D.B.; De Pater, C.J.; Romijn, R. Analysis of hydraulic fracture closure in laboratory experiments. *Spe Prod. Facil.* **2000**, *15*, 151–158. [CrossRef]
19. Seth, P.; Kumar, A.; Manchanda, R.; Shrivastava, K.; Sharma, M.M. Hydraulic Fracture Closure in a Poroelastic Medium and its Implications on Productivity. In Proceedings of the 52nd U.S. Rock Mechanics/Geomechanics Symposium, Seattle, Washington, DC, USA, 17–20 June 2018. ARMA-2018-695.
20. Wang, H.; Sharma, M.M. Modeling of hydraulic fracture closure on proppants with proppant settling. *J. Pet. Sci. Eng.* **2018**, *171*, 636–645. [CrossRef]
21. Warren, J.E.; Root, P.J. The behavior of naturally fractured reservoirs. *Soc. Pet. Eng.* **1963**, *3*, 245–255. [CrossRef]

22. Zeng, Q.; Yao, J. Numerical simulation of fluid-solid coupling in fractured porous media with discrete fracture model and extended finite element method. *Computation* **2015**, *3*, 541–557. [CrossRef]

23. Ho, C. Dual porosity vs. dual permeability models of matrix diffusion in fractured rock. In Proceedings of the the International High-Level Radioactive Waste Conference, Las Vegas, CA, USA, 29 April–3 May 2001. SAND2000-2336C.

24. Karra, S.; Makedonska, N.; Viswanathan, H.S.; Painter, S.L.; Hyman, J.D. Effect of advective flow in fractures and matrix diffusion on natural gas production. *Water Resour. Res.* **2015**, *51*, 8646–8657. [CrossRef]

25. Mudunuru, M.K.; Karra, S.; Makedonska, N.; Chen, T. Sequential geophysical and flow inversion to characterize fracture networks in subsurface systems. *Stat. Anal. Data Min. Asa Data Sci. J.* **2017**, *10*, 326–342. [CrossRef]

26. Cipolla, C.L.; Fitzpatrick, T.; Williams, M.J.; Ganguly, U.K. Seismic-to-simulation for unconventional reservoir development. In Proceedings of the SPE Reservoir Characterisation and Simulation Conference and Exhibition, Abu Dhabi, UAE, 9–11 October 2011. SPE-146876-MS. [CrossRef]

27. Nguyen, D.H.; Cramer, D.D. Diagnostic fracture injection testing tactics in unconventional reservoirs. In Proceedings of the SPE Hydraulic Fracturing Technology Conference, The Woodlands, TX, USA, 4–6 February 2013. SPE-163863-MS. [CrossRef]

28. Yu, W.; Sepehrnoori, K. Optimization of well spacing for bakken tight oil reservoirs. In Proceedings of the SPE/AAPG/SEG Unconventional Resources Technology Conference, Denver, CO, USA, 25–27 August 2014. URTEC-1922108-MS. [CrossRef]

29. Alramahi, B.; Sundberg, M.I. Proppant embedment and conductivity of hydraulic fractures in shales. In Proceedings of the 46th U.S. Rock Mechanics/Geomechanics Symposium, Chicago, IL, USA, 24–27 June 2012. ARMA-2012-291.

30. Terracina, J.M.; Turner, J.M.; Collins, D.H.; Spillars, S. Proppant selection and its effect on the results of fracturing treatments performed in shale formations. In Proceedings of the SPE Annual Technical Conference and Exhibition, Florence, Italy, 19–22 September 2010. SPE-135502-MS. [CrossRef]

31. Ertekin, T.; Abou-Kassem, J.H.; King, G.R. References. In *Basic Applied Reservoir Simulation*; SPE Textbook Series; SPE: Sirsi, India, 2001; Volume 7, ISBN 9781555630898.

32. Carlson, M.R. References. In *Practical Reservoir Simulation: Using, Assessing, and Developing Results*; PennWell Books: Tulsa, OK, USA, 2003; ISBN 978-0-87814-803-5.

33. Kim, J.; Kim, D.; Lee, W.; Lee, Y.; Kim, H. Impact of total organic carbon and specific surface area on the adsorption capacity in Horn River shale. *J. Pet. Sci. Eng.* **2017**, *149*, 331–339. [CrossRef]

34. Romm, E.S. References. In *Fluid Flow in Fractured Rocks*; English translation by W.R. Blake; Phillips Petroleum Company: Bartleville, OK, USA, 1972.

35. Pieters, D.A.; Graves, R.M. Fracture relative permeability: Linear or non-linear function of saturation. In Proceedings of the International Petroleum Conference and Exhibition of Mexico, Veracruz, Mexico, 10–13 October 1994. SPE-28701-MS. [CrossRef]

36. Fourar, M.; Bories, S. Experimental study of air-water two-phase flow through a fracture (narrow channel). *Int. J. Multiph. Flow* **1995**, *21*, 621–637. [CrossRef]

37. Diomampo, G. Relative permeability through fractures. *Stanf. Univ.* **2001**, SGP-TR-170. [CrossRef]

38. Speyer, N.; Li, K.; Horne, R. Experimental measurement of two-phase relative permeability in vertical fractures. In Proceedings of the Thirty-Second Workshop on Geothermal Reservoir Engineering, Stanford University, Stanford, CA, USA, 22–24 January 2007. SGP-TR-183.

39. Chima, A.; Geiger, S. An analytical equation to predict gas/water relative permeability curves in fractures. In Proceedings of the SPE Latin America and Caribbean Petroleum Engineering Conference, Mexico City, Mexico, 16–18 April 2012. SPE-152252-MS. [CrossRef]

40. Byun, J.H.; Jin, W.S.; Yi, J.S. Case study for effective stimulated reservoir volume identification in unconventional reservoir. *J. Korean Soc. Miner. Energy Resour. Eng.* **2018**, *55*, 127–146. [CrossRef]

41. Clarkson, C.R.; Jensen, J.L.; BLASTINGAME, T. Reservoir engineering for unconventional gas reservoirs: What do we have to consider? In Proceedings of the North American Unconventional Gas Conference and Exhibition, The Woodlands, TX, USA, 14–16 June 2011. SPE-145080-MS. [CrossRef]

42. Suliman, B.; Meek, R.; Hull, R.; Bello, H.; Portis, D.; Richmond, P. Variable stimulated reservoir volume (SRV) simulation: Eagle ford shale case study. In Proceedings of the SPE Unconventional Resources Conference-USA, The Woodlands, Texas, USA, 10–12 April 2013. SPE-164546-MS. [CrossRef]

43. Novlesky, A.; Kumar, A.; Merkle, S. Shale gas modeling workflow: From microseismic to simulation-a horn river case study. In Proceedings of the Canadian Unconventional Resources Conference, Calgary, AB, Canada, 15–17 November 2011. SPE-148710-MS. [CrossRef]

44. Forchheimer, P. Wasserbewegung durch boden. *Zeits V. Dtsch. Ing.* **1901**, *45*, 1782–1788.

45. Hagoort, J. References. In *Fundamentals of Gas Reservoir Engineering*; Elsevier: New York, NY, USA, 1988; Volume 23, ISBN 0-444-42991-3.

46. Rubin, B. Accurate simulation of non Darcy flow in stimulated fractured shale reservoirs. In Proceeding of the SPE Western Regional Meeting, Anaheim, CA, USA, 27–29 May 2010. SPE-132093-MS. [CrossRef]

47. Hubbert, M.K.; Willis, D.G. Mechanics of hydraulic fracturing. *Trans AIME* **1957**, *210*, 153–168.

48. Kam, P.; Nadeem, M.; Omatsone, E.N.; Novlesky, A.; Kumar, A. Integrated geoscience and reservoir simulation approach to understanding fluid flow in multi-well pad shale gas reservoirs. In Proceedings of the SPE/CSUR Unconventional Resources Conference–Canada, Calgary, AB, Canada, 30 September–2 October 2014. SPE-171611-MS. [CrossRef]

49. Li, Q.; Xing, H.; Liu, J.; Liu, X. A review on hydraulic fracturing of unconventional reservoir. *Petroleum* **2015**, *1*, 8–15. [CrossRef]

50. Craig, D.P.; Barree, R.D.; Warpinski, N.R.; Blasingame, T.A. Fracture Closure Stress: Reexamining Field and Laboratory Experiments of Fracture Closure Using Modern Interpretation Methodologies. In Proceedings of the SPE Annual Technical Conference and Exhibition, San Antonio, TX, USA, 9–11 October 2017. SPE-187038-MS. [CrossRef]

51. Palisch, T.T.; Duenckel, R.J.; Bazan, L.W.; Heidt, J.H.; Turk, G.A. Determining Realistic Fracture Conductivity and Understanding its Impact on Well Performance–Theory and Field Examples. In Proceedings of the SPE Hydraulic Fracturing Technology Conference, College Station, TX, USA, 29–31 January 2007. SPE-106301-MS. [CrossRef]

energies

MDPI

Article

Analysis of Pressure Communication between the Austin Chalk and Eagle Ford Reservoirs during a Zipper Fracturing Operation

Sriniketh Sukumar, Ruud Weijermars *, Ibere Alves and Sam Noynaert

Harold Vance Department of Petroleum Engineering, Texas A&M University, College Station, TX 77843, USA; sriniketh@tamu.edu (S.S.); ibere.alves@tamu.edu (I.A.); noynaert@tamu.edu (S.N.)
* Correspondence: r.weijermars@tamu.edu; Tel.: +1-979-845-4067

Received: 11 February 2019; Accepted: 29 March 2019; Published: 18 April 2019

Abstract: The recent interest in redeveloping the depleted Austin Chalk legacy field in Bryan (TX, USA) mandates that reservoir damage and subsurface trespassing between adjacent reservoirs be mitigated during hydraulic fracture treatments. Limiting unintended pressure communication across reservoir boundaries during hydraulic fracturing is important for operational efficiency. Our study presents field data collected in fall 2017 that measured the annular pressure changes that occurred in Austin Chalk wells during the zipper fracturing treatment of two new wells in the underlying Eagle Ford Formation. The data thereby obtained, along with associated Eagle Ford stimulation reports, was analyzed to establish the degree of pressure communication between the two reservoirs. A conceptual model for pressure communication is developed based on the pressure response pattern, duration, and intensity. Additionally, pressure depletion in the Austin Chalk reservoir is modeled based on historic production data. Pressure increases observed in the Austin Chalk wells were about 6% of the Eagle Ford injection pressures. The pressure communication during the fracture treatment was followed by a rapid decline of the pressure elevation in the Austin Chalk wells to pre-fracture reservoir pressure, once the Eagle Ford fracture operation ended. Significant production uplifts occurred in several offset Austin Chalk wells, coeval with the observed temporal pressure increase. Our study confirms that after the rapid pressure decline following the short-term pressure increase in the Austin Chalk, no residual pressure communication remained between the Austin Chalk and Eagle Ford reservoirs. Limiting pressure communication between adjacent reservoirs during hydraulic fracturing is important in order to minimize the loss of costly fracturing fluid and to avoid undue damage to the reservoir and nearby wells via unintended proppant pollution. We provide field data and a model that quantifies the degree of pressure communication between adjacent reservoirs (Austin Chalk and Eagle Ford) for the first time.

Keywords: Austin Chalk; Eagle Ford shale; hydraulic fracturing; pressure communication; production uplifts

---

1. Introduction

Understanding the pressure state in the Austin Chalk and Eagle Ford shale reservoirs and their possible communication is important for petroleum engineering operations in several technical and proprietary ways. First, the pressure depletion history in each of the reservoirs controls the production rate of its wells. Since the Austin Chalk has been producing several decades prior to the development of the Eagle Ford Formation, knowing the state of their respective pressure depletion remains important for production forecasting and future field development planning.

Second, limiting pressure communication between adjacent reservoirs during hydraulic fracturing is important in order to minimize the loss of costly frack fluid and to avoid undue damage to pumps of nearby wells via unintended proppant pollution, a problem commonly faced by operators (reported by the managing director of E2 Operating, via personal communications with the authors on 29 October 2017). During a hydraulic fracture treatment proppant pollution is the invasion of proppants into the stimulated rock volume of an offset well. The fracture treatment can also affect the downhole equipment of offset wells.

The main focus of this study is on the analysis of pressure response data of shut-in Austin Chalk wells during Diagnostic Fracture Injection Tests (DFIT) and subsequent zipper fracking of the two nearby Eagle Ford wells. Our study was conducted on a lease space beneath the RELLIS campus, a research facility that is administered by the Texas A&M University System in Brazos County (TX, USA). A physical image and schematic map of the RELLIS campus are displayed in Figure 1a,b. The aerial view of the RELLIS Campus (Figure 1a) highlights the relevant oil well site locations in relation to the schematic map (Figure 1b). The images show that the individual wells considered are noticeably spaced apart. Interwell distances vary between several hundreds of ft to over a thousand ft (see later).

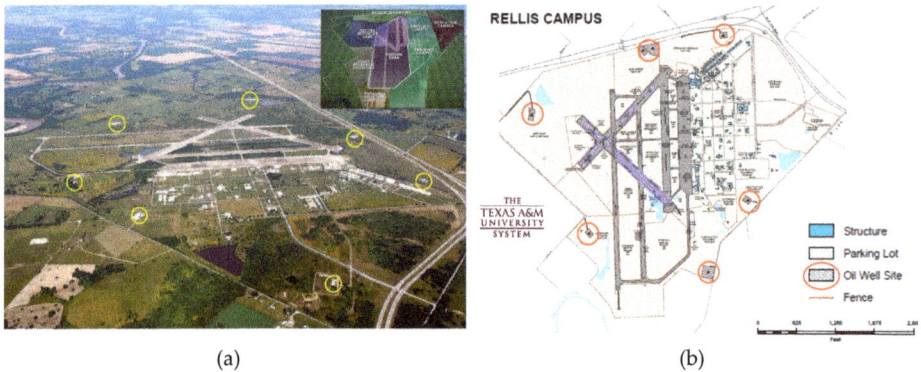

(a)   (b)

Figure 1. (**a**) Aerial view of the RELLIS area (left image) highlights oil well sites (yellow circles); (**b**) Map view (right image) shows a more abstract schematic of the campus, with oil sites highlighted. Our study presents evidence of pressure communication between wells hundreds to thousands of feet apart.

Our field study on the pressure communication between the wells of the individual companies was conducted using data provided by each of the operators (i.e., Austin Chalk and Eagle Ford leases, respectively). Well data are reported to the Texas A&M System in connection to their royalty share. Each reservoir is part of a split estate, which means that the mineral rights of the Austin Chalk and Eagle Ford Formation are leased to two different operators. Operators are pragmatic and have no incentive for judicial recourse in case of subsurface trespassing, which refers to the potential impact on mutual well productivity due to engineering interventions in adjacent petroleum reservoirs. Prior proceedings in the Texas High Court of Justice between adverse operators has declared the mutual responsibility to resolve any dispute lies with the individual operators [1].

This study explains the reservoir setting and well layout, initial pressure state in both reservoirs (Eagle Ford and Austin Chalk), and then proceeds to report the pressure data collected. Our study confirms there exists no pressure communication between the two reservoirs, either prior to, or after the fracture treatment. However, a significant temporal pressure response was measured in the Austin Chalk legacy wells during both the 2017 DFIT and the zipper frac operations in the Eagle Ford landing zone. We analyze the initial pressure state, temporal changes induced during, and the final pressure state in each reservoir after the interventions. The second part of the paper presents a conceptual model that can explain the physical process of the interwell pressure communication based on the field pressure data analyzed in the first part of our study.

2. Project Overview and Data Acquisition

We have collaborated extensively with several operators in the Eagle Ford Formation below the RELLIS Campus and used the provided field data to develop pressure depletion models [2,3] and production forecasts [4]. However, the overlying Austin Chalk Formation was developed in the early 1990s and although some logs are available from nearby wells in the formation, few details other than production data can be obtained for those older wells.

Six horizontal legacy wells, each with 4000 ft lateral length in the Austin Chalk reservoir landing zone beneath the RELLIS campus have either ceased to produce (3) or are marginal producers (3). These wells, named "Riverside 1 to 6" (or more simply R1 to R6) form the principal object of our study. The drilling and completion of two new wells in the Eagle Ford, with zipper fracking under extremely high hydraulic pressures used during fracture treatment of two Eagle Ford wells drilled in Nov/Dec 2017, provided a unique opportunity to gather pressure response data in the overlying Austin Chalk Formation via five pressure gauges, each mounted on a different Austin Chalk Riverside well.

2.1. Well Location and Trajectories

The Texas A&M University System administers the mineral rights of the RELLIS Campus in College Station, Texas, which includes the Eagle Ford shale, Austin Chalk and Buda Limestone plays that produce oil, and to a lesser degree, some associated natural gas. The development of the hydrocarbon plays (involving drilling, completion and necessary production operations such as shut-ins, artificial stimulations, including hydraulic fracturing and well workovers) is leased out to private operators. Our field study on the pressure communication between the wells of the individual companies was conducted using data provided by the various prior and current operators (i.e., for Austin Chalk and Eagle Ford leases, respectively). Well data are reported to the Texas A&M System in connection to their royalty share.

The RELLIS lease area hosts 12 wells drilled and completed during different epochs. Table 1 displays the names and parameters of the wells studied, with the well specifics based on data from the Texas Railroad Commission. The six Austin Chalk legacy wells are currently owned by E2 Operating, a subsidiary of Exponent Energy, which acquired the wells in 2014 from a bankruptcy sale. There have been many changes in ownership of the Austin Chalk wells, which were first completed in 1990s, not further elaborated here, as can be traced via the Texas Railroad Commission. The more recently developed six Eagle Ford wells are currently operated by Hawkwood Energy (Table 1), who bought the lease from Halcon Resources in 2017. Subsurface and production data were provided to us by various lease operators (i.e., E2 operation, Halcon Resources and Hawkwood Energy). All the companies mentioned in Table 1 are oil and gas operators in Brazos county, Texas, USA.

Table 1. Dates of completion of wells in the RELLIS lease area.

Symbol	Formation	Current Operator	Date of Completion	TVD (ft)	Lateral Length (ft)
R1	Austin Chalk	E2 Operating LLC	01 Jun 1991	7802	3258
R2[1]	Austin Chalk	E2 Operating LLC	01 Nov 1991	7628	4793
R4	Austin Chalk	E2 Operating LLC	01 Dec 1991	7628	4233
R3 (Not studied)	Austin Chalk	E2 Operating LLC	01 Apr 1992	7844	3566
R5[1]	Austin Chalk	E2 Operating LLC	01 Nov 1992	7630	2508
R6[1]	Austin Chalk	E2 Operating LLC	01 Oct 1992	7856	3904
R (Parent well)	Eagle Ford shale	Hawkwood Energy	01 Apr 2014	8240	8630
O (Parent well)	Eagle Ford shale	Hawkwood Energy	01 Apr 2014	8240	2942
M (Parent well)	Eagle Ford shale	Hawkwood Energy	01 Nov 2014	8240	6550
H1 (Parent well)	Eagle Ford shale	Hawkwood Energy	01 Nov 2014	8240	5950
H2 (Child well)	Eagle Ford shale	Hawkwood Energy	31 Dec 2017	8240	7905
H3 (Child well)	Eagle Ford shale	Hawkwood Energy	31 Dec 2017	8240	7359

[1] Riverside wells that were plugged and abandoned in 2018/2019.

A diagram of the well trajectories completed in the RELLIS lease area is shown in Figure 2a. Eagle Ford Wells H2 and H3 were completed most recently (2017) and can be considered the child wells of parent Wells R, O, M, H1, all of which were completed in 2014. Figure 2b illustrates the chronology of the development of the RELLIS lease area considered in our study. The dates of first production for the Eagle Ford Wells are the same as the dates of completion reported in Table 1.

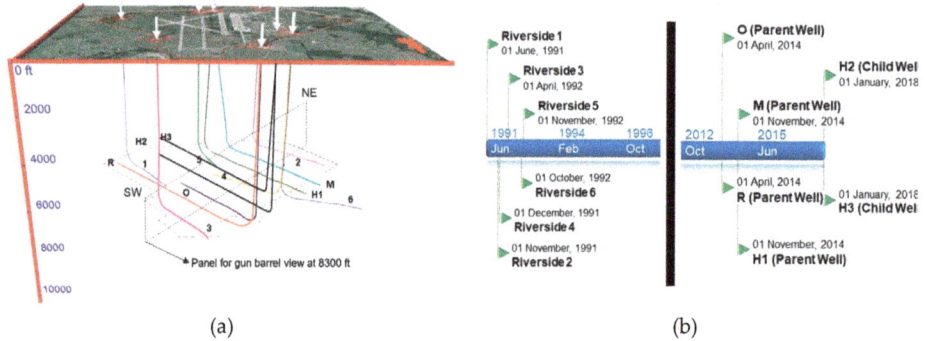

(a)   (b)

Figure 2. (**a**) RELLIS wellbore trajectories. The white arrows represent the surface location of each well. The dotted outline represents the landing zone. The rectangular panel shows the portion of the gun barrel view introduced in Section 2.3. Wells labeled R, O, M, H1, H2, H3 are completed in the Eagle Ford shale and wells labeled 1 to 6 are the Riverside wells completed in the Austin Chalk. The two Eagle Ford shale child wells, H2 and H3, are drilled from approximately the same location on the surface, and Wells H1, H2, H3 and O are mutually parallel. Wells H2 and H3 are 350 ft deeper at the toe (8450 ft) than at the heel side (8100 ft), due to a gentle slope of the producing landing zone of the wells. (**b**) Chronology of development of RELLIS oil and gas lease area. Dates of well completion are displayed. The black bar represents a time lapse from 1996 to 2012. The Eagle Ford Wells (H1, H2, H3, R, O, M) are much younger than the Austin Chalk Wells (R1–R6) which have been operational for over 25 years.

Prior to the recent rush to develop the Eagle Ford shale with modern multistage hydraulic fracturing techniques, only the Austin Chalk was developed in the RELLIS lease, because it is naturally fractured and production required only little well stimulation. Production for all the Austin Chalk wells started nearly three decades ago, first reported as of 1 July 1991, which is when the common production facility was completed for use by Well R1 initially. Each of the six Austin Chalk wells was fractured as a single stage with 7-inch casing and 30,000 bbl water, 11,000 lbs of diverter, and 18,000 gal of 15% hydrochloric acid. Additional completion data was not available. In 1992, the Austin Chalk Formation in Texas had a total of 4425 wells completed, which produced 330 million bbl of cumulative oil [5]. A more recent well count gives the 9500 wells in total and a cumulative production of 1.7 billion BOE [6]. The Austin Chalk, however still contains a large amount of unrecovered hydrocarbon resources, so the expansion of exploration in this formation could prove to be very profitable [6].

2.2. Initial Pressures of the Austin Chalk and Eagle Ford Hydrocarbon Reservoirs

Three of the six Austin Chalk wells have been recently plugged and abandoned (R2 and R5 in January 2018; R6 in spring 2019) by the operator to make room for building operations. Over the course of their lifespan from July 1991 to January 2018 (28 years of production), the six Austin Chalk wells have cumulatively produced 1 million bbl of oil and 3.5 bcf of natural gas. Wells R2, R5 and R6 were already not producing for several years and remaining producers R1, R3 and R4 were shut-in during the fracture treatment of Wells H2 and H3. Currently, of the three remaining Austin Chalk wells, one is inactive (not pumping) and the two active ones only produce a marginal 2–3 bbl/day.

Energies **2019**, *12*, 1469

Knowing the pressure of the Austin Chalk reservoir space immediately prior to the fracturing operation on Nov/Dec 2017 is relevant in order to better understand how the hydraulic pressure of Eagle Ford well stimulation communicated with the ambient pressure in the Austin Chalk reservoir space. The pressure depletion in the Austin Chalk reservoir just before the fracturing of Wells H2 and H3 can be estimated based on historic production and decline curves using production data from Texas RRC online.

2.2.1. Initial Pressure in the Austin Chalk Formation

All six original Austin Chalk wells (R1–R6) were connected to a single production gathering system. The cumulative hydrocarbon output of the aggregated production system since first production started is graphed in Figure 3a. The monthly decline of the hydrocarbon production over the 27-year well-life is separately plotted in Figure 3b. Note that all the gas produced in this formation is dissolved gas. For most of its production history, there existed no free gas under reservoir conditions since the reservoir pressure was above its bubble point pressure such that there was only liquid in the formation. Further, low productivity of Austin Chalk can be attributed to reduced reservoir pressures and dissolved-gas-drive mechanisms [7]. The Riverside wells were operated by pump jack for most of their production histories.

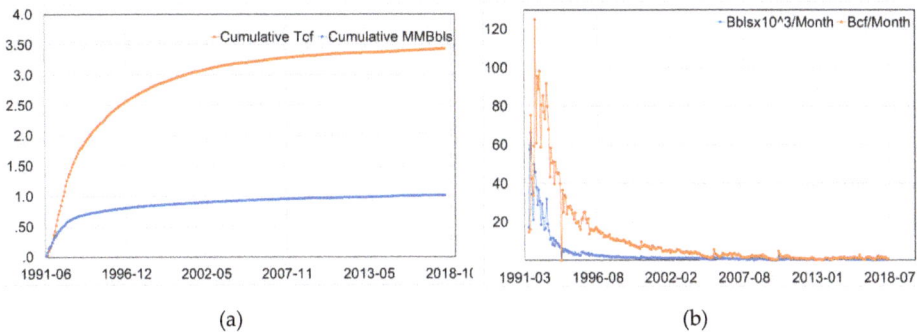

Figure 3. (**a**) Cumulative production of the six Riverside wells (R1-R6). (**b**) Monthly production decline curves for six Riverside wells. Oil (blue curve) is measured in bbl and gas (orange curve) in Mcf.

Using monthly production data, the reservoir pressure near the Austin Chalk wells at the time of the fracture treatment in Wells H2 and H3 was modeled based on the material balance technique outlined in [8], reproduced in Equation (1). The detailed methodology and parameters used are explained in Appendix A. The pressure depletion curves obtained are shown in Figure 4a. Keeping all other variables constant, a sensitivity analysis for the drainage area is presented in Figure 4b, which shows that the effect of depletion is stronger for small drainage areas. The logarithmic correlation obtained indicates that the pressure depletion effect is dependent on the amount of hydrocarbon in place, which can be represented through drainage area, (with all other variables kept constant). The equation for original oil in place, (N) is presented in Equation (2). Details of the nomenclature/parameters used in Equations (1) and (2), assumptions made and method of the depletion calculation are further discussed in Appendix A:

$$P_{ri} - P = \frac{N_p B_o + W_p B_w}{c_t (N B_{oi} + W B_{wi})} \tag{1}$$

$$N = 7758 A h \phi (1 - S_w) \tag{2}$$

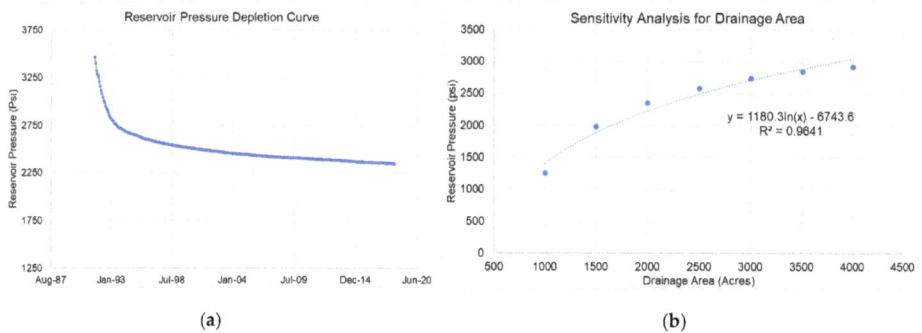

Figure 4. (**a**) Pressure depletion curve for the Austin Chalk Formation with an assumed drainage area of 2000 acres. The depletion rate declines after long term production. (**b**) Sensitivity analysis for the effect of drainage area on reservoir pressure depletion. A strong logarithmic correlation is obtained.

Based on Figure 4a, the current reservoir pressure of the Austin Chalk is estimated at 2354 psi, corresponding to an assumed drainage area of 2000 acres, which is the approximate acreage of the RELLIS campus lease area [9]. This value will be used in building a pressure response model (See Section 4.4).

2.2.2. Initial Pressure Eagle Ford Formation

Although the Eagle Ford shale is an ultra-low permeability formation with negligible natural fractures in the area studied, the occurrence of pressure communication between Eagle Ford shale and the naturally fractured Austin Chalk would mean the fracture stimulation pump schedule may need adjustment when optimizing the fracking process.

The wells recently completed in the Eagle Ford Formation confirmed that the initial reservoir pressure remained intact [2,3], despite nearly three decades of oil and gas extraction in the overlying Austin Chalk Formation. The initial reservoir pressure of the Eagle Ford prior to first well completion in 2014 (Table 1) was estimated based on history matching to be 4891 psi [2].

The initial pressure in the Eagle Ford of 4891 psi is higher than the depleted state of the Austin Chalk, at 2354 psi. Interestingly, the lower pressure allows fluid to migrate to the Austin Chalk Formation during the fracking of wells in the Eagle Ford. We will further analyze the pressure communication between the Austin Chalk and Eagle Ford reservoirs during the 2017 fracture treatment.

2.3. Austin Chalk Pressure Gauges Monitoring Eagle Ford Zipper Fracking Operation

The main focus of this study is on the analysis of pressure response data of shut-in Austin Chalk wells during zipper fracking of the two nearby Eagle Ford wells. A gun barrel view of all the wells below the RELLIS lease area is shown in Figure 5a to display well spacing and pressure gauge placements. Well spacing estimates are based on each well's trajectory. The Eagle Ford wells were drilled such that all wellbore trajectories were mutually parallel and in the direction of minimum horizontal stress of the region, which is assumed to coincide with the direction of the regional dip towards the Gulf of Mexico. Eagle Ford well spacings could therefore be easily measured from a wellbore trajectory map. The Austin Chalk legacy wells spacings were estimated on a line of best fit perpendicular for each wellbore, extrapolating for R3 and R4. While reasonably accurate, the well spacings should therefore be taken as only estimates, as an uncertainty of ±100 ft exists.

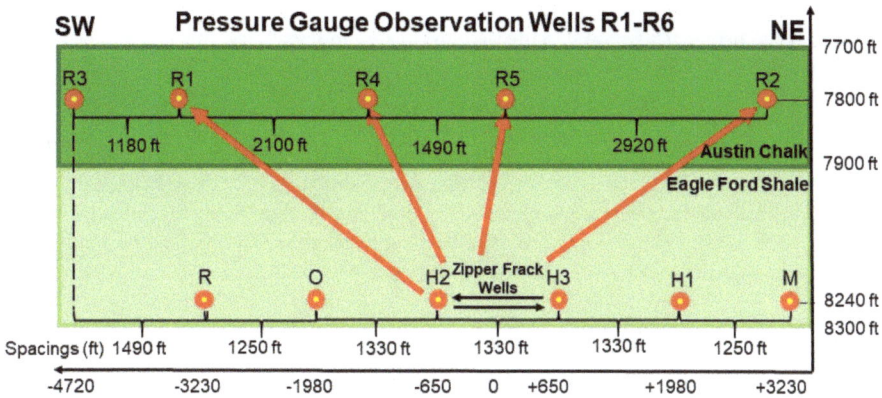

(a)

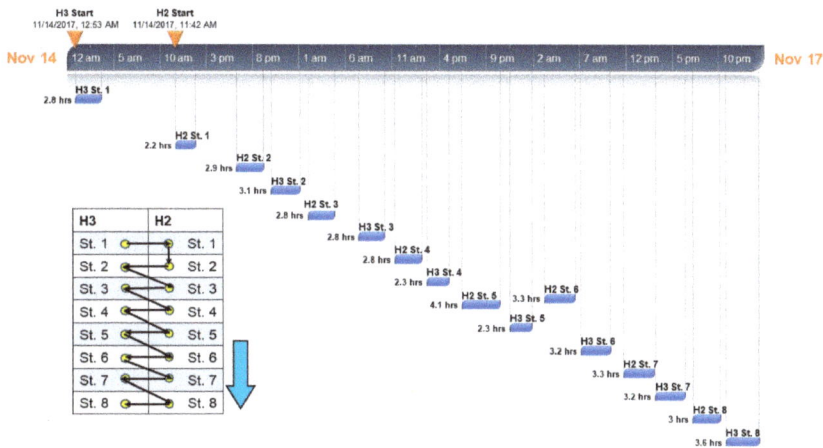

(b)

Figure 5. (**a**) Pressure sources and positions. Gun barrel view of all the hydrocarbon wells completed in the RELLIS lease area used for our pressure communication study. Red arrows indicate possible connections between pressure signal source and the observation pressure gauges, which monitored the annulus pressures on wellheads of Wells R1-R6. No pressure gauge was mounted on Well R3. Section is taken from panel in Figure 2a from South West to North East as outlined. Well R6 is outside the section and is therefore not shown in gun barrel. Well spacings are estimated using Well R3 as a reference line. The horizontal axis represents spacing relative to the midpoint of Wells H2 and H3. Vertical axis represents true vertical depth, and is exaggerated 6.6×. (**b**) Pressure signal timeline. Example of timeline depicting the relative durations of the first 8 stages of the zipper frack operation for Wells H2 and H3 (around the clock). The inset table represents the order of events in the operation. There was a slight delay at the start of the project and Stage 1 in Well H2 began almost half a day later than in H3. The remainder of the procedure experienced no delays and the zipper frack pattern occurred with no incidents reported.

Eagle Ford parent wells (R, O, M, H1) were drilled in 2014 (Table 1). Eagle Ford child Wells H2 and H3 were drilled and completed in fall 2017, and were closely monitored for response in neighboring wells. In the case studied here, the operators adopted an optimized fracking approach called zipper fracking, which involved the staggered/alternating stimulation of the two wells on a stage by stage basis from toe to heel [10], as shown in the timeline drawn in Figure 5b. This is not to be confused with simultaneous fracking ("Simulfrac"), a similar technique in which the two wells are fractured simultaneously, saving valuable time for operators [11]. In both cases, a primary aim (in addition to saving operation time and cost) is to create a network of complex hydraulic fractures, which can maximize stimulated rock volume, instead of fracturing linearly as with the traditional method [10]. Figure 5b further shows that there is no overlap between the durations of any two stages, so each stage acts as a distinct source for pressure response. Although Figure 5b shows only the timeline of the first 8 stages of Wells H2 and H3, we used and analyzed the pressure signals of all of the combined 101 stages involved in the fracturing operation (see Sections 2.4 and 3.1).

2.4. Acquisition of Eagle Ford Pressure Data

The proprietary fracture treatment files for the Eagle Ford shale Wells H2 and H3 were supplied for our study by the operator. The files include stage by stage post stimulation reports, and data on all relevant fracture treatment parameters such as treating, wellhead, pump side, wellhead and surface casing pressures, slurry flowrates, proppant and mesh size, and additive concentrations, with respect to absolute time, at a frequency of one measurement per second for each quantity. Table 2 shows the number of stages placed during the fracture treatment in each of the Eagle Ford well completions, along with the associated stage and cluster spacings. Wells H2 and H3, the subjects of our study, have the highest number of stages (51 and 50 stages respectively), and were fractured with an average of 9 clusters per stage. Figure 5b showed the timeline for the 2017 fracture treatment progress for Wells H2 and H3. The base, start peak and end peak pressures are the three most important events for each stage of the fracture treatment and were therefore summarized and plotted (see Section 3.1) to serve as a basis for correlation with the Austin Chalk response. In so doing, the voluminous data set supplied by the operator was significantly condensed, making it more suitable for further analysis. Eagle Ford well pumping schedules that were more prevalent in the recent past (2014), as well as common fracture treatment terminology used are discussed in Appendix B.

Table 2. Number of stages, stage spacing and perforation cluster spacing used in fracking operations for six Eagle Ford Wells, Brazos County, Texas.

Well Type	Well Name	No. of Stages (ft)	Stage Spacing (ft)	Perforation/Cluster Spacing (ft)
Parent	M	20	300	50
Parent	H1	22	300	50
Parent	O	13	240	60
Parent	R	35	250	63
Child	H2	51	56–177	6–19
Child	H3	50	45–180	6–20

2.5. Acquisition of Austin Chalk Pressure Data

The six Riverside wells were shut-in during the fracture treatment of Wells H2 and H3. Data logging pressure gauges were installed on the annuli of the Austin Chalk wellheads (Riverside 1–6) to measure any changes in the pressure. Processed data was logged at fifteen readings per minute. The time periods of successful and reliable data measurement for each well are reported in Table 3. A chronology of the pressure data acquisition in the monitored wells is developed in Figure 6. Diagnostic Fracture Injection Tests (DFIT) were conducted in both Wells H2 and H3 prior to the fracture operation in early November 2017, also shown in Figure 6.

Table 3. Observations and interpretations for Riverside 1.

Symbol	Comments/Interpretations
A1	Pressure variations due to well and offset wells production, which fluctuate due to rod-pump artificial lift.
B1	Pressure begins to spike due to Diagnostic Fracture Injection Test (DFIT™) conducted on 11/02/2017, for both Wells H2 and H3 almost simultaneously. The highlighted region is the full period of the pressure spike in Eagle ford Wells during the DFIT test.
C1	Pressure response spike was strongest on 11/03/2017 around 5:50 AM to 6:00 AM. Response increased up to 4000 psi for about one minute during this interval. Most likely an anomaly, but could also be an after effect of the DFIT™ test conducted the day before in which injection pressures increased up to 10,000 psi. About one minute of data is excluded from the response profile in Figure 9 due to outlier values (around 4000 psi) associated with the DFIT test.
D1	More distinct pressure spikes could be from the operator restarting the pumping schedule after a pause
E1	Same as A1
F1	This region is during 11/16/2017 and 11/17/2018, which corresponds to stages 7 to 9 on both Well H2 and H3.
G1	Large increase in pressure during 11/17/2017 20:02 to 11/19/2017 19:50 corresponding to H2 stages 10,11 and 12 and H3 stages 10 and 11 (possibly also stage 12 but stage 12 data is unavailable)
H1	The selected region is from 11/20/2017 11:22 AM to 12/01/2017 02:02 AM, which corresponds to H2 stages 17 to 43 and H3 stages 17 to 49.
I1	The selected region is from 12/01/2017 00:00 to 12/04/2017 00:00. This corresponds to H2 stage 45 to 51 (end of job) to H3 stage 50 (last stage). Pressure response increases towards the end of the job in H3.
J1	The peak pressure response of 265.8 psi occurs on 12/04/2017 07:33 AM and occurs after the operation is completed in both wells. Also, this is the maximum pressure reached mentioned in I1.
K1	This surge occurs monetarily on 12/07/2017 01:47 AM after the operation is completed in both wells, and is attributed to anomalous data, potentially due to equipment failure
L1	Anomalous data in this region (between 12/06/2017 to 12/07/2017) is attributed to equipment failure

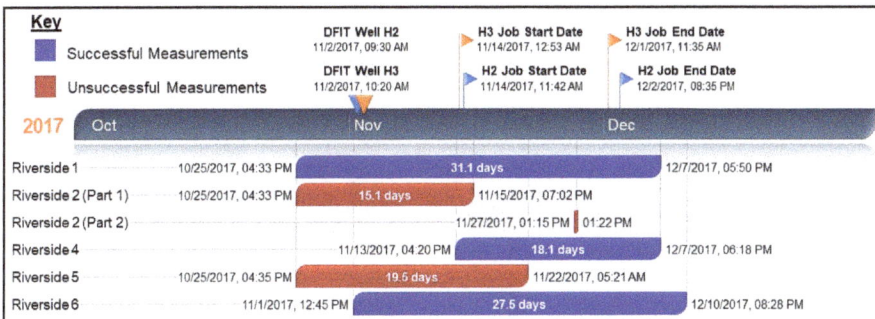

Figure 6. Chronology for Austin Chalk Pressure Data Collection. High quality pressure readings obtained from Wells R1, R4 and R6.

DFITs and other well testing procedures are more effectively investigated in more specialized work [12]. In our study, we simplify the effect of this complicated procedure by considering the DFIT pressure rise in the toe of the two Eagle Ford wells (H2, H3) as a distinct source of potential pressure communication with the Austin Chalk wells.

The pressure response readings for Wells R1, R4 and R6 are continuous during the DFIT and subsequent fracture treatment of the Eagle Ford wells and are therefore considered more extensively in developing our models and formulating conclusions. Data from Well R2 is discontinuous and is limited to just two brief entries (Part 1 and Part 2) on Figure 6. The pressure rise in Well R5 apparently killed the gauge early in the operation, so while the data set is continuous, it is not reliable data. The data collection period for Wells R2 and R5 ended permanently during the frack job. We attribute this to either memory overload or battery failure of the gauges. In spite of these technical issues, we were able to piece together a significant pressure response pattern by analysis of both the source and the response signals (Section 3).

3. Analysis of Results

The well head pressure data for the Eagle Ford fracture treatment were condensed to obtain a simplified input pressure signal (Section 3.1) that could then be used to visualize the correlation with the Austin Chalk pressure responses. The analysis of a fracture stage in the Eagle Ford is given in Section 3.1.1 and the combined pressure signal is discussed in Section 3.1.2. The correlated pressure response profiles are shown in Section 3.2 for Wells Riverside 1, 4 and 6, while those of Riverside 2 and 5 are discussed separately in Appendix C.

3.1. Pressure Analysis of Eagle Ford Wells H2 and H3

3.1.1. Analysis of Raw Data

Wellhead pressures for the fracture stages of the Eagle Ford wells provided by operators were used for later correlation with our pressure gauge measurements for Austin Chalk wells. The pressure build-up and subsequent pressure dissipation for Well H2 Stage 1 are shown in Figure 7. Treatment graphs show wellhead pressure variations plotted along time for a given stage, along with other relevant quantities like slurry rate and proppant concentration on the same axes. The base, start peak and end peak pressures are the three most important events in each stage of the fracture treatment. The three primary pressure states during the fracture treatment of each individual stage are labeled on Figure 7. Apart from minor operational differences, the treatment graphs for each stage in Wells H2 and H3 follow the same general pattern/shape of Figure 7.

- *Start Peak Pressure*: Highest pressure peak, which occurs at the very start of the plateau region of the pressure-time graph and corresponds to formation break down. Circulation fluid is pumped with no proppant to ensure the fractures are wide enough to accept the proppants, which is called creating a "pad". Proppant circulation then typically commences at 100 mesh and low concentrations (20–50 ppg) and increases over time (terminology explained in Table A2, Appendix B). Sometimes during this process, a viscous proppant-free solution called "sweep" is used to remove any solid residuals and clean the well before circulating more proppant.
- *Base Pressure*: Pressure that persists for a longer time, and is represented by the lowest pressure that occurred between the starting and ending peak pressures, which is the stable pressure required for injection of the constant rate of the slurry. Base pressure is attributed to fracture propagation in all directions away from the perforation, although preferential fracture growth occurs in the direction of maximum horizontal stress, perpendicular to the wellbore in the lateral direction [13]. Wells H2 and H3 were fracked with 27 perforations per stage (on average). This being the first stage for Well H2 fracturing, acid was circulated after formation breakdown, after which proppants of increasing concentrations are circulated up to 100 mesh. In the region between the starting peak and base pressure, the fractures propagate in all directions, confined between assumed lower and upper frack barriers. Lateral growth is assumed for the period where the pressure is stable, and subsequently increasing with respect to time, that is, between the base pressure and ending peak. In a typical fracture treatment, operators seek to maximize lateral fracture propagation to maximize the stimulated rock volume, by orienting wellbores and initiating fractures accordingly.
- *End Peak Pressure*: This corresponds to the time when pumping ceases and a pronounced end peak pressure occurs due to the highest proppant concentrations at the tip of the fracture ("screenout"). In order to avoid further pressure rise for Well H2 Stage 1 the operator cuts proppant supply. Operators need to be careful about pressure surges when ceasing pumping to end a stage. The goal is to regulate the proppant concentration precisely enough to minimize pressure build up towards the end of the stage. The final proppant concentration value for this well was 1.80 ppg at 100 mesh. Once this value was reached, the well was flushed with a fluid with no proppants to remove any residual acid and the stage was completed.

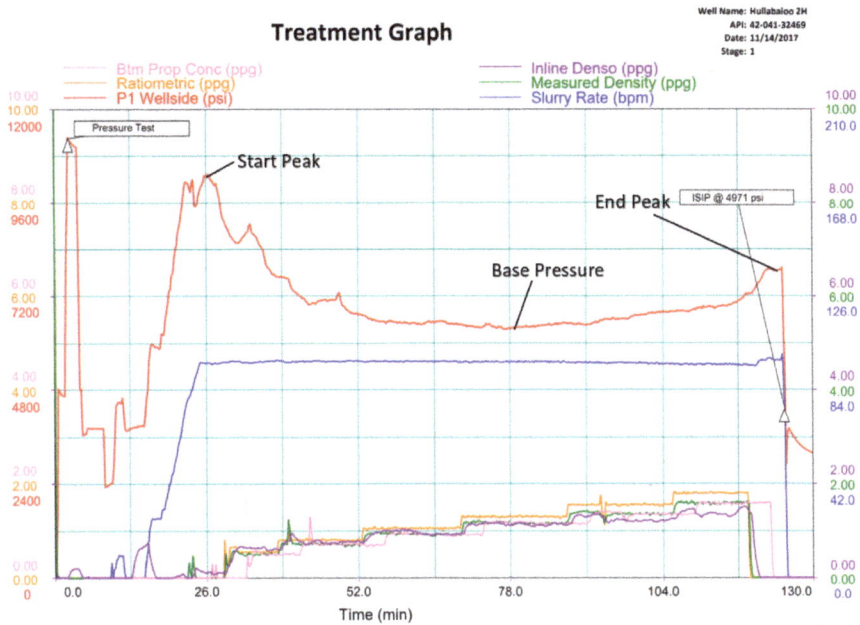

Figure 7. Fracture treatment graph for Stage 1 of Eagle Ford Well H2. All stages had a similar fracture treatment schedule.

3.1.2. Pressure Summary for all Stages of Eagle Ford Fracture Treatment

The post-stimulation reports provided by the operator were condensed by preparing summarized stage reports. The magnitudes of starting peak, base pressure and ending peak pressures of each stage in Wells H2 and H3 provide a first insight into the pressure profile, plotted in Figure 8a–c respectively. Figure 8d combines the starting peak, base peak and ending peak pressures in a combined plot for both wells. The plots provide an overview of the condensed Eagle Ford frack job pressure data against their relative timing. Next, the pressure signal of Figure 8a–d will be used to explain the nature of the pressure communication with the Austin Chalk Formation.

3.2. Pressure Response of Austin Chalk (Wells R1–R6)

Pressure responses of the five monitored Austin Chalk wells (R1 through R6, except R3) are discussed in detail in our study. The Eagle Ford pressure signal in the plots produced in this section consists of the combined pressure sources for Wells H2 and H3 as individually condensed in Figure 8d, but stage numbers are omitted in the correlated plots for the sake of clarity.

The following plots of Austin Chalk pressure response are based on high frequency pressure recordings (every 2 seconds) by the pressure gauges at the Austin Chalk wells. Given the difference in magnitudes between Eagle Ford and Austin Chalk pressures, the latter are plotted on a secondary axis (right-hand vertical scale in Figures 9–11), which produces one plot per well. Observations and interpretations made are displayed below each graph (Tables 3–5). Pressure response profiles and interpretations of Wells R2 and R5 are discussed in Appendix C.

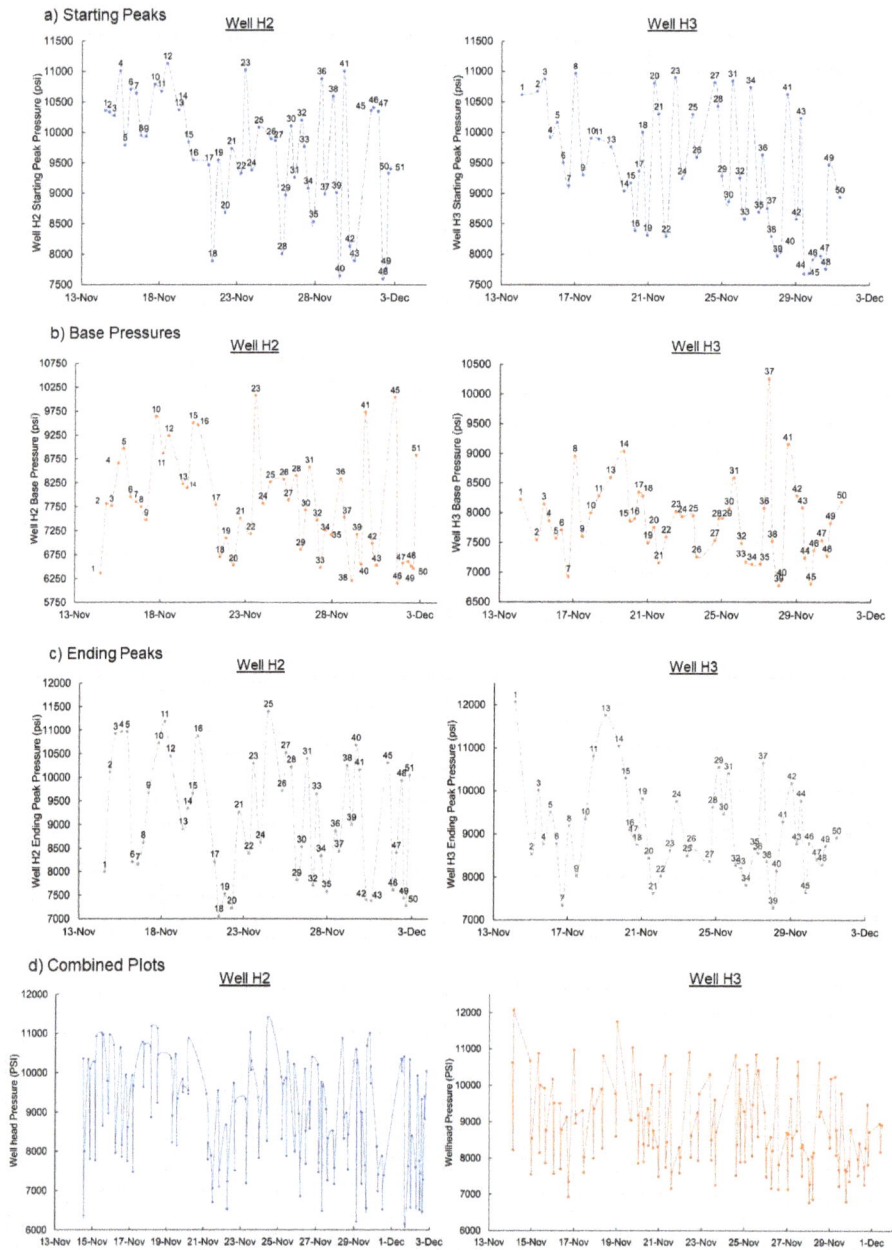

Figure 8. (**a**) Starting peak pressure (**b**) Base pressure, and (**c**) Ending peak pressure for each stage in Eagle Ford Wells H2 and H3, interpolated to more clearly show variation. Data points are labeled with corresponding stage number. Stage 1 is at the toe end and stage 51 and 50 are the final stages of the treatment, at the heel end of Wells H2 and H3, respectively. (**d**) Summary of Pressure change over time for 51 stage fracturing of Well H2 and 50 stage fracturing of H3. Data points represent discrete measurements and therefore the connections presented between data points are interpolations.

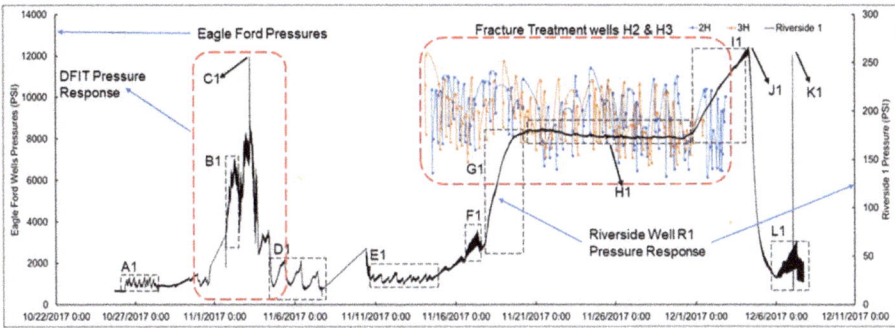

Figure 9. Correlated plots for Riverside 1. The left vertical axis is the pressure in Wells H2 and H3. The right axis is the pressure in the annulus of the Riverside wellhead. The significance of each labeled box is discussed in Table 3.

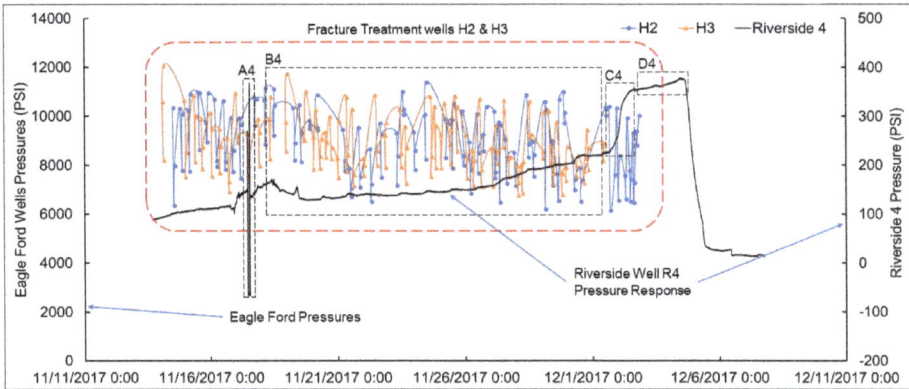

Figure 10. Correlated plots for Riverside 4. The left vertical axis is the pressure in Wells H2 and H3. The right axis is the pressure in the annulus of the Riverside wellhead. The significance of each labeled box is discussed below.

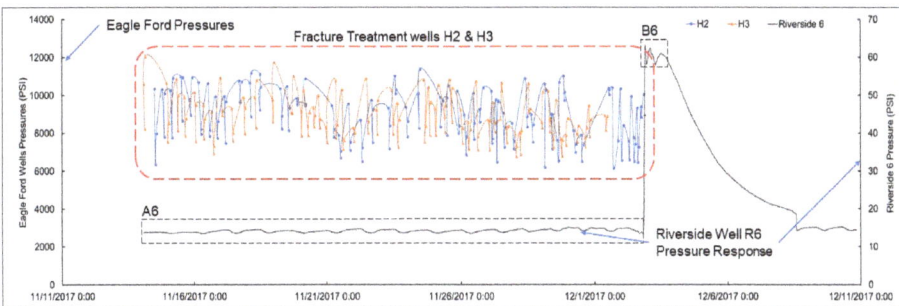

Figure 11. Correlated plots for Riverside 6. The left vertical axis is the pressure in Wells H2 and H3. The right axis is the pressure in the annulus of the Riverside wellhead. Significance of each labeled box is discussed below.

Table 4. Observations and interpretations for Riverside 4.

Symbol	Comments/Interpretations
A4	Pressure spike occurs on 11/17/2017 between 13:07 and 13:25. The surge momentarily stuns the gauge. The highlighted region corresponds to the interval between stages 9 and 10 on both Wells H2 and H3.
B4	The selected region is from 11/18/2017 on 12:04 to 12/01/2017 14:29, corresponding to H2 stages 12 to 45 and H3 stages 13 to 50 (end of the job in H3). It also includes H3 stage 12, but the data for this stage is unavailable so it remains uncertain.
C4	The region starts from 12/01/2017 17:51 to 12/02/2017 17:11, which corresponds to H2 stages 46 to 50. Pressure response increases almost immediately after the end of the job in H3.
D4	The region starts from 12/02/2017 17:11 (same time as when region C4 ends) and ends at 12/04/2017 11:01, which corresponds to H2 stage 51 (last stage) but is after the job in Well H3.

Table 5. Observations and interpretations for Riverside 6.

Symbol	Comments/Interpretations
A6	The plateau region starts from H2 stage 1 (11/14/2017 12:04) to H2 stage 51 (12/02/2017 20:21) and encompasses the entirety of the fracture treatment.
B6	Similar to R1 and R4, the pressure peaks at the end of the operation, and rapidly declines after reaching this maximum value. The highlighted region is from 12/02/2017 20:21 to 12/03/2017 02:17AM.

3.2.1. Riverside 1

Figure 9 shows the correlations between the signal of the pressure sources of fracture treatment stages in Well H2 and H3, and the responses in Well R1 on an absolute time scale. The plot for Well R1 is annotated with more detail than for the other wells to establish the causes for pressure variations before and after the job. Little wriggles can be noticed in the flat trend of the R1 pressure response curve in region H1. These wriggles loosely correlate with the start of the stages, most likely due to formation breakdown or starting peak pressure. This trend holds for most of the Eagle Ford operation. Towards the end of the operation in region I1, there is a large increase in Well R1 pressure response that persists for a few days after the Eagle Ford frack job ceases for a certain time period, and then rapidly decreases. The steep drop of the pressure response curve is attributed to leakoff and final closure of the pressure conduit between Well R1 and the Eagle Ford pressure source, and can be observed to some degrees in all the wells. Table 3 presents observations and interpretations for each region of the plot highlighted in Figure 9. The table also specifies approximate date/time values for the selected regions. These observations will all be useful in developing a conceptual model for pressure communication in later sections, and recording time intervals will help in making further correlations.

3.2.2. Riverside 4

The correlated plot for Riverside 4 is shown in Figure 10. The data for this well spans from just before the start of the fracturing in H2 and H3 and was collected until 12/8/2017, almost a week after the job in H2 and H3 ends. The plot highlights the important features of the pressure response profile (as in Figure 9) whose durations are shown in Table 4. Since the pre-frack data was not available, we cannot comment on the effect of the DFIT™ test conducted on 11/02/2017 on Riverside 4. Even so, the data set obtained for R4 is continuous, and is the most reliable out of all the wells studied (see Figure 6) and shows strong response to the fracture treatment on the same time scale. Table 6 also notes observations and durations associated with the regions of response highlighted in Figure 10.

Table 6. Offset production data. Changes in production are based on 3-month averages before and after November 2017.

Name	API Number	Direction Relative to H2/H3	Aug-17	Sep-17	Oct-17	Avg before	Nov (during)	Dec-17	Jan-18	Feb-18	Avg after	% Change
					Oil Production (bbl)							
Riverside Production Unit	04131502	Direct South	191	688	79	319	236	754	783	128	555	73.80
Fazzino Omni	04131486	South West	162	156	183	167	510	915	826	699	813	387
Brazos Farm	04131474	North West	6	0	0	2	284	449	259	205	304	15117
Fazzino-Penicka	04131530	West	137	124	150	137	169	189	173	153	171	25.3
Willie Kuder	04131464	South West	178	0	98	92	1	253	284	254	264	186.6
Magnolia Oil & Gas	04131307	North West	151	173	185	170	155	125	120	163	136	−19.8
Brazos Farm Ltd.	04131462	South West	0	0	12	4	14	14	18	49	27	575
					Gas Production (Mcf)							
Riverside Production Unit	04131502	Direct South	969	2185	1682	1427	871	1768	1539	1157	1488	4.29
Fazzino Omni	04131486	South West	368	460	684	602	897	1105	1234	967	1102	82.98
Brazos Farm	04131474	North West	32	0	0	153	580	735	606	372	571	273.20
Fazzino-Penicka	04131530	West	1217	1127	1180	1163	1129	1193	1137	999	1110	−4.61
Willie Kuder	04131464	South West	1811	1993	1143	1560	1296	995	478	692	722	−53.76
Magnolia Oil & Gas	04131307	North West	254	223	274	250	279	240	260	262	254	1.46
Brazos Farm Ltd.	04131462	South West	19	15	12	15.3	12	0	0	0	0	−100

Similar to the response profile of Riverside 1 (Figure 9), the little bumps on Figure 10 correspond to the start of stages (most likely due to formation breakdown or starting peak pressure as we defined it) in region B4. Not all fracking stages produce visibly large bumps in the Austin Chalk. This trend holds for most of the Eagle Ford operation. Another similarity is in region C4 and region I1, where there is an increase in pressure towards the end of the operation. Region D4 marks the maximum pressure reached. However, unlike in Well R1 (region J1 labeled on Figure 9) this maximum occurs over a plateau instead of one discrete point. We know that the maximum must be a plateau because the gradient of the plot decreases sharply between regions C4 and D4. Additionally, the pressure rapidly decreases in both Wells R1 and R4 after this maximum pressure point/plateau is reached. This is another important difference between R1 and R4, since in R4, the region of highest response occurs as a second plateau (with a noticeable, but minor positive gradient) instead of a discrete point, (as in Well R1), even though the maxima occur in both wells after the treatment is completed in Well H2. Further, the maxima in both Riverside wells occur at around the same absolute time (12/04/2017).

3.2.3. Riverside 6

The pressure response profile of Riverside 6 is shown on Figure 11, and the observations and time durations were recorded in Table 5. The key feature of this well is that there is only one significant pressure response in R6 that almost instantaneously occurs towards the end of the fracturing in Eagle Ford. Additionally, the magnitude of response is less than those of Riverside 1 and 4 (see Figures 9 and 10 respectively) and takes significantly longer duration after the treatment ends to return to its initial reservoir state. Data for this well was collected until after the pressure returned to its initial state in about one week after the job ended. Similar to the highlighted region H1 (on Figure 9) and B4 (on Figure 10), the response for the majority of the frack job is characterized by a plateau region in pressure with wriggles loosely correlating with the starting times of stages. The selected region A6 (on Figure 11) covers the entire frack operation. R6 follows the plateau trend of a steady pressure response for the almost the entire operation, unlike in Wells R1 and R4. Similar to R1 and R4, the pressure response of Well R6 peaks at the end of the operation and rapidly declines after reaching this maximum value. The difference in Well R6 is that the rise to this pressure occurs almost instantaneously, which could be attributed to equipment failure, but could also be a strong pressure communication between the wells. The significant pressure response commences almost instantly after the frack job in H2 is completed, unlike in R1 and R4 which began growing to a maximum after the job in H3 was complete, while the fracking of the final stages of Well H2 were still ongoing (recall the job for H2 was finished a day after H3 as shown in Figure 8d). The pressure response maxima are reached at around the same time for Wells R1, R4 and R6 (i.e., at different times on 12/04/2017).

4. Interpretation of Results

The principal purpose of our study is to develop a conceptual model for the observed pressure communication between the two reservoirs (Eagle Ford and Austin Chalk). The estimated pressure acting on the boundary between the two reservoirs during the fracking of the Eagle Ford wells is modelled based on the pressure responses observed in the Austin Chalk wells discussed in Section 3.2. Our analysis will quantify (and qualify) the correlation of the pressure response profiles using the following observations:

- The relative lateral spacings between Eagle Ford Wells H2 and H3 and the Austin Chalk observation wells (Section 4.1)
- Changes in average production in wells in the vicinity of the H2–H3 pair, including impact on the Riverside production unit (Section 4.2)

We then develop a conceptual model using the results of our analysis (Sections 4.3 and 4.4) that serves to explain the principal mechanisms responsible for the observed pressure communication across the reservoir boundary between the Eagle Ford and the Austin Chalk Formations.

4.1. Vertical Communication

The reservoir pressures in the Eagle Ford an Austin Chalk reservoirs immediately prior to the the fracking operations in Wells H2 and H3 (Section 2.2), and the Austin Chalk pressure responses during the fracture treatment (Section 3.2 and Appendix C) are used to better understand the detailed nature of the pressure communication between the Eagle Ford and Austin Chalk reservoirs. The principal pressure response magnitude, rate and durations for each of the observation wells are calculated in Appendix D. Pressure response magnitude is highlighted by the thickness of arrows in the gun barrel view of Figure 12a, which shows that the pressure communication intensifies from SW to NE.

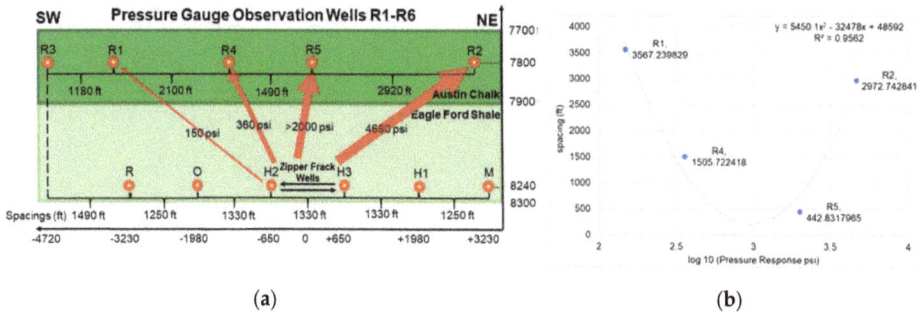

(a)

(b)

Figure 12. (a) Gun barrel view with pressure response intensity in monitored Austin Chalk wells emphasized by arrow width. Well spacings are estimated using Well R3 as a reference line. The horizontal axis represents spacing relative to the midpoint of Wells H2 and H3. Vertical axis represents true vertical depth, and is exaggerated 6.6× relative to the horizontal length scale. (b) Plot of logarithm of pressure response against diagonal well spacings calculated from Figure 12a. Data labels are the Austin Chalk well names (R1, R4, R5, R2) with measured diagonal spacings in ft. A strong parabolic correlation is obtained.

Based on the observations from Figure 12a, one may suggest that the magnitude of pressure response is a function of distance to the fracked Wells H2 and H3. Diagonal spacings of Austin Chalk wells relative to the midpoint of H2–H3 pair are calculated and plotted against the logarithm of the pressure response in Figure 12b, correlating the information presented in Figure 12a. Diagonal well spacing and logarithm of pressure response therefore have a parabolic relationship. One possible explanation for the pressure communication intensifying from SW to NE is that a denser natural fracture network occurs in the NE part of the Austin Chalk, which establishes a better connection with the hydraulic fractures from the Eagle Ford. We assume that some hydraulic fractures in the vertical direction in the Eagle Ford wells will connect with a natural fracture, which will ultimately lead up to a Riverside wellbore. Natural fractures farther away from the wellbore have a better chance of activation if they fall within the influence zone of a long hydraulic fracture [14]. Alternatively, some of the observed pressure communication may occur by fluid transmission through the primary pore network of the Austin Chalk, which has a 12% average porosity (in a potential range of 10% to 22%) and an average permeability of 0.12 mD (in a range of 0.01 md to 15 mD), according to local field studies [7].

One may also speculate that the hydraulic fractures of Well H3 have wider apertures, because that well was fractured with coarser proppants (40/70 mesh heavy) much more frequently than Well H2 (100 mesh heavy), which would allow transmission of more frack-fluid and energy, and therefore could have resulted in a stronger series of connections with the natural fracture network in the Austin Chalk. In any case, Wells R2 and R5 show the highest intensity pressure response because these wells are relatively close to Well H3.

4.2. Observed Production Uplifts

Independent evidence for temporary pressure uplift in the Austin Chalk due to the fracking operation in the Eagle Ford Wells H2 and H3 is provided by increases in nearby Austin Chalk well productivity. For example, the Fazzino Well (API 04131486) operated by Wild Horse showed a distinct production uplift (Figure 13), which more than doubled its production of both oil and natural gas on the time scale of the fracture operation, and persisted for several months afterwards. We reason that the natural fracture networks being activated through the intensity of the hydraulic fractures from the Eagle Ford is responsible for the uplift. The temporary pressure uplift observed resembles a fracture hit [6,15] which in our case did not result in permanent interwell communication. Earlier production uplifts seen in Figure 13 can be attributed to well workovers and shut-ins. Typically, such periods of zero production are followed by brief episodes of enhanced production. However, no shut-in preceded the latest rise in the Fazzino well, which the operator therefore attributed to the nearby fracking operation in the Eagle Ford Wells H2 and H3.

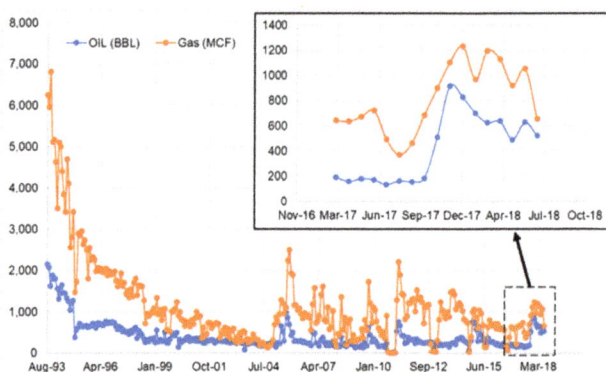

Figure 13. Production profile of Fazzino Well, including a zoomed in section showing with greater resolution the production uplift attributed to the fracturing of Wells H2 and H3. The vertical axes for both plots have the same units. That is, bbl for oil and Mcf for gas. The production data used was obtained from the Texas Railroad Commission's public sources.

Further, although Texas RRC reports that the Fazzino Well of approximately 3000 ft from the pressure signal (i.e., the H2–H3 Well pad), it is entirely possible that the production uplift observed was caused as a result of pressure communication due to the fracking of Wells H2 and H3. A similar incident occurred in 2014, in which a significant frack hit was observed in two other Eagle Ford wells in the same area. In that instance, Well O responded considerably to the fracking of Well H1, which is located 4000 ft away (in horizontal direction to the wells) from Well O.

Additionally, production data from the Texas Railroad Commission was considered in computing average changes in production for offset wells in the vicinity of the H2–H3 pair in the three months prior to and succeeding the fracture operation (which started on 14 November 2018). Three-month averages were computed for both oil and natural gas. The most significant production uplifts are summarized in Table 6. Some of the production uplifts may have occured as a result of other refracks and workover operations taking place in the county at about the same time. Additionally, whether or not a well shows an average increase or decrease in production depends on how the average values are computed. For Table 6, average before includes monthly data from August, September and October 2017 and average after includes December 2017, January and February 2018. However, the data in Table 6 shows that the production uplift of the Fazzino Well emphasized in Figure 13 is attributable to the fracture operations, given that Table 6 reports 387% and 83% increases in oil and gas production, respectively.

4.3. Interpretations of Pressure Response Profiles and Conceptual Model

To build a conceptual model that can explain the temporary nature of and the physical process of pressure communication, we first consider the following additional details from the pressure response profiles presented in Section 3.2. Our reference is the response of Riverside 1 (Figure 9), but similar pressure response patterns and inferences were observed in the other Riverside wells (see Figures 10, 11, A2 and A3)

The typical DFIT response [16] is characterized by a sudden surge in pressure that is dominated by the input pressure signal, followed by a rapid release of pressure, known as the reservoir dominated region in which the reservoir returns to its original pressure state. Interestingly, the response patterns for the main fracture treatments of all five observation wells (see Figures 9–11, Figures A2 and A3) also shows this pattern, to varying degrees. We identify four phases that characterize the pressure response pattern of Well R1, as follows:

(1)	The first increase in pressure (regions F1 and G1 labeled on Figure 9) is due to the hydraulic fractures propagating outwards in the vertical direction and connecting to the naturally fractured Austin Chalk system. As frack fluid enters the natural fracture network, the reservoir pressure of the Austin Chalk increases, which causes fluids to migrate towards Riverside wellbores, which were shut-in for the period of the fracking operation, but resumed production soon after the operation concluded. (Correspondingly, Table 6 showed the later increase in both oil and gas production of the Riverside production pad, as a result of the fracture treatment, in the months following the operation).

(2)	The plateau region (region H1 labeled on Figure 9) can be attributed to the zipper fracturing nature of the operation. In a zipper fracking operation, the fractures propagate towards each other so that the induced stresses near the tips force fracture propagation in a direction perpendicular to the wellbores [10]. The lateral fracture propagation prevents further vertical fracture growth, which results in a nearly constant pressure response in the Riverside well, given that each stage of the Eagle Ford fracture treatment was conducted in a similar way. The small bumps on the pressure response (region H1 labeled on Figure 9) suggest that the Austin Chalk reservoir pressure would increase further, if not restricted by the induced stresses at the fracture tips caused by the zipper fracking operation. This method of zipper fracturing is also highly effective in creating an altered zone within the Eagle Ford itself (in the horizontal direction).

(3)	The second increase in pressure (region I1 labeled on Figure 9) begins towards the end of the treatment of Well H3. Since there is no more interference from an additional fracture treatment, the pressure increases until after the treatment of Well H2 ends, and pressure declines to its original state.

(4)	The rapid decline in pressure occurs after a small time delay due to the residual effect of fracture treatment. If we model the entire response phenomena as a single fracture stage, the maximum response point (point J1 labeled on Figure 9) would be analogous to a closure pressure after which the flow becomes reservoir dominated. Stress shadows are then able to close the induced/connected fractures and fracture fluid is no longer forced into the Austin Chalk. Correspondingly, the production uplift effect for the Riverside well pad declined over a longer time span, as was shown by production data for three months after the fracking operation ended (Table 6).

The above pressure response observations for Well R1 largely apply to the other Riverside wells as follows:

•	*Riverside 4*: Although there is no early pressure increase in Well R4, the pressure is slowly increasing throughout the frack job (regions B4, C4 and D4 labeled on Figure 10), and the magnitude of pressure communication is higher than for Riverside 1 (Figure 9), suggesting that Riverside 4 has better communication with a naturally fractured zone, allowing for greater extent of fracture network development. The sudden spike in pressure in the middle of the fracture treatment job (Region A4 on Figure 10) could mean that a direct fracture connection was formed, allowing for pressure to surge, that even momentarily stuns the gauge as shown by a momentary negative pressure value (region A4, labeled on Figure 10). Similar to Riverside 1 (region I1 labeled on Figure 9), Riverside 4 shows a second pressure increase (region C4 labeled on Figure 10) at the end of the treatment of Well H3 that ends when the treatment of H2 ends, after which there is a slight increase for a short duration, due to the residual effect of the treatment. Reservoir dominated portion subsequently returns the reservoir to its original state.

- *Riverside 6*: Similar to Wells Riverside 1 and 4 (region H1 labeled on Figure 9, and region B4 labeled on Figure 10 respectively), Well R6 shows a plateau region (region A6 labeled on Figure 11) for the entirety of the frack job. The difference is that for Well R6, the increase in pressure (region B6 labeled on Figure 11) occurs immediately after the end of the treatment of H2. The pressure subsequently declines in the same way as in Wells R1 and R4. The magnitude of communication is significantly lower in this well because it is spaced further apart from wells H2 and H3 than the other Riverside wells (Well R6 does not appear on the gun barrel cross section in Figures 5a and 12a) Further, the lateral section of this well could have a weak connection to natural fractures, which means fewer hydraulic fractures can develop a network.

- *Riverside 2*: Even though there is not enough data available, a region of pressure communication can be observed (Point B2 on Figure A2), although it is followed by a sudden decline in pressure.

- *Riverside 5*: Strong pressure surges are seen during and after the DFIT tests (conducted on Eagle Ford Wells H2 and H3 on 2 November 2017), in addition to observed response in middle of the fracture treatment (regions A5 and B5 respectively, labeled on Figure A3). The surges once again can be attributed to events similar to those assumed for Riverside 1 (see above), except in this case in lieu of higher quality data, we conclude that the surges correspond to two discrete, strong connections being formed.

4.4. Conceptual Model

We assume that the Eagle Ford shale and Austin Chalk form a single hydrocarbon system, in which the Eagle Ford shale is the underlying source rock [17]. Section 2.2 showed that there exists a significant pressure (2354 psi) difference between the Eagle Ford and Austin Chalk Formations. The pressure difference could cause oil from the hydrocarbon-rich Eagle Ford shale to naturally migrate upwards into the lower pressured Austin Chalk Formation. However, without any form of human intervention, this migration must happen over geologic time, since both reservoirs have very low permeabilities. The hydraulic fracture treatment in the Eagle Ford pressurized the reservoir significantly and temporarily increased the pressure difference with the Austin Chalk. The hydraulic fracture treatment creates a fracture pattern in the Eagle Ford, which may connect to the existing natural fracture network in the Austin Chalk. Frack fluid used in the stage stimulation of Wells H2 and H3 is the ultimate source of the temporary pressure communication across the two fracture networks.

Figure 14a shows a schematic wellbore diagram of the vertical section of the Austin Chalk observation wells, highlighting the annular location of the wellhead where the pressure gauges were mounted. Figure 14b shows a conceptual model for the observed pressure communication, with a vertical cross section of the reservoirs, taken perpendicular to the panel (of the gun barrel view) shown in Figure 5a. Figure 14b shows the base pressure signal introduced in Figure 8b superimposed onto the lateral section to show the pressure experienced for the longest duration on each section of the wellbore over the course of the fracture treatment. Arrows indicate pressure transmission which ultimately causes pressure responses in Riverside wells. Pressure communication and fluid mobility in the model are both facilitated by the network of natural fractures in the Austin Chalk.

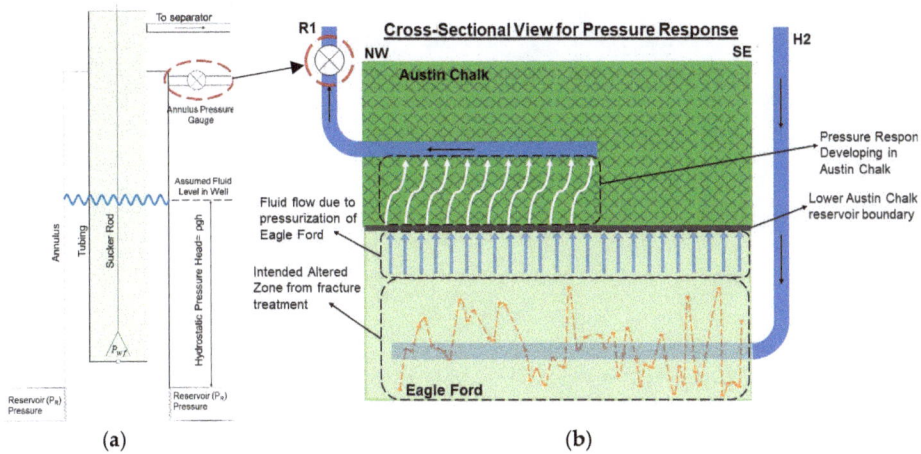

Figure 14. (a) Schematic wellbore diagram for the vertical section of an Austin Chalk observation well, showing static fluid column for a shut-in well and highlights (in red) the annular location on the wellhead where the pressure gauges were mounted. (b) Conceptual model illustrating fluid flow and pressure communication that ultimately cause production uplifts in the Riverside production unit.

Often when fracture pressure communication is discussed, the observed pressure communication effects are attributed to changes in the in-situ stress of monitored well fractures, due to the propagation of hydraulic fractures in an offset well in the same reservoir, described further in [18]. Poroelastic interactions between monitor fractures and propagating hydraulic fractures, for a general unconventional reservoir and well configuration, are also modeled in cases where the wells are in the same, typically shale, formation [19]. These effects are not considered in our work since the Austin Chalk is a naturally fractured carbonate, which allows it to act as a conduit for actual physical fluid-based communication.

5. Conclusions

Our study analyzed empirical evidence for pressure communication with the Austin Chalk Formation during the stimulation of two new (2017 child) wells in the Eagle Ford shale, which caused the pressure to locally surge in both reservoirs. The conductive fracture network formed in the Austin Chalk is assumed to have transmitted some fluid pressure from the Eagle Ford to the annulus of five monitoring wells in the Austin Chalk. Pressure gauge responses in the Austin Chalk wells were measured during the fracture treatment in the Eagle Ford. The magnitude of pressure response in the Austin Chalk wells is only a fraction of the Eagle Ford injection pressures. The pressure stimulation of the two Eagle Ford wells occurred in alternating stages (zipper fracking).

The Riverside wells began responding to the Eagle Ford frack job through an initial increase in pressure as hydraulic fractures begin to propagate outwards and connect to the naturally fractured network of the Austin Chalk. The initial rise in pressure is followed by a plateau region, which is limited by the stresses induced in a zipper fracking operation that causes fractures to propagate towards each other, in a direction perpendicular to the wellbore. Immediately after the fracture treatment in Well H3 was completed, a second pressure rise occurred, which persisted until after the operation in Well H2 was completed. The pressure response rapidly declined to its pre-treatment state, which confirms that pressure communication was temporary in nature. The observed time delay was a result of stress shadows closing the induced/connected fractures, after which frack fluid is no longer forced into the Austin Chalk.

Our conceptual model of pressure communication takes into account the pressure depletion of the Austin Chalk reservoir due to decades of production, prior to the fracture treatment in the nearby Eagle Ford wells. The depleted, average reservoir pressure near the Austin Chalk wells is estimated to be 2354 psi. Based on history matching of earlier Eagle Ford (2014 parent) wells, the pressure in the Eagle Ford landing zone immediately prior to the fracture treatment was 4891 psi. The initial pressure in the annulus of the five Austin Chalk observation wells was approximately 20 psi immediately prior to the Eagle Ford fracture treatment. However, the pressure rose to 265.8 psi in Riverside 1, 378 psi in Riverside 4, and 63.3 psi in Riverside 6. Pressure response profiles of wells Riverside 2 and 5 show similar trends.

We summarize conclusions, based on our interpretations of the pressure response profiles in wells R1–R6, as follows:

(1) Pressure communication between the two well sets (Eagle Ford-Austin Chalk) is a temporary phenomenon, taking approximately up to three days (from the start of the zipper fracking of the H2–H3 Well pair) to establish, lasting approximately 11 to 12 days to reach a plateau, which is followed by a brief final screen out peak that drops off nearly instantaneously. In all the pressure response profiles considered, the pressure rise in the Austin Chalk wells rapidly declines back to pre-treatment annulus pressures.

(2) The magnitude of pressure rise in the Austin Chalk is significantly lower than Eagle Ford fracture treatment injection pressures (about 6%)

(3) Pressure communication is thought to occur due to pressurization of isolated fracture stages in the Eagle Ford wells, which temporarily increases the pressure differential between the Austin Chalk and Eagle Ford shale.

(4) Coeval production uplifts in the months following the fracking of the Eagle Ford wells were observed in the offset Austin Chalk wells, in addition to the Riverside wells themselves, and are associated with the natural fractured network of the Austin Chalk, that is further activated through the fracking of the Eagle Ford wells.

(5) Hydraulic fractures from the Eagle Ford open due to fluid injection during fracture treatment and are assumed to temporarily connect with the natural fracture system in the Austin Chalk reservoir. Poroelastic effects are not considered in our work due to the nature of the Austin Chalk.

(6) The pressure response in the annulus of the Austin Chalk wells is characterized by a time delay, because the pressure surges in Wells Riverside 1 and 4 ceased shortly after (35.2 h in Riverside 1 and 38.7 h in Riverside 4) the zipper fracking operation of Eagle Ford Wells H2 and H3 was completed.

Author Contributions: Data curation, S.S.; Investigation, S.S.; Methodology, R.W., I.A. and S.N.; Supervision, R.W.; Writing—review & editing, R.W. and S.S.

Funding: This project was sponsored by startup funds of the senior author (R.W.) from the Texas A&M Engineering Experiment Station (TEES).

Acknowledgments: We would like to thank Hawkwood Energy, Exponent Energy and the Texas A&M University System for providing access to the RELLIS lease wells and data sets, which facilitated our study. Tyler Moehlman is thanked for help as an undergraduate student collecting pressure data in the field.

Conflicts of Interest: The authors declare no conflict of interest.

Appendix A. Pressure Depletion Calculation in Austin Chalk

When the pressure in the Austin Chalk is below its bubble point pressure, the oil formation volume factor is given as a function of pressure as shown in Equation (A1). The formation volume factor B_o is an important parameter in calculating pressure depletion. Equation (A2) shows the calculation of pressure depletion from initial reservoir pressures based on production data. Equation (A2) is derived from material balance [8]. The pressure in the reservoir can be iteratively calculated using monthly production data by solving Equation (A3) which is obtained from substituting Equation (A1) into

Equation (A2) and solving for pressure P. Gas production is not a part of Equation (A2) (and hence Equation (A3)) since there is no free gas initially in the reservoir:

$$B_o = 1 + \frac{(B_{ob} - 1)P}{P_b} \tag{A1}$$

$$P_{ri} - P = \frac{N_p B_o + W_p B_w}{c_t(NB_{oi} + WB_{wi})} \tag{A2}$$

$$P = \frac{P_{ri}c_t(NB_{oi} + WB_{wi}) - W_p B_w - N_p}{\frac{N_p}{P_b} * (B_{ob} - 1) + c_t(NB_{oi} + WB_{wi})} \tag{A3}$$

When the reservoir pressure is above the bubble point, the expression for formation volume factor changes, requiring the use of Equation (A4), which calculates the formation volume factor in an undersaturated reservoir. As in the below bubble point case, we substitute Equation (A2) to obtain the expression for pressure depletion, shown in Equation (A5), which needs to be solved using iterative techniques. The bisection method, which gives results to any desired level of precision (we use 0.01%) [20] was used to solve Equation (A5):

$$B_o = B_{ob} \exp(c_o(P_b - P)) \tag{A4}$$

$$P_{ri} - P = \frac{N_p B_{ob} \exp(c_o(P_b - P)) + W_p B_w}{c_t(NB_{oi} + WB_{wi})} \tag{A5}$$

The methodology adopted was to first calculate pressure depletion using Equation (A5) from monthly production data, replacing the initial reservoir pressure term P_{ri} with the calculated pressure P after every interval. When pressure falls below the bubble point, Equation (A3) would need to be used instead. Using monthly production data (from Texas RRC online), we can calculate the resulting reservoir pressure at the time of the fracture treatment in Wells H2 and H3. All the parameters required for the above calculations (with nomenclature) are shown in Table A1. The original reservoir pressure is first calculated from pressure gradient of 0.45 psi/ft and the average true vertical depth of the six Riverside wells. Petrophysical parameters are needed to perform original oil and water in place calculation. Although there is understandably a large uncertainty in values, we consider a scenario, in which the water saturation is the highest and therefore the percentage of water produced (to that of oil produced) is an arbitrarily chosen high value. The bubble point pressure and composition of the reservoir fluids present in the Eagle Ford are the same as that for the Austin Chalk vertically above, which is reasonable since the Austin Chalk and Eagle Ford shale likely form a single hydrocarbon production system [17].

The pressure depletion curves obtained are shown in Figure 4a Keeping all other variables constant, we perform a sensitivity analysis for the drainage area and present the results in Figure 4b, which shows that the effect of depletion is stronger for small drainage areas. For small drainage areas, P_{ri} falls below the bubble point pressure, requiring use of Equation (A3) (see Section 2.2.1 for Figure 4a,b). The initial reservoir pressures plotted refer to the average of the reservoir pressures calculated for the duration of the fracture treatment. When the drainage area becomes infinitely large, the pressure depletion effect becomes negligibly small. Likewise, the reservoir pressure goes to zero for small drainage areas. The relationship is logarithmic in nature, with the correlations shown on the plot. Given that the RELLIS field spans over an area of 2000 acres [9], we will use the corresponding reservoir pressure from Figure 4a (2354 psi) in building our pressure response model.

Table A1. Variables used for pressure depletion model for Austin Chalk.

Symbol	Parameter	Value	Unit
	Pressure Gradient	0.45	psi/ft
	True Vertical Depth (TVD)	7731.3	ft
P_{ri}	Reservoir pressure before production	3479.1	psi
	Percentage of water produced	50	%
c_t	Rock Compressibility	4.0×10^{-6}	1/psi
	Drainage Area	4000	acres
	Thickness	200	ft
	Water Saturation	0.8	No Unit
	Oil Saturation	0.2	No Unit
B_{oi}	Initial Oil Formation Volume Factor	1.15	RB/STB
B_{wi}	Initial Water Formation Volume Factor	1.0	RB/STB
	Porosity	0.12	No Unit
	Unit Conversion (for OOIP and OWIP calculation)	7758	Conversion
N	Original Oil in Place	1.30×10^8	bbl
W	Original Water in Place	5.96×10^8	bbl
P_b	Bubble Point Pressure	2398	psi
c_o	Undersaturated Oil compressibility	1.0×10^{-5}	1/psi
B_{ob}	Maximum Oil Formation Volume Factor	1.16	RB/STB

Appendix B. Pumping Schedule of Well H1 in Fracture Treatment of 2014

This appendix presents pumping schedule used for hydraulic fracture treatment in Well H1, which is representative for all present wells (Table 1) of the RELLIS area. Common terminology used in stimulation reports is also shown. The pumping process of 2014 fracture treatment of Well H1 was also similar to that used in 2017 fracture treatment of Wells H2 and H3. Wells were acidized, padded, and circulated with increasing levels of proppant concentration and decreasing proppant size as the stage progressed. Figure A1 shows the treatment graph for Well H1 Stage 7 for comparison with pumping schedules of Wells H2 and H3, in Figure 7. (shown in Section 3.1). The most common terms used in a post-stimulation report are summarized in Table A2.

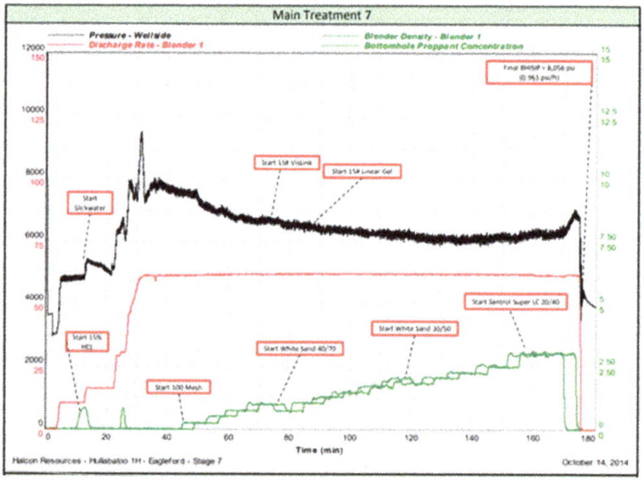

Figure A1. Well H1 Stage 7. The same properties apply as in Wells H2 and H3 (Figure 7) except there is much more variation in Well H1, with regards to the proppants used, both in terms of grain and mesh size, which explains the high increase in proppant concentration (red line). In H2 and H3, 100 mesh is mostly used, with 40/70 being employed for select stages, so the corresponding graph would have a lower gradient or be flatter. Type of proppant and gels used are labeled.

Table A2. Common fracturing terminology.

Term	Meaning
Start Peak Pressure	Wellhead pressure peak at the start of the job
Base Pressure	Minimum pressure that persisted for the longest time (in the plateau region)
End Peak Pressure	Wellhead pressure peak at the end of the job
Breakdown	The pressure/applied stress at which the formation breaks down
Acid	When acid is being circulated to loosen the formation
Shut Down	When pressure pumps shut down
Displace acid/Pump Ball	Stop circulating acid and start circulating slurry
Pad	Circulate fluid with no solid in it until the fracture is wide enough to accept proppant
Sweep	Circulate small volume of viscous fluid called a carrier gel to clean/remove solid residue from the well
Flush	Circulate a fluid that removes any remaining acid
End Stage	Indicates the end of stage. We may still take measurements for a few more minutes
x ppg 100 Mesh	When slurry at that proppant weight is circulated. Mesh is inversely proportional to grain size. 100 Mesh proppants are therefore fine-grained
x ppg 40/70 Mesh	When slurry at that proppant weight is circulated. 40/70 mesh proppants are coarser than 100 mesh.
ppg	A common unit of density; lb/gal

Appendix C. Pressure Response Profiles for Riverside 2 and Riverside 5

This appendix presents the pressure response profiles of Wells R2 and R5 in the same fashion as the profiles for Wells R1, R4 and R6 (shown in Section 3.2) Not enough reliable data was collected for Wells R2 and R5 to produce a complete correlation, but results are still included since valuable insights can be gained from even the limited data available.

Riverside 2

The data collected for Riverside 2 is very limited (as discussed in Section 2.5 and shown in Figure 6). As a result, the pressure response profile shown in Figure A2 consists mostly of interpolations (indicated by thin black lines). The data is sparse due to equipment failure, however Figure A2 still shows that the gauges did detect an increased pressure in R2 during the treatment, even if for a short time. The data for Riverside 2 was unavailable after 11/27/2017, so B2 does not necessarily represent a maximum pressure. This would be since the gauge ran out of battery at that time or was otherwise unable to continue taking accurate readings. Table A3 discusses the estimated point of increase and the approximate timings of events A2 and B2. The important observation from Well R2 response lies not in the correlation trend, but rather in the magnitude. Even for the minimal data connected, the figure shows the data in region B2 has magnitudes in regions of 3000 to 5000 psi, which is significantly more than Wells R1 and R4. Interestingly, there was no pressure response detected for the duration of the DFIT Test, even though pressure data was collected during this time (11/02/2017 to 11/03/2017).

Table A3. Observations and interpretations for Riverside 2.

Symbol	Comments/Interpretations
A2	The point selected occurs on 11/15/2017 19:02 and corresponds to Wells H2 and H3 between stages 4 and 5. This point indicates the assumed start of pressure increase. For Well R2, the pressure increase approximates to a single straight line and takes place early on in the frac job.
B2	11/15/2017 13:15 H2 between stages 2 and 3, and H3 between stages 3 and 4. This point is the end of the pressure increase and what appears to be an instantaneous pressure drop.

Riverside 5

The response profile for Riverside 5 presents almost no viable continuous data set, as shown in Figure A3. While it is easy to assume this was caused by faulty gauges, the fact remains that there still does exist a pressure communication signal on the time scale of major events. Data is unavailable for Riverside 5 after 11/22/2017 05:21 AM. This well does not show any of the trends as the other Riverside wells studied. This would be since the gauge ran out of battery at that time or were otherwise unable to continue taking accurate readings.

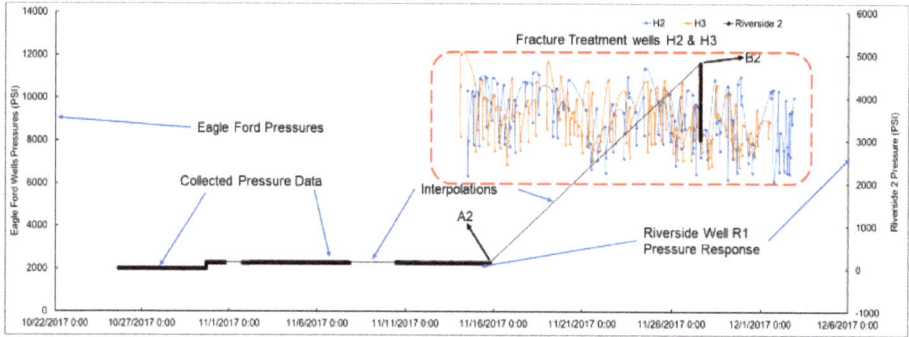

Figure A2. Correlated plots for Riverside 2. The left vertical axis is the pressure in Wells H2 and H3. The right axis is the pressure in the annulus of the Riverside wellhead. The significance of each labeled box is discussed in Table 1. The lighter black lines in the plot are interpolations that connect between the missing data.

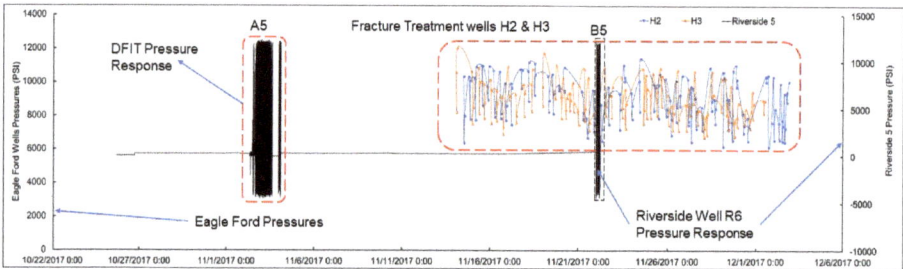

Figure A3. Correlated plots for Riverside 5. The left vertical axis is the pressure in Wells H2 and H3. The right axis is the pressure in the annulus of the Riverside wellhead. The significance of each labeled box is discussed in Table A3.

Table A4. Observations and interpretations for Riverside 5.

Symbol	Comments/Interpretations
A5	The highlighted region is from 11/02/2017 10:13 AM to 11/04/2017 05:32 AM and corresponds to the DFIT™ test conducted on both wells on 11/02/2017 around 09:30 AM in Well H2 and 10:20 AM to 10:40 AM in H3. This should be the cause of the sudden rises and drops in the pressure. The pressure spike suggests that the pressure response was a surge strong and sudden enough to kill the gauge.
B5	The region is from 11/21/2017 18:28 to 11/22/2017 05:21 AM, which corresponds to H2 between stages 19 and 20 and H3 between stages 22 and 23. The stimulation reports for these stages mention nothing unusual. The gauge picked up this spike even though it was potentially killed from the DFIT Test pressure response.

The pressure gauges installed could read a maximum value of 2000 psi, which means the responses recorded for this well could exceed this value, since the highlighted regions indicate that the gauges were recording values beyond its maximum, which is what killed the gauge to begin with. As identified in Table A4, Riverside 5 shows the strongest response to the DFIT test and fracture treatment of all the wells studied, which is identified even in the case of faulty equipment and an incomplete data set.

Appendix D. Austin Chalk Pressure Response Calculations

Table A5 describes in detail the pressure response calculations made for each Riverside observation well. Magnitude, duration and rate of pressure response in each of the Austin Chalk wells were determined from observations of the pressure response profiles of each well (shown in Figures 9–11, Figures A2 and A3 respectively)

Table A5. Rate, duration and intensity calculations for communications measured from available wellhead data. Timings are based on accurate date values.

Parameter	Value	Units	Comments
Riverside 1			
Average pressure	21.42	psi	Average wellhead pressure before DFIT™ test (until 10/31/2017 16:00, after that it increases sharply)
DFIT™ test intensity of response	232.48	psi	Peak value (C1 on Figure 9-Average. pressure
Duration of total response	473.65	hours	Duration of the total shape; start at 11/14/2017 14:15, 27 psi, end at 12/04/2017 07:54
Rate of 1st increase	1.073	psi/hour	Average slope (pressure rise/duration); start at 11/14/2017 14:15, 27 psi, 11/20/2017 11:22, 178.4 psi
Duration of Plateau	254.67	hours	Duration of H1 block
Average response of plateau	153.68	psi	Average of plateau minus average pressure (175.1-21.43) =153.68 psi
Max response	244.38	psi	J1 Peak occurs on 12/04/2017 07:33 minus average
Rate of 2nd increase	1.17	psi/hour	Gradient of line from end of H1 block to J1 peak; use average response value of plateau to calculate; 77.52 hrs;
Duration of clustered region at the end	41.92	hours	Taken from 12/05/2017 23:55 to 12/07/2017 17:50 (end of data set)
Time delay to start response	Negligible	hours	Negligible time delay from the first stage data in Well H3
Time delay after the job	35.20	hours	Between end of job in Well H2 and point J1
Riverside 4			
Average pressure	19.30	psi	Taken from the end of the data set
Initial response	81.80	psi	This is the first pressure response including time delay on 11/16/2017 22:43; value-average
Time delay to start response	Negligible	hours	Time from the first Eagle Ford data point to initial response
Plateau region duration	314.42	hours	Consider region B4
Plateau region avg response	141.64	psi	Plateau avg minus Avg. pressure 160.95-19.30
Plateau region rate	0.315	psi/hour	99.20/314.42 max-min/duration
Rate of 2nd increase	5.19	psi/hour	121.1 psi/23.33 hrs; consider C4
Max response	358.70	psi	Signifies end of response; Max-Average
Time delay end point	38.67	hours	Job end time to max point time
Rate of 2nd plateau	0.6	psi/hour	(378-352.9)/41.83; Consider region D4
Riverside 6			
Average pressure	14.18	psi	Average pressure
Max response	49.09	psi	Max point minus average; 63.27-14.18; Other meaningful quantities cannot be calculated since the pressure rises almost instantaneously
Rate of decrease	0.32	psi/hour	This should be a negative value;(60.64-18.07)/131.67
Riverside 2			
Average pressure	148.48	psi	Average pressure
Max response	4676.82	psi	Max point minus average; 4825.3-148.48
Duration of response	282.22	hours	Time from A2 to B2; Time of interpolated line
Rate of response	16.57	psi/hour	Rate of increase for interpolated line; 4676.82/282.22
Riverside 5			
Average pressure	250.52	psi	Taken from the flat region between the two responses
Duration of flat region between responses	429.04	hours	Taken from the flat region between the two responses
Duration of DFIT™ response	43.32	hours	First response; duration of A5
Duration of start job response	10.88	hours	Second response; duration of B5
Time delay for start job response	184.25	hours	Time when B5 region started minus time when the job started in H3
Rate of response (gradient of flat region)	0.409	psi/hour	175.3/429.04

References

1. Rodgers, L. Subsurface Trespass by Hydraulic Fracturing: Escaping Coastal v. Garza's Disparate Jurisprudence through Equitable Compromise. *Tex. Tech Law Rev. Online Ed. 99.* **2013**, *45*. Available online: http://texastechlawreview.org/volume-45/ (accessed on 2 November 2018).
2. Khanal, A.; Weijermars, R. Pressure depletion and drained rock volume near hydraulically fractured parent and child wells. *J. Pet. Sci. Eng.* **2019**, *172*, 607–626. [CrossRef]
3. Weijermars, R.; Alves, I.N. High-resolution visualization of flow velocities near frac-tips and flow interference of multi-fracked Eagle Ford wells, Brazos County, Texas. *J. Pet. Sci. Eng.* **2018**, *165*, 946–961. [CrossRef]
4. Hu, Y.; Weijermars, R.; Zuo, L.; Yu, W. Benchmarking EUR estimates for hydraulically fractured wells with and without fracture hits using various DCA methods. *J. Pet. Sci. Eng.* **2018**, *162*, 617–632. [CrossRef]
5. McGowen, H.E.; Krauhs, J. Development and Application of an Integrated Petroleum Engineering and Geologic Information System in the Giddings Austin Chalk Field. *Soc. Pet. Eng.* **1992**. [CrossRef]
6. Parshall, J. Can Austin Chalk Expansion Lead to Revival? *Soc. Pet. Eng.* **2017**. [CrossRef]
7. Hester, C.T.; Walker, J.W.; Sawyer, G.H. Oil Recovery by Imbibition Water Flooding in the Austin and Buda Formations. *Soc. Pet. Eng.* **1965**. [CrossRef]
8. Dake, L.P. *Fundamentals of Reservoir Engineering*, 8th ed.; Elsevier: Amsterdam, The Netherlands, 2001.
9. Weijermars, R.; Burnett, D.; Claridge, D.; Noynaert, S.; Pate, M.; Westphal, D.; Yu, W.; Zuo, L. Redeveloping depleted hydrocarbon wells in an enhanced geothermal system (EGS) for a university Campus: Progress report of a real-asset-based feasibility study. *Energy Strategy Rev.* **2018**, *21*, 191–203. [CrossRef]
10. Rafiee, M.; Soliman, M.Y.; Pirayesh, E. Hydraulic Fracturing Design and Optimization: A Modification to Zipper Frac. *Soc. Pet. Eng.* **2012**. [CrossRef]
11. Vermylen, J.; Zoback, M.D. Hydraulic Fracturing, Microseismic Magnitudes, and Stress Evolution in the Barnett shale, Texas, USA. *Soc. Pet. Eng.* **2011**. [CrossRef]
12. Craig, D.P.; Barree, R.D.; Warpinski, N.R.; Blasingame, T.A. Fracture Closure Stress: Reexamining Field and Laboratory Experiments of Fracture Closure Using Modern Interpretation Methodologies. *Soc. Pet. Eng.* **2017**. [CrossRef]
13. Economides, M.J.; Hill, A.D.; Ehlig-Economides, C.; Zhu, D. *Petroleum Production Systems*, 2nd ed.; Prentice Hall: Upper Saddle River, NJ, USA, 2013; Available online: http://ezproxy.library.tamu.edu/login?url=http://search.ebscohost.com/login.aspx?direct=true&db=cat03318a&AN=tamug.4023753&site=eds-live (accessed on 2 November 2018).
14. Daneshy, A. Effect of Dynamic Active Fracture Interaction DAFI on Activation of Natural Fractures in Horizontal Wells. *Soc. Pet. Eng.* **2016**. [CrossRef]
15. Yu, W.; Wu, K.; Zuo, L.; Tan, X.; Weijermars, R. Physical Models for Inter-Well Interference in shale Reservoirs: Relative Impacts of Fracture Hits and Matrix Permeability. In Proceedings of the 4th Unconventional Resources Technology Conference, San Antonio, TX, USA, 1–3 August 2016. [CrossRef]
16. Valko, P.P. Hydraulic Fracturing Short Course: Fracture Design, Fracture Dimensions, Fracture Modeling. 2005. Available online: http://slideplayer.com/slide/4948692/ (accessed on 10 November 2018).
17. Martin, R.; Baihly, J.D.; Malpani, R.; Lindsay, G.J.; Atwood, W.K. Understanding Production from Eagle Ford-Austin Chalk System. *Soc. Pet. Eng.* **2011**. [CrossRef]
18. Seth, P.; Manchanda, R.; Kumar, A.; Sharma, M. Estimating Hydraulic Fracture Geometry by Analyzing the Pressure Interference Between Fractured Horizontal Wells. *Spe Annu. Tech. Conf. Exhib.* **2018**. [CrossRef]
19. Seth, P.; Manchanda, R.; Zheng, S.; Gala, D.; Sharma, M. Poroelastic Pressure Transient Analysis: A New Method for Interpretation of Pressure Communication Between Wells During Hydraulic Fracturing. *Soc. Pet. Eng.* **2019**. [CrossRef]
20. Chapra, S.C.; Canale, R.P. *Numerical Methods for Engineers*; McGraw-Hill: New York, NY, USA, 2015.

energies

MDPI

Article

Numerical Study of Simultaneous Multiple Fracture Propagation in Changning Shale Gas Field

Jun Xie [1], Haoyong Huang [1,2,*], Yu Sang [1], Yu Fan [1], Juan Chen [1], Kan Wu [3,*] and Wei Yu [3,4]

[1] Petrochina Southwest Oil&Gasfield Company, Chengdu 610017, China; xiejun01@petrochina.com.cn (J.X.); sangy@petrochina.com.cn (Y.S.); fanyu@petrochina.com.cn (Y.F.); c_juan@petrochina.com.cn (J.C.)
[2] School of Petroleum Engineering, China University of Petroleum, Qingdao 266555, China
[3] Harold Vance Department of Petroleum Engineering, Texas A&M University, College Station, TX 75254, USA; yuwei127@gmail.com
[4] Hildebrand Department of Petroleum and Geosystems Engineering, The University of Texas at Austin, Austin, TX 78712, USA
* Correspondence: huang_hy@petrochina.com.cn (H.H.); kan.wu@pe.tamu.edu (K.W.); Tel.: +86-028-86018593 (H.H.); +86-010-5125652857 (K.W.)

Received: 28 February 2019; Accepted: 4 April 2019; Published: 8 April 2019

Abstract: Recently, the Changning shale gas field has been one of the most outstanding shale plays in China for unconventional gas exploitation. Based on the more practical experience of hydraulic fracturing, the economic gas production from this field can be optimized and gradually improved. However, further optimization of the fracture design requires a deeper understanding of the effects of engineering parameters on simultaneous multiple fracture propagation. It can increase the effective fracture number and the well performance. In this paper, based on the Changning field data, a complex fracture propagation model was established. A series of case studies were investigated to analyze the effects of engineering parameters on simultaneous multiple fracture propagation. The fracture spacing, perforating number, injection rate, fluid viscosity and number of fractures within one stage were considered. The simulation results show that smaller fracture spacing implies stronger stress shadow effects, which significantly reduces the perforating efficiency. The perforating number is a critical parameter that has a big impact on the cluster efficiency. In addition, one cluster with a smaller perforating number can more easily generate a uniform fracture geometry. A higher injection rate is better for promoting uniform fluid volume distribution, with each cluster growing more evenly. An increasing fluid viscosity increases the variation of fluid distribution between perforation clusters, resulting in the increasing gap between the interior fracture and outer fractures. An increasing number of fractures within the stage increases the stress shadow among fractures, resulting in a larger total fracture length and a smaller average fracture width. This work provides key guidelines for improving the effectiveness of hydraulic fracture treatments.

Keywords: cluster efficiency; perforating number; Changning shale gas; multiple fracture propagation

1. Introduction

The Changning shale gas field in Sichuan Basin is well known as the main shale gas production area in China, and has begun commercial production since 2012. With less than a decade of production, there is still much to learn about the most efficient way to produce shale gas. With the increasing practical experience of hydraulic fracturing, the economic shale gas production from this field is able to be optimized and gradually improved [1]. The production of single shale-gas wells has been continuously improved and the average daily production has increased from 11.1×10^4 m^3 to 28×10^4 m^3. However, the average well production rate and estimated ultimate recovery (EUR) are significantly lower than shale gas production in North America [2–4].

The technique of multi-stage hydraulic fracturing is the key to develop unconventional gas reservoirs [5–8]. The production in the Haynesville shale demonstrates that one of the most effective ways to increase production is to maximize the number of fracture initiation points along the lateral. Because of the limited drainage radius of the created fractures, the well production increases while the spacing between each perforated cluster interval decreases. The recent completions in the Haynesville shale show that many operators are completing wells with tighter cluster spacing than previously attempted, and this trend has continued [9]. However, when the completions for increasing the number of clusters in one stage are used in the Changning field, the well production is not significantly increased.

The complex fracture geometry is often generated and predicted in shale gas reservoirs rather than simple planar fractures through advanced fracture diagnostic and microseismic monitoring results [10]. One usually considers that increasing the perforation clusters in one stage can generate a similar number of fractures after hydraulic fracturing. However, production logging and tracer detection demonstrated that not all fractures along the horizontal wellbore can effectively propagate [11–15]. Fracturing fluids and proppants do not enter into each cluster evenly. Some clusters have a large proportion of the intended liquid and proppant and generate "super" fractures, resulting in other clusters accepting very little fluid to grow. As a result, the difference of shale gas production between different clusters along the lateral is very big. The data of the production logs from the wells of the Sichuan Basin with 4–5 perforation clusters in one stage indicated that some clusters may be ineffective and do not contribute to production. However, the further optimization of the fracture design requires a better understanding of the effects of engineering parameters on simultaneous multiple fracture propagation. The fracture propagation model has been widely used in unconventional reservoirs for the completion design, such as the Permian Basin and Eagle Ford Shale [16–18]. Through the modeling research, optimization strategies have been achieved, which support the improvement of single well production. The optimal fracture design can materially increase the effective fracture number and enhance the well productivity. However, the rock mechanics parameters used in the simulation are the actual data of Changning, which are significantly different from US fields. The minimum horizontal stress and the Young's modulus are the different parameters used for this specific field. The minimum horizontal stress gradient of Changning is 0.0249 MPa/m, while the minimum horizontal stress gradient of US fields is 0.0199 MPa/m. The Young's modulus is about twice as much as that in US fields. Consequently, the multistage fracturing completion of the Changing field should be optimized to achieve a high cluster efficiency and increase the opportunities to distribute fluid and proppant evenly across all targeted clusters. In this paper, based on the complex fracture propagation model (XFRAC) and the Changning field data, a series of case studies were performed to investigate the effects of multiple engineering parameters on multiple fracture propagation. The fracture spacing, perforating number, injection rate, fluid viscosity and number of fractures within one stage were studied. For a deeper understanding of the complex physics related to simultaneous multiple fracture propagation and evaluating the uniformity of the fracture length, three perforation clusters in a stage were simulated, and the deviation of the normalized fracture length was calculated. The description of the model is presented in the following section.

2. Methodology

A complex fracture propagation model, developed by Wu [19], was used to simulate simultaneous multiple fracture propagation in shale gas formation. The rock deformation and fluid flow were iteratively coupled in the model. The rock deformation was modeled by a simplified 3D displacement discontinuity method [20]. The shear and normal displacement discontinuities were calculated for each fracture element. The normal displacement discontinuity is the opening of fractures, and the shear displacement discontinuity is used to predict the fracture propagation path at each time step. A non-planar fracture geometry will be induced if the shear displacement discontinuity is nonzero. To improve the computation efficiency, the simplified displacement discontinuity method eliminated the discretization in the vertical (fracture height) direction. The solution of the method can be made explicit as follows:

$$\sigma_{sL}^i = \sum_{j=1}^{N} A_{sL,sL}^{ij} D_{sL}^j + \sum_{j=1}^{N} A_{sL,nn}^{ij} D_n^j$$

$$\sigma_{nn}^i = \sum_{j=1}^{N} A_{nn,sL}^{ij} D_{sL}^j + \sum_{j=1}^{N} A_{nn,nn}^{ij} D_n^j$$

(1)

where i and j represents elements i and j, N is the total element number, D_n^j is a normal displacement discontinuity on element j, and D_{sL}^j is a shear displacement discontinuity on element j. σ_{sL}^i and σ_{nn}^i are given traction boundary conditions. The distribution of pressure along the fracture path can be computed by the fluid flow model, which can provide these tractions. The constitutive model is based on the assumption of the plane-strain and elastic deformation. $A_{nn,sL}^{ij}$ is the coefficient matrix that can give the normal stress at element i because of a shear displacement discontinuity at element j. $A_{nn,nn}^{ij}$ represents the normal stress at element i induced by an opening displacement discontinuity at element j. Analogous meanings can be attributed to $A_{sL,sL}^{ij}$ and $A_{sL,nn}^{ij}$. The detailed derivation of the model can be found from the work by Wu [19].

The fluid flow in the shale gas wellbore and each fracture are fully coupled, similar to the electric circuit network. The flow rate of every fracture is similar to the current, and the pressure is analogous to the electric potential. We applied Kirchoff's first and second laws to compute the flow rate distribution among every fracture within a stage. The total volumetric injection rate, Q_T, is given, and the injection rates into each fracture, Q_i, are dynamically calculated by the model. The wellbore storage effect was ignored in the model. The sum of the injection rates of all the fractures is equal to the total injection rate,

$$Q_T = \sum_{i=1}^{N} Q_i$$

(2)

Kirchoff's second law described the continuousness of the pressure along the horizontal wellbore, considering the pressure drop of the wellbore friction and the perforation friction [21]. The sum of the pressure in the first element of a fracture branch, perforation friction pressure drop, and wellbore friction pressure drop together is equal to the pressure in the wellbore heel. The equation is given by:

$$p_o = p_{pf,i} + p_{cf,i} + p_{w,i}$$

(3)

where p_o is the total pressure of the wellbore heel, $p_{w,i}$ is pressure of the first element of the fracture, $p_{pf,i}$ is the pressure loss of the perforation friction pressure loss, and $p_{cf,i}$ is the pressure loss of the horizontal wellbore. The identification number of the fracture branches is represented by 'i'. The pressure drop of the perforation friction can be calculated by a function of the square of the flow rate and perforation friction. The lubrication theory was applied to describe the fluid flow in the fracture and the associated pressure drop. The model assumed that the fracture is a slot between parallel plates. Multiple fracture propagation has been simulated by the model and compared with a numerical model [22] to benchmark the accuracy of capturing the physical process of stress shadow effects.

3. Case Study

3.1. Base Case

In this section, we demonstrate the phenomenon of uneven fracture growth and how to facilitate a more uniform fracture propagation. The base case has three fractures propagating simultaneously in a single stage (Figure 1), which has a uniform cluster spacing of 23.3 m. All parameters were selected from the Longmaxi formation of the Changning shale gas field in China and are listed in Table 1. We assume that one perforation cluster induces only one hydraulic fracture. Hence, the perforation-cluster spacing is the same as the initial-fracture spacing. The effects of the natural fractures and near-wellbore tortuosity are not taken into account. It is assumed that the reservoir is homogeneous in regard to slight differences of the in-situ stress state and rock mechanical properties. The final fracture geometry and flow volume distribution of the base case are shown in Figures 2 and 3, respectively. Because of the strong stress shadow effects, the middle fracture is much shorter, while the two exterior fractures are much longer. The average percentage of the flow rate into every cluster is 33%. The middle fracture only received 19.6%, which is much less than the intended percentage, while the exterior fractures received about 40.2% of the total fluid. The stress shadow effects and the friction pressure drop along the wellbore result in the curves of the interior and exterior fractures diverging. Based on the base case, we modified the values of the fracture spacing, perforating number, injection rate, fluid viscosity and number of fractures within the stage to analyze how these factors affect the effectiveness in promoting a uniform fracture growth. These parameters were changed one at a time from the base case.

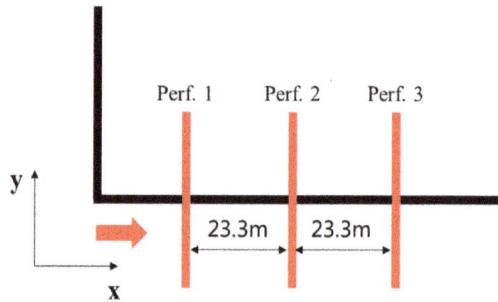

Figure 1. Three transverse fractures with a uniform spacing of 23.3 m in a single stage.

Table 1. Input parameters for simulation cases in this study.

Properties	Case 1	Base Case	Case 2	Case 3	Unit
Fracture spacing	10	23.3	15	30	m
Perforation density/cluster	12	16	20	24	-
Injection rate	10	12	14	16	m^3/min
Fluid viscosity	2.0	3.5	10	24	mPa·s
Number of fractures within the stage	2	3	4	5	-

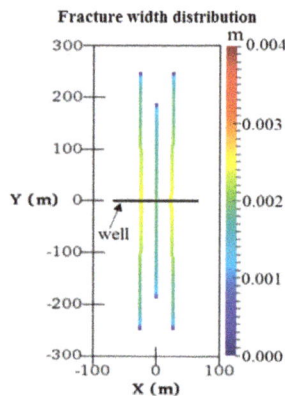

Figure 2. Three transverse fractures propagating simultaneously in a single stage.

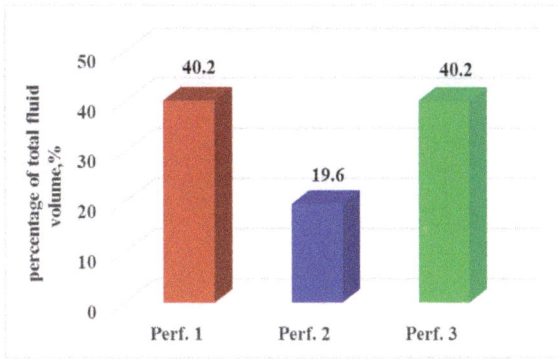

Figure 3. Percentage of total flow volume entering into each perforation cluster.

3.2. Effect of Fracture Spacing

With ultralow matrix permeability, one of the most effective ways to increase shale gas production is to optimize the number of fracture initiation points along the lateral. However, stress shadow effects can result from overly closely spaced fractures, resulting in an inefficient completion. Hence, we investigated three different fracture spacing effects on the fracture geometry and compared this with the base case. Each stage consists of three clusters, and the fracture spacings are 10 m, 15 m, 23.3 m, and 30 m (Figure 4), respectively. The simulation results show that the stress shadow effects increase with the decreasing fracture spacing, resulting in two longer outer fractures and a shorter middle fracture, as shown in Figure 5 and Table 2. The non-uniform fracture growth will significantly reduce the perforation efficiency. This is because larger stress shadow effects would increase the flow resistance of the middle fracture; less fluid enters into the middle fracture.

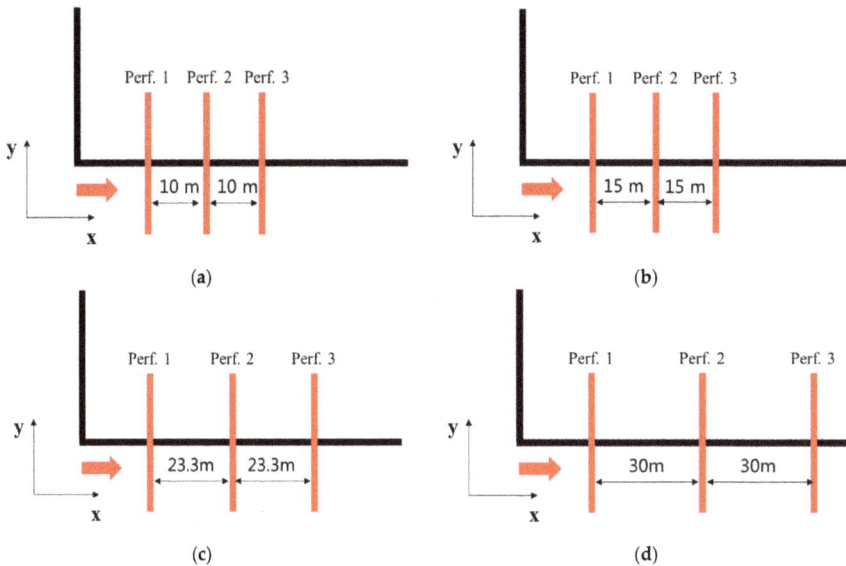

Figure 4. Different fracture spacings in a single stage: (**a**) 10 m; (**b**) 15 m; (**c**) 23.3 m; and (**d**) 30 m.

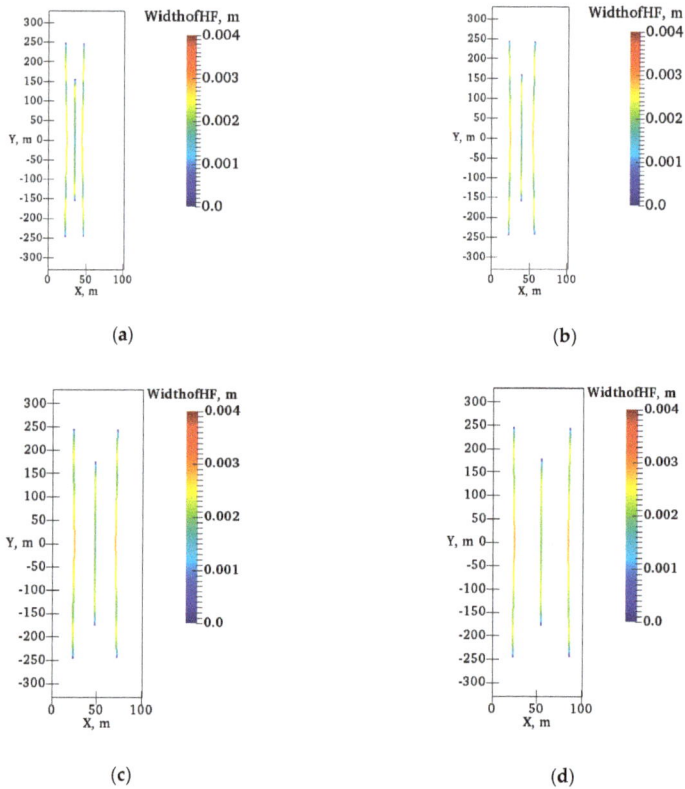

Figure 5. Effects of the different values of fracture spacing on the fracture geometry: (**a**) 10 m; (**b**) 15 m; (**c**) 23.3 m; and (**d**) 30 m.

Table 2. The results of fracture length affected by different cluster spacings.

Fracture Spacing, m	15	23.3	10	30
The length of fracture 1, m	490.1	492.6	495	492.6
The length of fracture 2, m	320.3	351.3	312.6	358.1
The length of fracture 3, m	488.3	489	494.1	487.3

3.3. Effect of Perforating Number

Perforation friction is a function of perforation density. The base case has a uniform perforation design with 16 perforations for each cluster. In this subsection, three different cases were investigated: two of them increase to 20 and 24 perforations for each cluster respectively, and another case uses only 12 perforations for each cluster. Figure 6 and Table 3 illustrate that fractures grow more non-uniformly with the increasing perforation density for each cluster. The larger the perforation density, the shorter the middle fracture and the longer the two outer fractures. In addition, it can be found that 12 perforations per cluster for three clusters in one stage are the optimal design in the Changning shale gas field. While the perforation density for each cluster increases from 12 to 24, the length of the middle fracture is reduced by 150% and the width is reduced by 25%, which significantly decreases the cluster efficiency.

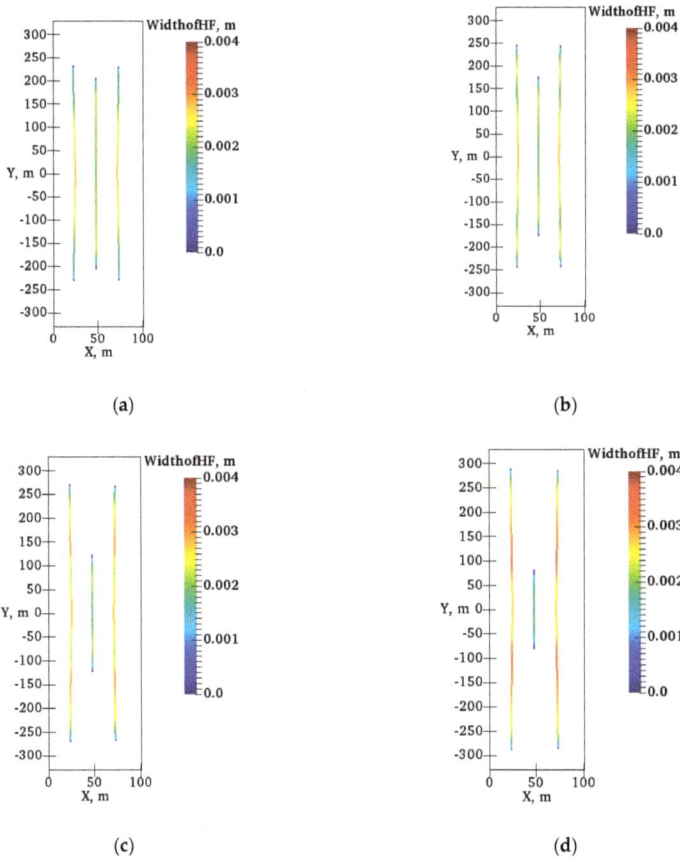

Figure 6. Effects of different perforations per cluster on the fracture geometry: (a) 12; (b) 16; (c) 20; and (d) 24.

Table 3. The results of fracture length affected by different perforating numbers.

Perforating Number	12	16	20	24
The length of fracture 1, m	463.3	492.6	541.3	580.3
The length of fracture 2, m	410.1	351.3	246.5	163.1
The length of fracture 3, m	460.9	489	536.6	573.8

3.4. Effect of Injection Rate

The injection rate is another important factor for affecting hydraulic fracturing treatments. Additionally, we kept other parameters the same as the base case and investigated the effects of different injection rates on the fracture geometry (Figure 7 and Table 4). Four injection rates were considered: 10 m^3/min, 12 m^3/min, 14 m^3/min and 16 m^3/min. Since the injection time for all of the cases was the same, more fluid volume was injected for the larger injection rate. Figure 7 shows that a more uniform fracture geometry was achieved for the larger injection rate. The simulation results demonstrate that a higher injection rate is better for promoting a uniform fluid volume distribution and an even growth for each cluster in the Changning shale gas field. This is because that larger injection rate can mitigate stress shadow effects and generate a higher perforation friction pressure drop. Consequently, the injection rate of the Changning shale gas field should be increased to improve the cluster efficiency.

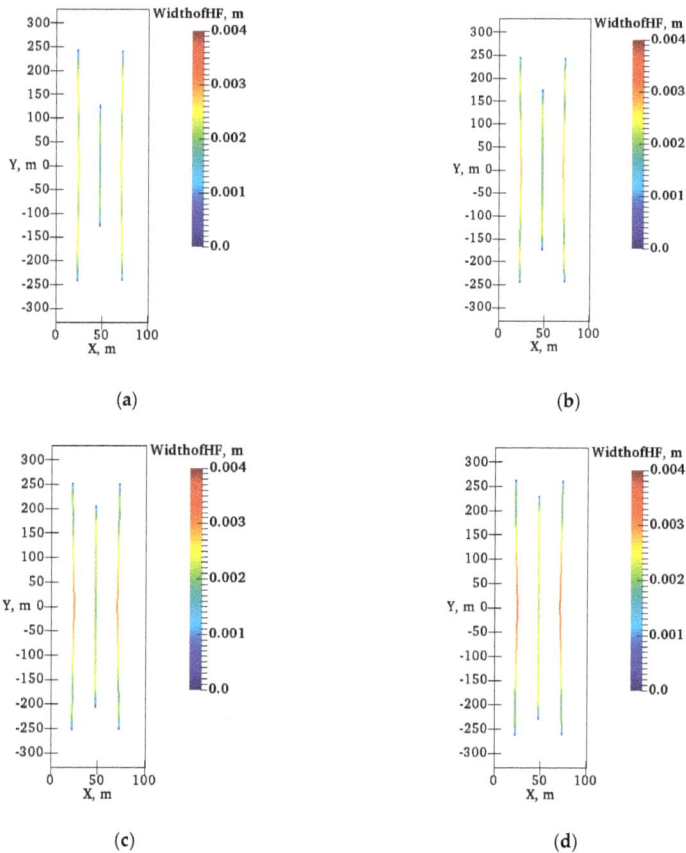

Figure 7. Effects of different injection rates on the fracture geometry: (**a**) 10 m³/min; (**b**) 12 m³/min; (**c**) 14 m³/min; and (**d**) 16 m³/min.

Table 4. The results of fracture length affected by different injection rates.

Injection Rate, m³/min	10	12	14	16
The length of fracture 1, m	487.7	492.6	507.2	526.7
The length of fracture 2, m	255.1	351.3	413.8	460.4
The length of fracture 3, m	484	489	503.6	523.3

3.5. Effect of Fluid Viscosity

We studied the effects of different fluid viscosities on the fracture geometry, and the simulation results are shown in Figure 8. The other parameters are the same as those of the base case. The three different viscosities of the injection fluid are, respectively, 2.0 mPa·s, 10 mPa·s and 24 mPa·s. An injection fluid with a larger viscosity created a higher fluid pressure within the fracture and a wider fracture width (Figure 8 and Table 5). A higher fluid pressure generated stronger stress shadow effects, resulting in a larger variation of fluid distribution between perforation clusters. Figure 8 illustrates that the length of the middle fracture is reduced by 40.6%, and the width increases by 76%, when the fluid viscosity increases from 2.0 mPa·s to 24 mPa·s. For that reason, the viscosity of the injection fluid should be decreased to 2.0 mPa·s in the Changning shale gas field.

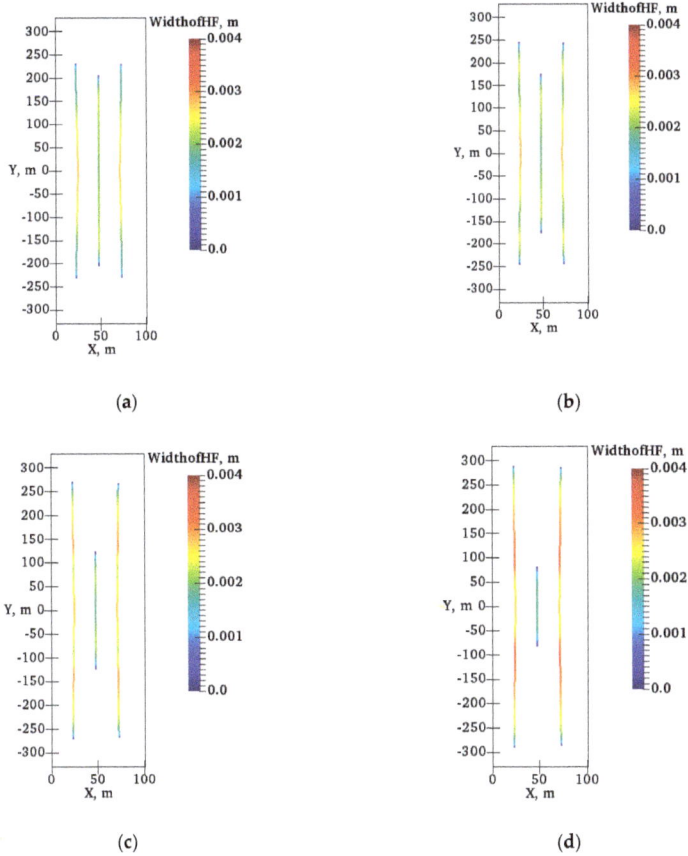

(a)   (b)

(c)   (d)

Figure 8. Effect of fluid viscosity on the fracture geometry: (a) 2 mPa·s; (b) 3.5 mPa·s; (c) 10 mPa·s; and (d) 24 mPa·s.

Table 5. The results of fracture length affected by different fluid viscosities.

Fluid Viscosity, mPa·s	2.0	3.5	10	24
The length of fracture 1, m	487.7	492.6	512.1	531.6
The length of fracture 2, m	377.4	351.3	288.5	224.3
The length of fracture 3, m	485.5	489	503.0	511.7

3.6. Effect of Number of Fractures Within the Stage

The number of fractures within the stage is an important factor of hydraulic fracturing treatments, which is the most effective way to increase production in the Haynesville [14]. Therefore, we studied the impacts of different cluster numbers on the fracture geometry in a single stage. Under the condition of the fixed stage length of 70 m, the cluster number of four cases is, respectively, 2, 3, 4 and 5, while the other parameters are the same as the base case. The simulation results are shown in Figure 9 and Table 6. They illustrate that as the cluster number within the stage increases, the cluster spacing decreases, and the stress shadow effects increase, leading to a longer total fracture length and shorter average fracture width. The optimal number of clusters in a single stage needs to be determined in combination with the production simulation and economic evaluation. However, according to the simulation results, if more than 4 clusters within the stage are used, one needs to utilize the intrastage diversion techniques [23,24] to enhance cluster efficiency in the Changning shale gas field.

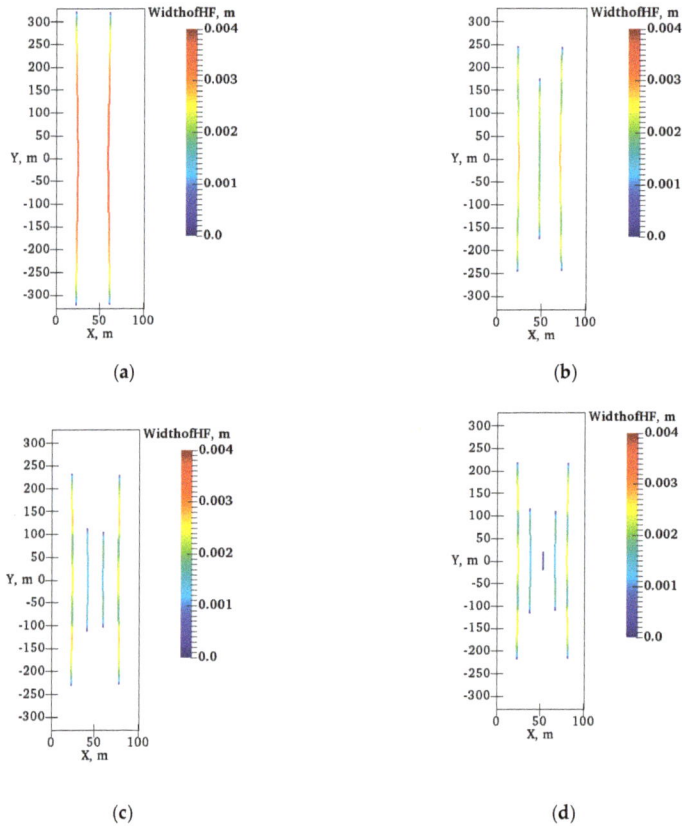

Figure 9. Effects of different numbers of fracture within the stage on the fracture geometry: (**a**) 2; (**b**) 3; (**c**) 4; and (**d**) 5.

Table 6. The results of fracture length affected by different number of fractures within the stage.

Number of Fractures within the Stage	2	3	4	5
The length of fracture 1, m	643.7	492.6	463.3	438.5
The length of fracture 2, m	641.7	351.3	226.5	232.7
The length of fracture 3, m		489	210.9	39.4
The length of fracture 4, m			457.6	220.9
The length of fracture 5, m				434.3

4. Discussions

In order to evaluate the effects of the fracture spacing, perforating number, injection rate, fluid viscosity, and number of fractures within the stage on the fracture geometry in the Changning shale gas field, we defined a deviation of the normalized fracture length [25]. This sequence can indicate the main controlling factors for the effectiveness of fracture treatments. First, according to the basic input parameters, the average fracture length in the example is calculated. Then, we calculated the deviation of the three fractures. In the same way, we calculated the maximum and minimum deviation corresponding to the maximum and minimum values of each uncertain parameter. Finally, we sorted each uncertain parameter according to the maximum and minimum deviation values. Based on the sorting result and the deviation of the normalized fracture length, the Tornado plot (Figure 10) was obtained. The x-axis is the calculated deviation of the normalized fracture length and represents the effects of uncertain parameters on the uniformity of the fracture growth. The order of uncertain parameters in the y-axis was determined by the absolute difference between the maximum and minimum deviation of the normalized fracture length. The green bar represented a positive effect and the black bar represented a negative effect. The middle mark represented the deviation of the normalized fracture length of the base case. The Tornado plot shows that the number of fractures within the stage is the most important parameter for affecting the fracture geometry in the Changning shale gas field. A larger variation of fracture geometry will be created with either the increasing number of fractures, the decreasing flow rate, the increasing perforating number or the increasing fluid viscosity. The fracture spacing has a relatively smaller impact on the fracture geometry. It should be mentioned that the spatial variations of the stress state, natural fractures and near wellbore tortuosity are not considered in this study, but will be examined in our future work. Therefore, we should improve the number of fractures in the stage with the intrastage diversion techniques. In addition, 16 m^3/min of flow rate, 12 perforations for each cluster and an injection fluid with 2.0 mPa·s are better for improving the effectiveness of the stimulation treatments in the Changning shale gas field.

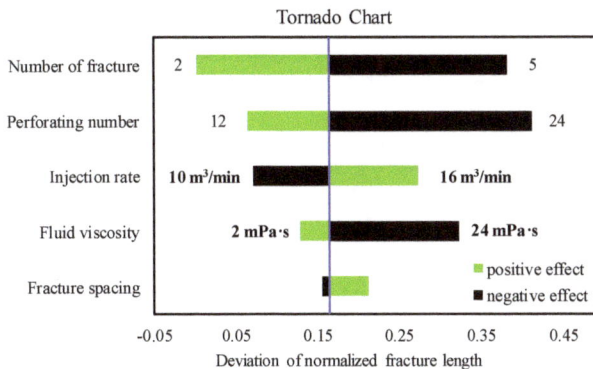

Figure 10. Rank of five uncertain parameters on the deviation of the normalized fracture length.

5. Conclusions

We applied a complex fracture propagation model to simulate multiple fracture propagation in the Changning shale gas field. The effects of the fracture spacing, perforating number, injection rate, fluid viscosity, and number of fractures within the stage on the fracture geometry were investigated based on field data from the Longmaxi shale formation in the Changning shale gas reservoir. The following conclusions can be drawn from this study:

(1) The main factors for controlling the cluster efficiency in the Changning shale gas field are the cluster numbers, the perforation density, the injection rate, and the liquid viscosity.

(2) Hydraulic fracture treatments with more than four clusters per stage, a lower injection rate, larger perforating number, larger viscosity fluid, and closer fracture spacing can result in an increasing gap between the inner fracture and outer fractures, and they will likely exhibit a bad production performance.

(3) This study provides a better understanding of the way to appropriately optimize a hydraulic fracturing treatment design which can increase the effective fracture number and promote the shale gas well performance in Changning.

Author Contributions: Conceptualization, J.X. and H.H.; methodology and investigation W.Y. and K.W.; writing—original draft preparation, H.H.; writing—review and editing, Y.S. and Y.F.; project administration, J.C.; funding acquisition, Y.S. and Y.F.

Funding: This research was funded by "National Science and Technology Major Project of China, grant number 2016ZX05023-005-00" and "The Chinese Academy of Engineering Key Consulting Research Project, grant number 2018-XZ-20" and "Major Science and Technology Special Project of CNPC, grant number 2016E-0612".

Conflicts of Interest: The authors declare no conflict of interest.

References

1. Xie, J. Rapid shale gas development accelerated by the progress in key technologies: A case study of the Changning-Weiyuan national shale gas demonstration zone. *J. Nat. Gas Ind.* **2018**, *5*, 283–292. [CrossRef]
2. Cadotte, R.J.; Whitsett, A.; Sorrell, M.; Hunter, B. Modern Completion Optimization in the Haynesville Shale. In Proceedings of the SPE Annual Technical Conference and Exhibition, San Antonio, TX, USA, 9–11 October 2017; Society of Petroleum Engineers: Richardson, TX, USA, 2017.
3. Engelder, T.; Lash, G.G.; Uzcategui, R.S. Joint sets that enhance production from Middle and Upper Devonian gas shales of the Appalachian Basin. *AAPG Bull.* **2009**, *93*, 857–889. [CrossRef]
4. Yu, W.; Wu, K.; Liu, M.; Sepehrnoori, K.; Miao, J. Production forecasting for shale gas reservoirs with nanopores and complex fracture geometries using an innovative non-intrusive EDFM method. In Proceedings of the SPE Annual Technical Conference and Exhibition, Dallas, TX, USA, 24–26 September 2018.
5. Shou, K.J. A High Order Three-Dimensional Displacement Discontinuity Method with Application to Bonded Half-Space Problems. Ph.D. Dissertation, University of Minnesota, Minneapolis, MN, USA, 1993.
6. Siriwardane, H.J.; Layne, A.W. Improved Model for Predicting Multiple Hydraulic Fracture Propagation from a Horizontal Well. Presented at the SPE Eastern Regional Meeting, Lexington, KY, USA, 22–25 October 1991. [CrossRef]
7. Weng, X. Modeling of Complex Hydraulic Fractures in Naturally Fractured Formation. *J. Unconv. Oil Gas Resour.* **2015**, *9*, 114–135. [CrossRef]
8. Yan, C.; Deng, J.; Hu, L.; Chen, Z.; Yan, X.; Lin, H.; Tan, Q.; Yu, B. Brittle failure of shale under uniaxial compression. *Arab. J. Geosci.* **2015**, *8*, 2467–2475.
9. Dong, Z.; Holditch, S.A.; Mcvay, D.A. Resource evaluation for shale gas reservoirs. In Proceedings of the SPE Hydraulic Fracturing Technology Conference, The Woodlands, TX, USA, 6–8 February 2012.
10. Thompson, J.; Fan, L.; Grant, D.; Martin, R.B.; Kanneganti, K.T.; Lindsay, G.J. An overview of horizontal-well completions in the Haynesville Shale. *J. Can. Pet. Technol.* **2011**, *50*, 22–35. [CrossRef]
11. Warpinski, N.R.; Mayerhofer, M.J.; Davis, E.J.; Holley, E.H. Integrating fracturing diagnostics for improved microseismic interpretation and stimulation modeling. In Proceedings of the URTeC 1917906, the Unconventional Resources Technology Conference, Denver, CO, USA, 25–27 August 2014.
12. Olson, J.E.; Wu, K. Sequential versus Simultaneous simultaneous Multimulti-zone Fracturing fracturing in Horizontal horizontal Wellswells: Insights insights from a Nonnon-planar, Multimulti-frac Numerical numerical Modelmodel. Presented at the SPE Hydraulic Fracturing Technology Conference, The Woodlands, TX, USA, 6–8 February 2012.
13. Kan, W.; Anusarn, S.; Tang, J. Numerical study of flow rate distribution for simultaneous multiple fracture propagation in horizontal wells. In Proceedings of the 50th US Rock Mechanics/Geomechanics Symposium, Houston, TX, USA, 26–29 June 2016; American Rock Mechanics Association: Alexandria, VA, USA, 2016.
14. Sani, A.M.; Podhoretz, S.B.; Chambers, B.D. The Use of Completion Diagnostics in Haynesville Shale Horizontal Wells to Monitor Fracture Propagation, Well Communication, and Production Impact. In Proceedings of the SPE/CSUR Unconventional Resources Conference, Calgary, AB, Canada, 20–22 October 2015; Society of Petroleum Engineers: Richardson, TX, USA, 2015.
15. Ugueto, C.; Gustavo, A.; Huckabee, P.T.; Molenaar, M.M.; Wyker, B.; Somanchi, K. Perforation cluster efficiency of cemented plug and perf limited entry completions; Insights from fiber optics diagnostics. In Proceedings of the SPE Hydraulic Fracturing Technology Conference, The Woodlands, TX, USA, 9–11 February 2016; Society of Petroleum Engineers: Richardson, TX, USA, 2016.

16. Guo, X.; Wu, K.; Killough, J.; Tang, J. Understanding the mechanism of interwell fracturing interference based on reservoir-geomechanics-fracturing modeling in Eagle Ford Shale. In Proceedings of the Unconventional Resources Technology Conference, Houston, TX, USA, 23–25 July 2018; Society of Exploration Geophysicists, American Association of Petroleum Geologists, Society of Petroleum Engineers: Richardson, TX, USA, 2018; pp. 4097–4117.

17. Li, J.; Yu, W.; Wu, K. Analyzing the impact of fracture complexity on well performance and wettability alteration in Eagle Ford shale. In Proceedings of the Unconventional Resources Technology Conference, Houston, TX, USA, 23–25 July 2018; Society of Exploration Geophysicists, American Association of Petroleum Geologists, Society of Petroleum Engineers: Richardson, TX, USA, 2018; pp. 2072–2086.

18. Patterson, R.; Yu, W.; Wu, K. Integration of microseismic data, completion data, and production data to characterize fracture geometry in the Permian Basin. *J. Nat. Gas Sci. Eng.* **2018**, *56*, 62–71. [CrossRef]

19. Wu, K. Numerical Modeling of Complex Hydraulic Fracture Development in Unconventional Reservoirs. Ph.D. Dissertation, The University of Texas, Austin, TX, USA, 2014.

20. Wu, K.; Olson, J.E. Simultaneous Multi-Frac Treatments: Fully Coupled Fluid Flow and Fracture Mechanics for Horizontal Wells. *SPE J.* **2015**, *20*, 337–346. [CrossRef]

21. Elbel, J.L.; Piggott, A.R.; Mack, M.G. Numerical modeling of multilayer fracture treatments. In Proceedings of the SPE Permian Basin Oil and Gas Recovery Conference, Midland, TX, USA, 18–20 March 1992.

22. Wu, R.; Kresse, O.; Weng, X.; Cohen, C.; Gu, H. Modeling of interaction of hydraulic fractures in complex fracture networks. In Proceedings of the SPE Hydraulic Fracture Technology Conference, The Woodlands, TX, USA, 6–8 February 2012.

23. Xiong, H.; Wu, W.; Gao, S. Optimizing Well Completion Design and Well Spacing with Integration of Advanced Multi-Stage Fracture Modeling & Reservoir Simulation-A Permian Basin Case Study. In Proceedings of the SPE Hydraulic Fracturing Technology Conference and Exhibition, Woodlands, TX, USA, 23–25 January 2018.

24. Huang, J.; Datta-Gupta, A.; Augustine, J.R. Optimization of hydraulic fracture development and well performance using limited entry perforations. In Proceedings of the SPE Oklahoma City Oil and Gas Symposium, Oklahoma City, OK, USA, 27–31 March 2017; Society of Petroleum Engineers: Richardson, TX, USA, 2017.

25. Wu, K.; Olson, J.E. Numerical Investigation of complex fracture networks in naturally fractured reservoirs. *SPE Prod. Oper.* **2016**, *31*, 300–309.

energies

MDPI

Article

Elementary Pore Network Models Based on Complex Analysis Methods (CAM): Fundamental Insights for Shale Field Development

Ruud Weijermars * and Aadi Khanal

Harold Vance Department of Petroleum Engineering, Texas A&M University, 3116 TAMU, College Station, TX 77843-3116, USA; akhanal@tamu.edu
* Correspondence: R.Weijermars@tamu.edu

Received: 18 February 2019; Accepted: 22 March 2019; Published: 1 April 2019

Abstract: This paper presents insights on flow in porous media from a model tool based on complex analysis methods (CAM) that is grid-less and therefore can visualize fluid flow through pores at high resolution. Elementary pore network models were constructed to visualize flow and the corresponding dynamic bottomhole pressure (BHP) profiles in a well at reservoir outflow points. The pore networks provide the flow paths in shale for transferring hydrocarbons to the wellbore. For the base case model, we constructed a single flow path made up of an array of pores and throats of variable diameter. A passive ganglion (tracer) of an incompressible fluid was introduced to demonstrate the deformation of such ganglions when moving through the pores. The simplified micro-flow channel model was then expanded by stacking flow elements vertically and horizontally to create complex flow paths representing a small section of a porous reservoir. With these model elements in place, the flow transition from the porous reservoir fluid to the wellbore was modeled for typical stages in a well life. The dynamic component of the bottomhole pressure (BHP) was modeled not only during production but also during the drilling of a formation (with either balanced, underbalanced or overbalanced wellbore pressure). In a final set of simulations, the movement of an active ganglion (with surface tension) through the pore space was simulated by introducing a dipole element (which resisted deformation during the movement through the pores). Such movement is of special interest in shale, because of the possible delay in the onset of bubble point pressure due to capillarity. Capillary forces may delay the reservoir to reach the bubble point pressure, which postpones the pressure-drop trigger that would lead to an increase of the gas–oil ratio. The calculation of the estimated ultimate recovery (EUR) with an erroneous assumption of an early increase in the gas–oil ratio will result in a lower volume than when the bubble point delay is considered.

Keywords: pore network; flow models; bottomhole pressure; bubble point pressure

1. Introduction

Pore sizes in shale basins range from a few nm to several μm [1–4]. Pore size and pore networks control the porosity of a rock and connectivity between the pores controls the bulk permeability. Pores in shale are so narrow that the already reduced permeability of shale formations is further affected by the flow behavior in the reservoir, deviating from Darcy flow. Hydrocarbon reservoir fluids typically have molecular diameters on the order of 1 Angstrom or 0.1 nm. Various conceptual models exist to describe and quantify the behavior of fluids in nano-pores [5]. The effects include so-called molecular size filtration and sievage, which become particularly relevant when the pore diameter drops from 10 to 5 nm [6,7]. For multi-component hydrocarbon phases (e.g., CO_2, CH_4, C_{2-4}, C_{5-6}, and C_{7+}) the gas–oil ratio will generally increase when the reservoir pressure drops due to the pressure decline to below the bubble point pressure as a consequence of continued production. The phase behavior can be modeled, which has an iterative impact on the reservoir pressure and compressibility needs to be taken into account too. Calculations of the equation of state and vapor-liquid equilibrium must be linked with the velocity and pressure field solutions to investigate the fundamental nature of the interaction between the bubble point pressure and the capillary pressure due to interfacial tension and pore geometry.

Rock–fluid interaction in variably sized pores may cause interfacial tension changes that influence phase behavior via capillary pressure effects. Such thermodynamic processes result in so-called non-Darcy flow, which refers to molecular effects that influence the fluid flow behavior other than—or in addition to—viscous forces. The networks of pores form important flow paths in shale for transferring hydrocarbons to fractures and finally to the wellbore. Recovery rates of oil and gas from unconventional source rocks such as shale are still dismal (<10% for oil; <20% for gas). The low recovery rates are primarily due to the ultra-low permeability of the rocks, which retards the flow of hydrocarbons toward the hydraulic fractures that are connected to the production well [8,9]. The retarded flow of hydrocarbons via the shale pores is too slow for economic extraction, which can be partly alleviated by flow stimulation through the hydraulic fracture treatment. Studying the effect of the hydraulic fractures on the drainage efficiency of the well requires integration of field data with detailed models, not only of flow in the fractures, but also of the fluid exchange between the matrix pore space and the transecting fractures. Methods are needed to accelerate the drainage process from shale wells.

Advanced imaging techniques, such as scanning electron microscopy, and bulk characterization techniques, such as gas adsorption, mercury intrusion, or small-angle neutron scattering, have confirmed that shale consists of a complex network of pores, ranging from micro-pores (less than 2 nm diameter; actual nano-pores), via meso-pores (2–50 nm diameter) to macro-pores (larger than 50 nm) [10–12]. The specific nature of pores in porous media may affect fluid motion in several ways. The porosity is a measure of fluid storage capacity (storativity). A locally higher porosity slows down the flow, resulting in longer transit times [13]. The porosity distribution does not affect the detailed flow path in a porous medium with single-phase fluid flow and particle paths exclusively controlled by the permeability distribution as a measure of resistance to fluid passage. Many producing shale plays in the US have comparable permeability ranges between microDarcy to nanoDarcy, with corresponding pore size varying between nm and μm. Examples of the scale variations of pore structures are included in Figure 1.

When the reservoir fluid migration involves multi-phase flow behavior, particle paths will be affected not only by the local permeability tensor, but also by the interaction of the fluid phases with the pore space (molecular size filtration and sieving, capillarity delaying phase changes in narrow pore spaces). Currently, the flow mechanisms through multi-porosity systems with nm-scale pores are not well understood. Notwithstanding many open questions, numerous studies have shown that fundamental properties such as the phase behavior, viscosity, and interfacial tension for multi-phase flow through confined micro-channels in ultra-low permeability shale deviate from the conventional behavior due to the capillary walls and fluid interactions [14,15]. In the larger, unconfined pore spaces, fluid molecule–molecule interactions are the dominant transport mechanism and molecule-pore wall interactions are negligible [16]. However, in the nano-scale pores, the pore size becomes comparable to the dimensions of the molecular mean free path, which causes an increase in the mutual interaction of hydrocarbon molecules and with pore walls, due to Van der Waals forces [17,18]. In gas reservoirs, examples of nano-pore scale effects include gas slip and adsorption effects [19] and Knudsen diffusion, which requires correction for apparent gas permeability [7,20].

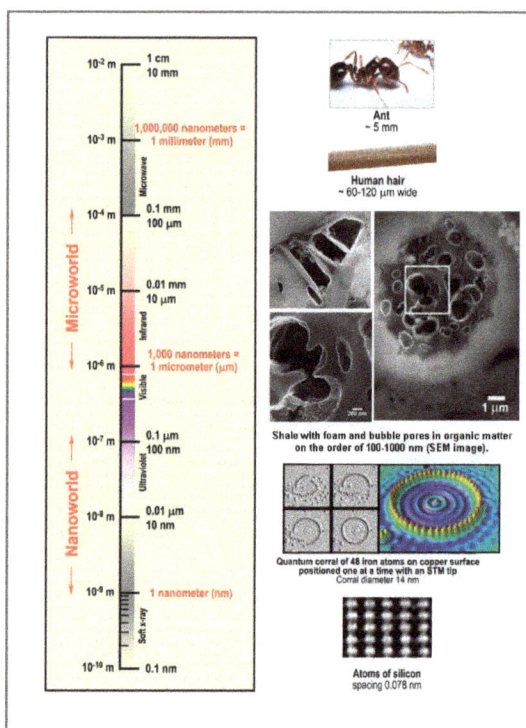

Figure 1. Augment scale of things with images of rock porosity diameters covering the same range. Adapted from [21]. SEM images from [11].

In liquid hydrocarbon maturity windows, non-Darcy flow effects may impact gas to oil ratios by lowering of the bubble point pressure due to the occurrence of elevated capillary forces in nano-pores [22]. In nano-pore rocks, bubble point effects are not only regulated by local pressure but also by rock–fluid interaction in the variably sized nano-pores, which will cause interfacial tension changes that influence phase behavior via capillary pressure dynamics. For example, the bubble point pressure may decrease due to capillary effects, resulting in two-phase flow being reached later in the well life than in the absence of such effects [23]. This phenomenon is also seen in so-called liquid rich shale (LRS) volatile oil reservoirs, where the bubble-point pressure is anomalously depressed due to the change in Pressure-Volume-Temperature (PVT) properties associated with pore-confinement [24,25]. Prediction of such behavior from simulation models is essential to improve the accuracy of EUR estimations for LRS reservoirs.

This study uses complex analysis methods (CAM) to construct elementary pore flow models. Although many pore flow models exist, CAM has never before been used to construct pore-scale network-models, until this study. The common aim of pore network models is to (1) improve our understanding of molecular flow effects at the pore scale and (2) use the pore flow models to upscale such effects for use at reservoir scale and in reservoir simulators. The ultimate aim is to successfully translate the theoretical and experimental results into practical insights, which may lead to (1) more accurate determination of estimated ultimate recovery (EUR), (2) improved predictions of the impact of enhanced oil recovery (EOR) interventions on the EUR, and (3) more accurate forecasts of monthly production profiles, which would result in more reliable evaluations of project economics, which must underpin reserves classification. Although fulfillment of the stated aims will require much more investigation, some early results of our new pore network flow model are presented here for the first time. What is particularly unique for CAM-based pore network models is the ability (a) to visualize particle paths at the pore scale and (b) to study the flow transition from the porous reservoir space to the wellbore space.

2. Modeling Approach

Many current investments and regulatory decisions in oil and gas projects are primarily based on models with constitutive equations and flux function on a continuum scale (i.e., the length scale larger than the pore scale). With investment focus shifting to shale plays, up-scaling of the pore-scale flow behavior to the continuum scale needs to properly account for all pore-scale effects, which is why flow studies at the pore-scale are relevant. This section reviews prior work (Section 2.1) and then introduces CAM-based pore network models (Sections 2.2 and 2.3). The pore network model of Section 2.3 is subsequently used to assemble a model of the transition zone between the porous reservoir and the wellbore (Section 3).

2.1. Prior Modeling Efforts

The pore structure of rock samples can be imaged at high resolution [26] but capturing the complex orientations and dimensions of the pores in actual flow models of porous media still remains a challenge. A crucial aspect of flow in porous media is the interaction between the matrix and hydraulic (and any natural) fractures near the treatment zone. Modeling flow in fractured porous media, such as rocks and unconsolidated sands, are among the most challenging problems both in hydrogeology [27,28] and in reservoir engineering [29,30]. One approach prevalent in simulations of fractured reservoirs uses the dual-porosity model [31] and its numerous expansions [29]. The reservoir is modeled by separating domains of connected fractures with 100% porosity and a certain assumed permeability with matrix pores acting as storage space. Transfer functions and shape factors were introduced to control the exchange of fluid between matrix and fractures [32–34].

Early mathematical models of flow through rock fractures are mainly limited to cubic law formulations for fluid flow transport through just a single fracture [35], expanded in later work with factors accounting for wall roughness [36,37]. The transient effect of stress on flow in a single fracture due to dilation effects has also been quantified [38–40], with further efforts to account for inertia effects due to turbulent fluid at high rates of injection [41]. More recent work takes into account solute transport [42,43] and chemical interaction [44,45].

Modeling advective displacement in fractured porous media is challenging enough for single-phase flow, but becomes more complicated when a multi-phase flow is considered. Capillary forces between immiscible wetting phase (e.g., gas) and non-wetting phase (e.g., oil) may cause entrapment of the oil in certain pores [46,47]. Prior research introduced theoretical concepts on the role of capillary forces in water-flooding efficiency [48,49], backed up with some experimental data that show the effect of capillarity on oil transport in pore interstices, including trapping and snapping of oil droplets [50]. Recent experimental work has also shown that oil–water interaction in porous media may result in oil-in-water emulsions, which affect the pressure field and thus the oil recovery speed [51]. When the fluid is miscible, the fluid mixture may be considered to move as a single phase modeled by Darcy's law assuming incompressible miscible displacement. However, flow mechanisms of emulsions through porous media are not well understood. When the multi-phase flow is immiscible, fluid transport can be described by a fractional flow model, using operator splitting methods [52].

More recently, multi-phase pore-scale simulations have become possible [53–56], which import the pore geometry obtained from micro-CT (computerized tomography) scans of real rocks. Such models avoid the simplifications adopted in mathematically generated pore-network models and allow pore size driven flow accounting for capillary or viscous dominate flow as appropriate. However, such so-called direct simulation is still very computationally intensive (and thus costly and cumbersome), which is why the use of simplified pore and pore network models is still merited. Pore models using single channels and with straight flow streams have been previously used [57]. Pore space network models are commonly based on a regular lattice assumption, typically a cubic lattice [58]. An in-depth review of pre-network models is given in [59].

2.2. Complex Analysis Method (CAM)

We introduce a new, simple two-dimensional approach to model flow paths in the pores and the pore-throats of the porous network. The method expands our recent research program that models the flow in porous media with a grid-less simulator based on complex analysis methods ((CAM); [60–65]). The method can handle permeability anisotropy and reservoir heterogeneity due to the presence of discrete fractures [66,67]. The method also allows for porosity variations to accelerate or decelerate the transit time along streamlines [13]. A prime motivation to use CAM, as a modeling method complementary to models based on finite difference methods, is the grid-less and continuously scalable nature of the code. The high resolution of CAM models may provide complementary insight to finite volume-based simulators. The capability to model flow in fractured porous media at high resolution is particularly useful for fracture treatment design optimization [65,67–69], in order to ultimately improve recovery factors of shale wells and maximize return on investment.

2.2.1. Single Pore Throat Channel Example

For the base case model, we use a single flow path made up of an array with an increasing number of pores and throats of variable diameter. The fluid flow in the reservoir is assumed to occur by a natural drive from an aquifer, which can be represented by a far-field flow with a constant velocity and, if necessary, a particular tilt angle [66]. When the flow in the porous medium is studied at a macroscopic scale, the streamlines for the far-field flow entering the reservoir space from the left in a homogeneous system remained completely unperturbed (all streamlines remained perfectly parallel). For a flow rate of 1.5 m/year (4.8×10^{-8} m/s) the fluid would travel 28.8×10^{-6} m in 600 s, as outlined by time-of-flight-contours (TOFCs, red) (Figure 2). The streamlines are represented by green curves and the transit time is marked by TOFCs, spaced for intervals of 60 s.

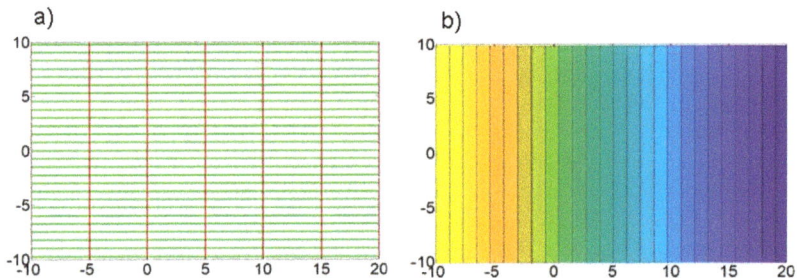

Figure 2. (**a**) Flow model space marked with streamlines (green) and the time-of-flight-contours (red). Flow comes in from the left. (**b**) Isobars for pressure field (warm colors indicate higher pressure) calculated by taking the real part of the complex potential for configuration in (**a**). Algorithms for the model are given in Appendix A. Length units are scaled in μm (both x-axis and y-axis).

Figure 3 shows an elementary pore flow model, using a constant far-field flow velocity of 1.5 m/year (4.8×10^{-8} m/s). The algorithms to simulate the flow in micro- and nano-pores are formulated in complex analysis calculus and details are given in Appendix A. A large number of simplifying assumptions are adopted. The model is 2D and all fluid flow occurs in the plane of study. We assume all flow tubes (micro-channels) are made up by one or more pores and throats and pore throat orientations stay horizontally aligned with the flow direction from the negative x-axis to the positive *x*-axis. Gravity forces and inertia effects are neglected, justified by the local scale and the slow motion of the fluid, respectively. Capillary forces due to multi-phase flow are initially neglected and heterogeneity due to spatial variations in water-wet and oil-wet rock surfaces are also ignored. So, the outset is a very simple pore network model, occupied by a Newtonian, incompressible fluid. The flow rate through the pores may either be steady or time-dependent and is a function of strength v for the channel element [66]. The flow strength of the pores/pore-throats can be scaled such that these vary, depending on the strength of the steady-state far-field flow and the dimensions of the pores.

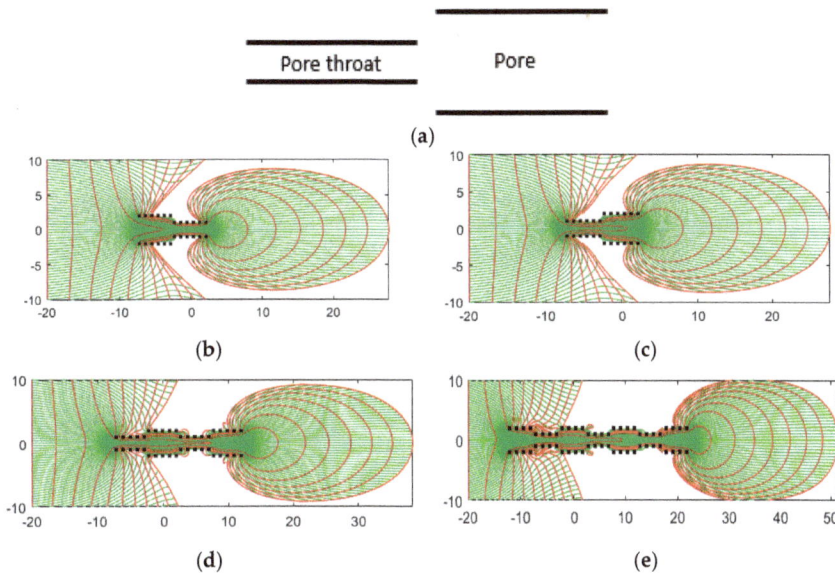

Figure 3. (**a**) Schematic of an elementary pore-model showing a pore and a pore throat. (**b**): Elementary pore-flow model comprised of array of various pores and pore-throats. Flow is represented by streamlines (green) and time-of-flight-contours (TOFC) (red), with spacing of 60 s. Total time-of-flight is 600 s, time step $\Delta t = 0.008$ s. Height and width for pore-throat(s) is 2×10^{-6} and 5×10^{-6} m, respectively and for pore-spaces 4×10^{-6} and 5×10^{-6} m, respectively. The far-field flow velocity is 3.2×10^{-6} m/s. The strength of pore-throat and pore space is 10^{-18} m^4/s, about 150 times the velocity of the far-field flow (after adjusting for unit dimensions). Length units are scaled in μm (both x-axis and y-axis). (**c–e**) Same as (**b**) but with different arrangements of pores and pore throats.

Figure 3a–d shows how fluid flow paths and local velocities are affected by the pores and pore-throats in different configurations. The effect of the pores on the streamlines is visualized using different combinations of the pores and pore-throats acting as a single horizontal micro-channel. We assume that the depth, width, and height of the pore-throats is 1×10^{-6} m, 5×10^{-6} m, and 2×10^{-6} m, respectively. Similarly, the pore dimensions are 1×10^{-6} m, 5×10^{-6} m, and 4×10^{-6} m, respectively. For both pores and pore-throats, the assigned strength is 7.2×10^{-7} m^4/s, and consequently due to a constant flux moving through a narrowing space. The highest flow rates occur in the narrower pore-throats. The total time of simulation is 600 s, using a time-step of 0.008 s. The relatively short duration of the simulation and time-step captures the movement of fluid across the micro-pores (Figure 3a–d). Inflow and outflow domains are shown to highlight constriction of streamlines and acceleration of the flight time in the pore space, as marked by the time-of-flight contours (red).

The pressure for the configurations of Figure 3c is given in Figure 4, which shows that the pressure differential at the pore-throat entrances is higher as compared to that of the wider pores. The pressure profile in such micro-pores varies with pore throat size and increases when capillary forces are involved (multi-phase flow). Local flow rates in the pore throats may be several times faster than in the wider pores (ignoring multi-phase flow effects).

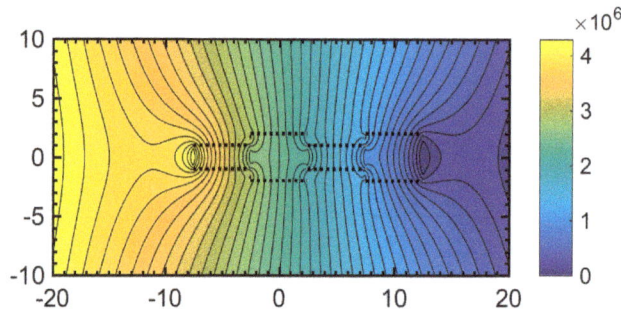

Figure 4. Pressure field (Pa) calculated by taking the real part of the complex potential for configuration in Figure 3c. The increase in pressure is localized in the pore throats. Narrower pore diameters show higher pressure gradients, as expected. The macroscopic permeability is 600 nD, viscosity as 1 cP, far field flow velocity is 1.5 m/year, and pore strengths is 225 m^4/year. Length units are scaled in µm (both x-axis and y-axis).

2.2.2. Movement of a Passive Ganglion through an Elementary Pore-Network

A passive ganglion, made up by tracers of incompressible fluid introduced in the far-field flow at time zero, was followed as it moved along the particle paths to study the effect of pores on fluid flow (Figure 5a). For the base case model, we assume that the capillary pressure around the ganglion iss negligible, hence the ganglion easily deforms when passing through the pores. Figure 5b shows that the ganglion starts to "feel" the effect of pores towards the flow front at a time as early as 60 s. At around 180 s (Figure 5c), the initially circle-shaped ganglion completely deforms into a dumbbell-shaped body (due to stretch) when the streamlines maze and converge inside the pore space. After a flow time of 300 s (Figure 5d), the ganglion reverts nearly to its original shape, albeit with some imperfections. Later in this study, we present a case with a ganglion with interfacial tension, which may resist to stretch and deform due to surface tension, slowing flow in (or even blocking) the pore space.

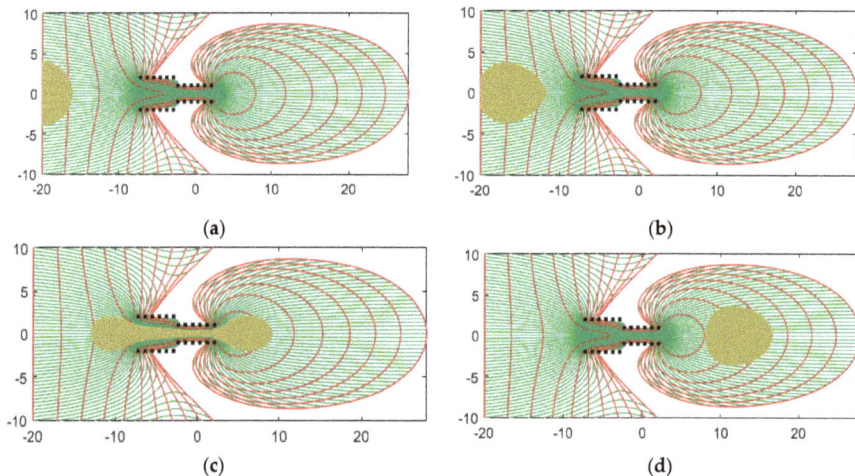

Figure 5. Movement of a passive ganglion (brown body) through elementary micro-pore domain in central part of the flow model. Dimensions of the pores, strength, and duration of simulation are the same as in Figure 3. Flow is steady and shows (**a**) initial position of the circular ganglion, with a radius of 4×10^{-6} m and (**b–d**) position and shapes of the ganglion after 60, 180, and 300 s, respectively. Length units are scaled in µm (both x-axis and y-axis).

2.3. Multi-Pore Network Model

The simplified micro-flow channel model (Figures 3 and 5) can be expanded with an assumption that several elementary models are stacked vertically and horizontally to create complex flow paths that are more representative for a small section of a porous reservoir. The pore-space of a representative reservoir section can be modeled superposing short channel segments with varying height and width in an array of pores and throats. The dimensions of the pores and throats in the pore network model are kept spatially constant, but heterogeneities in pore spaces would affect the flow paths and can be introduced by spatial variations in pore dimensions.

The motivation for constructing the micro-pore network model is to demonstrate how the pores interact and affect the overall flow path. Figure 6 shows that flow converges towards the pore network with each pore and pore-throat, given the same flow strength. The dimensions of the pores, the strength of the far-field flow and total simulation time are constant and identical to those used in the base case models of Figures 3 and 5. However, the strength of the pores is reduced to prevent fluid flow congregating only toward the central row of pores. Some boundary effects cannot be excluded: Outer pores receive slightly less fluid and remedying that situation would require a much larger pore network system to reduce outer boundary effects. The micro-pore network model (Figure 6), being well aware of the boundary effects experienced by the outer pores, provides a practical tool to explore fundamental questions regarding capillary effects and time-dependent changes in pore pressure due to PVT effects on fluid miscibility.

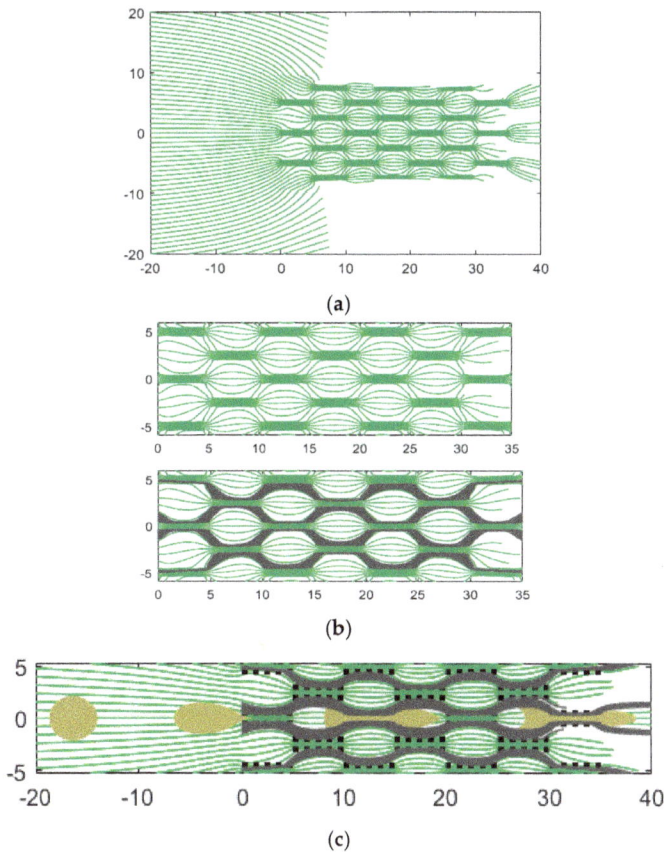

(a)

(b)

(c)

Figure 6. *Cont.*

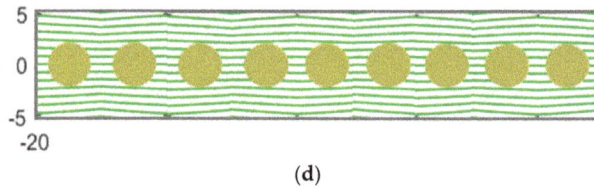

(d)

Figure 6. (a) Overview of multi-pore flow space showing streamlines. (b) Central section of the multi-pore model with the matrix between the pores filled with dark dye. The dimensions of the pores, the strength of the streamlines, and duration of simulation is the same as in Figures 3 and 5. The strength of the pores is 30 times the strength of the far-field flow. (c) Flow of passive ganglion (brown body) through the micro-pore network model. Snapshots of passive ganglion are given for 60, 240, 360, and 480 s. (d) Continuum model with dyed blob to study the movement of fluid particles would not change shape as in pore network model, and interfacial contact surface or length does not change, unlike that seen in case (c) due to deformation in the pore space. Length units are scaled in μm.

The explicit boundaries of the pore-spaces (Figure 6a) can be defined by delineating no-flow boundaries around the pores as shown in Figure 6b. Fluid particles that do not cross the pore space are grayed out and act as no-flow barriers between individual pore arrays. The path of a passive ganglion was tracked in the experimental pore-network model as seen in Figure 6c. A ganglion (brown circle) with a radius of 2.25×10^{-6} m moves for ten minutes through a pore system with a far-field velocity of 1.5 m/year (4.8×10^{-8} m/s). The strength of the pores is 1.44×10^{-6} m^4/s and the total time of flight is 600 s. The position and shapes of the passive ganglion passing through the pore network are shown at regular time steps of 60, 240, 300, 360, and 540 s, respectively. Figure 6d shows a continuum model with dyed blob moving with the flow, where the shape of the blob doesn't change, as in the pore network model. Interfacial contact surface or circumferential length does not change, unlike that seen in Figure 6c, where the interfacial outline of the ganglion shows substantial length increases due to deformation in the pore space.

The simulation of Figure 6 again neglects any capillary pressure. The mathematical model needs to be modified to represent the surface tension of a ganglion and to account for the capillary pressure. Before discussing such a modification with movement of an active ganglion through the pores affected by capillary pressure effects in a later section (Section 4), we first need to expand the pore network model to include the intersection of the reservoir space and the wellbore, which can be used to construct dynamic bottomhole pressure profiles for reservoir fluid entering the wellbore when leaving the pore space.

3. Dynamic Bottomhole Pressure Profiles

With the basic pore network model in place, we next construct flow models to determine the dynamic bottomhole pressure (BHP) due to the flow transition between the reservoir space and the wellbore. For a non-flowing well, the BHP is equal to the hydrostatic pressure due to the dead weight of the fluid column in the well. When the well is flowing, the BHP may change according to the pressure gradients associated with the fluid circulation. The below models are presented for the dynamic pressure gradient due to flow, which adds a dynamic BHP to the static BHP to obtain a total BHP for the flowing well. The effect of natural gas lift due to dissolution below the bubble pressure was initially ignored in the dynamic pressure model. While the release of dissolved gas in the production tube would contribute to increase the dynamic component of the BHP, the lighter gas would lower the static BHP and therefore tended to suppress the overall BHP for the flowing well.

3.1. Model Design

The model set-up is sketched in Figure 7a. The reservoir section is modeled by the pore network model of Figure 6. The wellbore section is modeled by an elementary model of the fluid circulation in the production tube (Figure 7b). Alternatively, the wellbore section can be modeled by fluid circulation via the drill string and the annulus (see later). The model design (Figure 7a) includes a mirror image of the well system to ensure stability in the CAM solution space. However, the mirror image well system is not necessarily required and we later realized that only a little distortion occurs by leaving out the mirror well. Nonetheless, the models presented in this study utilized the full system symmetry.

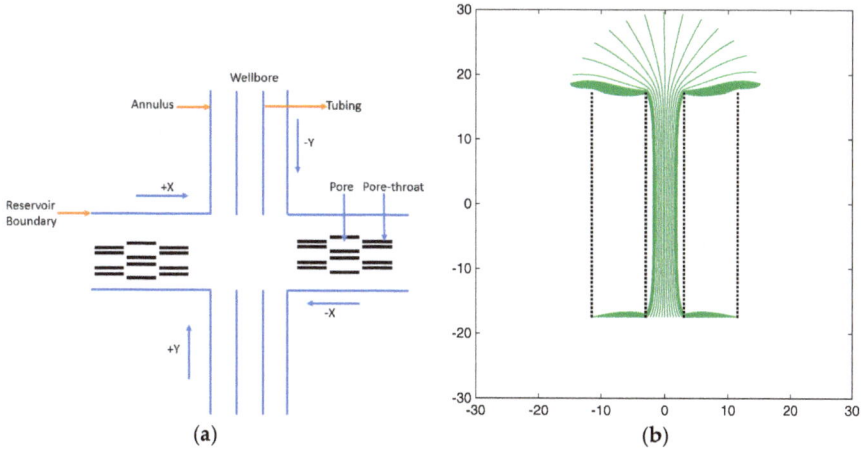

(a)   (b)

Figure 7. (**a**) Schematic of the reservoir–wellbore transition model with flow directions. (**b**) Production tube flow model with streamlines (green) for flow from reservoir to the wellbore obtained by solving the stream function using a particle tracking method based on incremental particle displacement in a Eulerian reference frame (as opposed to a Lagrangian reference frame). Length units are scaled in cm (both x-axis and y-axis).

3.2. Producer and Injection Wells

During production, the flowing well generally shows an increased BHP relative to the static BHP, simply because formation fluid moves into the well, which adds a dynamic pressure gradient to the static pressure gradient in the production tube. Figure 8a shows the streamlines for the producing well. The wellbore flow occurs by a natural pressure drive. We emphasize that the model only focuses on the flow dynamics near the base of the well and gravity is neglected for simplicity. The weight of the static fluid column in the wellbore adds a static BHP component which increases with depth. The total BHP equals the static BHP plus the dynamic BHP. The dynamic pressure distribution is contoured in Figure 8b. The assumed model parameters are given in Table 1. Pore sizes were upscaled by a factor of 10^{-6} for visualization.

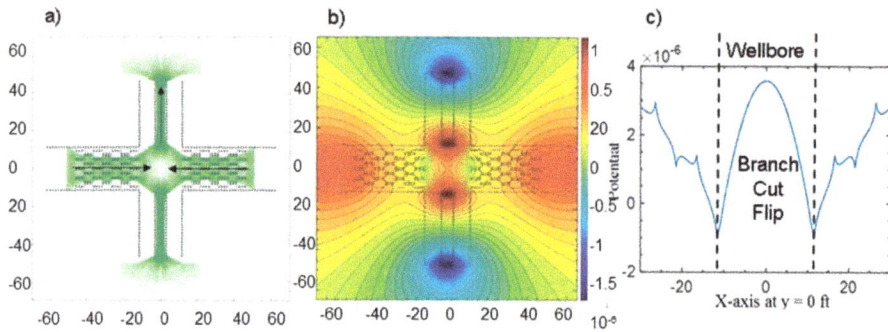

Figure 8. Producer well. (**a**) Streamlines (green) for flow from reservoir to the wellbore obtained by solving the stream function for the integrated system. (**b**) Pressure field obtained by solving the corresponding potential function scaled in units of $cm^2\ s^{-1}$ (Case 4, Table 1). (**c**) Dynamic bottomhole pressure profiles (along vertical axis in $cm^2\ s^{-1}$) for production well. Central portion of profile (in wellbore) needs to be folded downward to correct for branch cut effect (i.e., bottomhole pressure (BHP) minimum occurs in the wellbore center). Length units are scaled in cm (both x-axis and y-axis).

Table 1. Inputs for wellbore models.

Physical Quantity	Producer Case 4	Injector Case 5	Underbalanced Case 3	Overbalanced Case 6
Pore Strength $(cm^4 \cdot s^{-1})$	5×10^{-6}	5×10^{-6}	5×10^{-6}	5×10^{-6}
Pore-throat Strength $(cm^4 \cdot s^{-1})$	1×10^{-5}	1×10^{-5}	1×10^{-5}	1×10^{-5}
Pore Width (cm)	4	4	4	4
Pore-throat Width (cm)	1	1	1	1
Reservoir Boundary Strength $(cm^4 \cdot s^{-1})$	6×10^{-4}	6×10^{-4}	6×10^{-4}	6×10^{-4}
Annulus Strength $(cm^4 \cdot s^{-1})$	6×10^{-4}	6×10^{-4}	6×10^{-4}	1.20×10^{-3}
Tubing Strength $(cm^4 \cdot s^{-1})$	6×10^{-4}	6×10^{-4}	6×10^{-4}	6×10^{-4}
Reservoir Boundary Width (cm)	23	23	23	23

Several fundamental insights can be gleaned from the base production model of Figure 8a,b. First, in the reservoir, the pressure dropped toward the well, which led to fluid acceleration or velocity increase of fluid particles moving toward the well (see Appendix A and [70]). Streamlines initially converge, but then diverge as fluid moves from the reservoir into the well. The entire velocity field is governed by the equation of motion, as captured in the complex potential used to model the streamlines of Figure 8a. In our present study, the pressure plots (such as in Figure 8b) include regions of adverse pressure gradients, where flow appears to move counter to the pressure gradient. Removal of the so-called integral branch cuts leads to the local development of these apparent, adverse pressure gradients. Such regions are artifacts of the computational solution method that assigns a single value to a multi-valued complex elementary function [71,72]. For example, choosing the positive rather than the negative root when jumping over a potential branch cut boundary will swap the sign of the pressure gradient. Adverse pressure gradients occur in the peripheral quadrants of the simulated reservoir and in the central square space where fluid flowed from the porous reservoir into the wellbore (Figure 8b). The bottomhole pressure profile in Figure 8c can be corrected for the adverse pressure gradient by flipping down the central profile region between the branch cuts coinciding with the wellbore. The flip will result in a continuous V-shaped pressure profile across the wellbore an into the adjacent reservoir space.

A model for the injection well case is given in Figure 9a,b. The dynamic component of the BHP in the wellbore appears lower than in the directly adjacent reservoir space, which is, again, a purely mathematical branch cut effect. The pressure profile in Figure 9c marks the locations (dashed lines) where the branch cut causes apparent pressure gradient changes. The solution is accurate for the assumptions and initial conditions used and the dynamic BHP component can be corrected by simply folding up the central trough in Figure 9c to obtain a continuous Λ-shaped pressure profile for the injection well. A continuous Λ-shaped pressure profile appears as fluid leaves the porous reservoir space and enters the wellbore, detail of which has not been modeled in any prior study. The BHP gradient in the wellbore should alert production engineers that pressure readings from downhole pressure gauges will be greatly influenced by their actual lateral position in the bottomhole space.

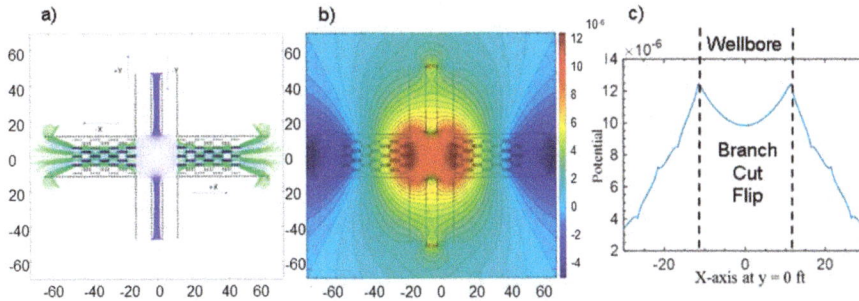

Figure 9. Injection well. (**a**) Blue streamlines are for fluid from the wellbore invading the reservoir space. Green streamlines are for fluid originally in the reservoir and displaced by the wellbore fluid. The result is obtained by solving the stream function of the composite flow field. (**b**) Pressure field obtained by solving the corresponding potential function scaled in units of cm^2 s^{-1} (Case 5, Table 1). (**c**) Dynamic BHP pressure profile (along vertical axis in cm^2 s^{-1}) at reservoir–wellbore transition. Length units are scaled in cm (both x-axis and y-axis).

3.3. Wellbore Pressures during Drilling

When drilling the well, a factor that influences the dynamic component of the BHP for a given mud weight is the viscous friction of the mud return circulation in the annulus. For a so-called pressure balanced well, there is no contribution of formation fluid to the circulation of the drilling fluid. All circulation is by drilling mud alone, which moves to the bottom of the well via the drill string pipe and then exits via the mud motor (in the bottomhole assembly, BHA) to return to the surface via the annulus (Figure 10a). Our simple model shows that, for a balanced well, the dynamic pressure gradient contribution to the BHP, due to the mud circulation directly below the drill pipe, is lower than for the nearby annulus (Figure 10b).

In well control, the primary concern is to keep the BHP close to the formation pressure to avoid the occurrence of either underbalanced or overbalanced wellbore pressures, essentially to minimize any dynamic contribution to the BHP due to fluid exchange with the penetrated formations. As long as the drilling fluid pressure is balanced or only slightly exceeds the local formation pressure at each depth, the well will not take in undue flow from escaping formation fluid after the drill penetrates the formation(s). However, when there is fluid exchange with the reservoir during drilling, due to mismatches in the balance of the mud pressure and the formation fluid pore pressure, two basic cases may occur: Underbalanced and overbalanced pressures at the wellbore wall.

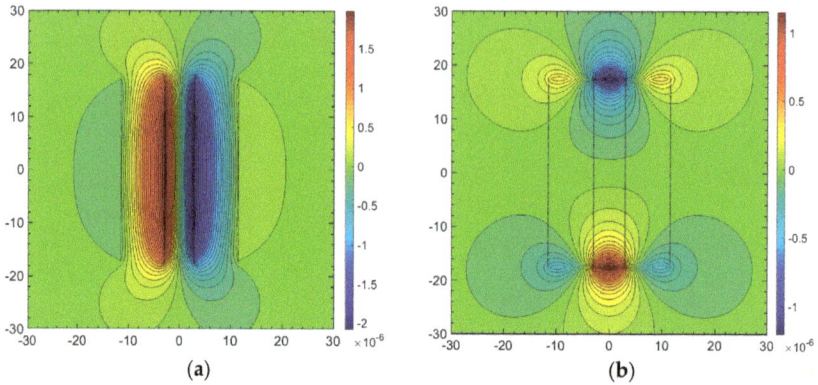

Figure 10. (**a**) Streamlines and stream function solution (color grades) for mud circulating down the drill string and up the annulus. Minor circulation occurs outside the annulus, which is an artifact of the model method. (**b**) Corresponding potential function solution. Length units are scaled in cm (both x-axis and y-axis).

Underbalanced wellbore pressure may lead to a premature flow of formation fluids into the well, with a risk of pressure escalation into a blowout. For such cases, the formation pressure adds a dynamic pressure component to the BHP when formation fluid flows into the well (Figure 11a,b). A dynamic pressure low developed at the wellbore wall, which establishes a favorable pressure gradient for the fluid moving into the wellbore from the far-field reservoir space. Pores that are filled with fluid near the wellbore wall have a higher pressure than the fluid in the adjacent wellbore. The potential occurrence of negative net pressures on the inner wellbore wall when the formation pressure is underbalanced was assumed in prior studies [73,74].

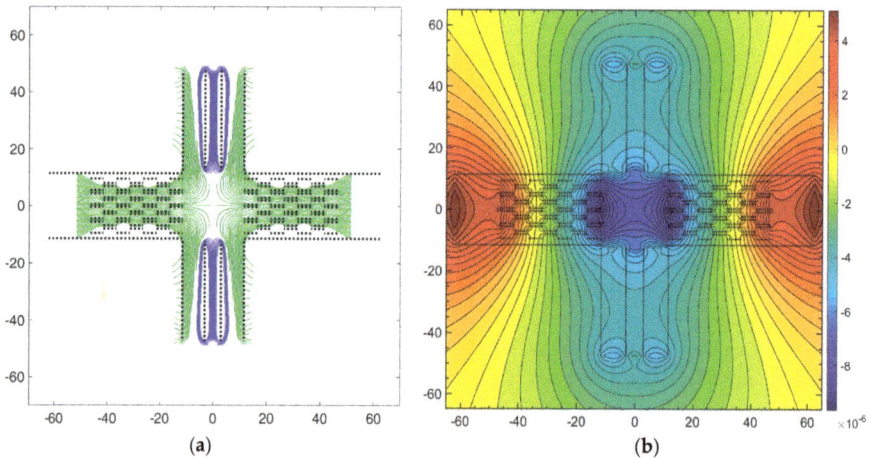

Figure 11. Underbalanced pressure while drilling. (**a**) Green streamlines are for fluid flowing from the penetrated formation to the fluid annulus between the drill string and the wellbore. Blue streamlines are for circulating drilling mud. Due to drilling mud pressure deficit (underbalance) native pore fluid from the formation can move up the annulus. Flow field is solved using a stream function describing the composite flow field. (**b**) Pressure field obtained by solving the corresponding potential function scaled in units of $cm^2 s^{-1}$ (Case 3, Table 1). Length units are scaled in cm (both x-axis and y-axis).

The final case modeled is for an overbalanced wellbore pressure, which may lead to a dynamic increase of the BHP when drilling fluid starts to invade the formation, leading to drilling mud losses (Figure 12a,b). Note tha the model for the overbalanced wellbore of Figure 12a,b is akin to the model for the injection well case of Figure 9a,b. For underbalanced formation pressure while drilling (Figure 11) and overbalanced formation pressure while drilling (Figure 12), dynamic bottomhole pressure profiles are given in Figure 13a,b. For the underbalanced pressure profile (Figure 13a), formation rock pores that are filled with fluid near the wellbore wall have a pressure lower than the fluid in the adjacent wellbore. Remember one must apply a branch cut flip to the central profile section in the underbalanced wellbore (Figure 13a) to obtain the correct V-shaped pressure profile. Likewise, for the overbalanced wellbore (Figure 13b), the dynamic BHP profile will have a Λ-shaped pressure profile after correction for the branch cut flip.

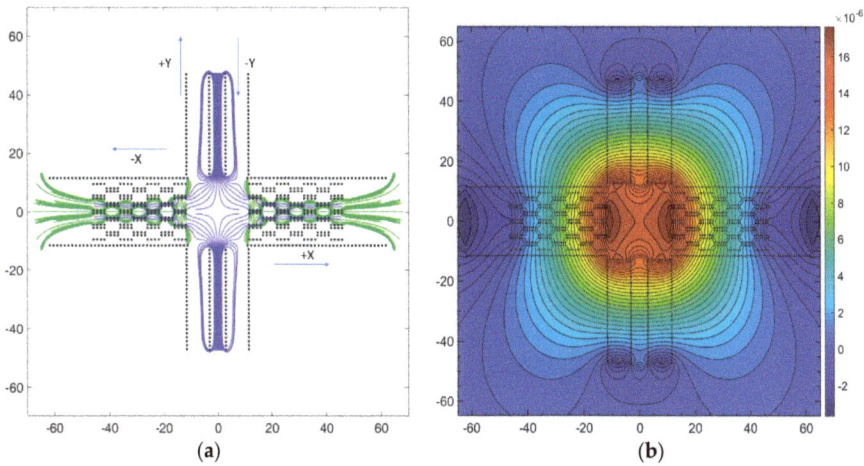

Figure 12. Overbalanced pressure while drilling. (**a**) Blue streamlines are for circulating drilling mud. Formation fluid cannot move into the annulus die to overbalanced drilling, but drilling fluid can invade the reservoir pore space. Green streamlines are for fluid originally in the penetrated formation, which is displaced deeper into the formation space due to drilling mud pressure excess (overbalance). Flow field is solved using a stream function describing the composite flow field. (**b**) Pressure field obtained by solving the corresponding potential function scaled in units of cm^2 s^{-1} (Case 6, Table 1). Length units are scaled in cm (both x-axis and y-axis).

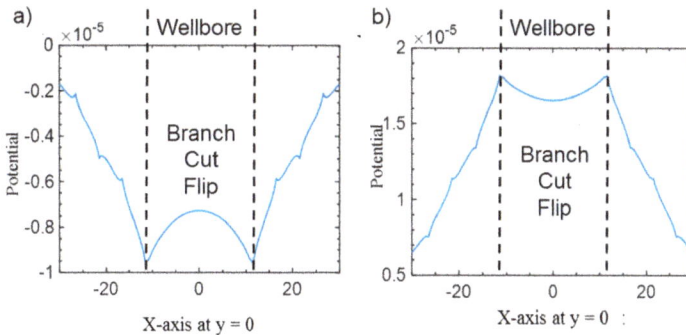

Figure 13. Dynamic bottomhole pressure profiles using the potential function (along vertical axis in cm^2 s^{-1}) for (**a**) underbalanced formation pressure while drilling (Figure 11) and (**b**) overbalanced formation pressure while drilling (Figure 12). Length units are scaled in cm (both x-axis and y-axis).

As noted in earlier in Section 3.2, the dynamic bottomhole pressure component in a horizontal section across the wellbores (halfway the formation) is not constant but varies laterally. The occurrence of such local pressure gradients should alert drilling engineers that pressure readings from the bottomhole assembly will be greatly influenced by the lateral position of the pressure gauge in the bottomhole space.

4. Multi-Phase Flow Effects

The simulations of Sections 2 and 3 neglected any capillary pressure. The pressure inside and outside of a passive ganglion passing through the pore models (Figures 5 and 6) remain the same when no interfacial tension exists between the dyed blob and the ambient fluid. Flow in a reservoir with pressure above bubble point occurs as single-phase oil displacement (Sections 2 and 3). If native water is present, it may be assumed to be drained by the piston-like movement of the non-wetting fluid (oil) as portrayed in Figure 14a. Locally, the flow may be accompanied by imbibition of the capillary space by the wetting fluid (native water) as in Figure 14b.

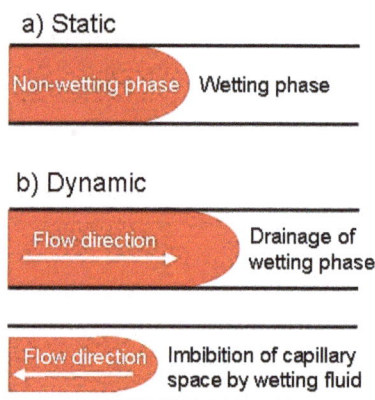

Figure 14. (**a**) Two-phase fluid contact with surface tension between non-wetting phase (e.g., oil) and wetting phase (e.g., water). (**b**) Piston-like movement occurs when non-wetting fluid (oil) sweeps the wetting fluid.

The oil ganglions must move through the pore space much like a single-phase flow, to leave behind as little residual oil as possible in the pore space of a reservoir drained by a well (Figures 5 and 6). However, when there is interfacial tension, a ganglion of non-wetting fluid from the bulk of the reservoir upon entering a narrow cylindrical pore tube (Figure 15a) will experience a capillary pressure, which will increase when pores become narrower until the pressure reaches the limiting value given by the Young–Laplace equation. There will be a pressure difference inside and outside the ganglion equal to the capillary pressure. When the (non-wetting) fluid exits a pore throat as shown in Figure 15b due to the pressure gradient, the capillary pressure will first increase when the interface expands. On further flow, the pressure will decrease and the volume of the sphere will increase while the capillary pressure decreases.

(a)

(b)

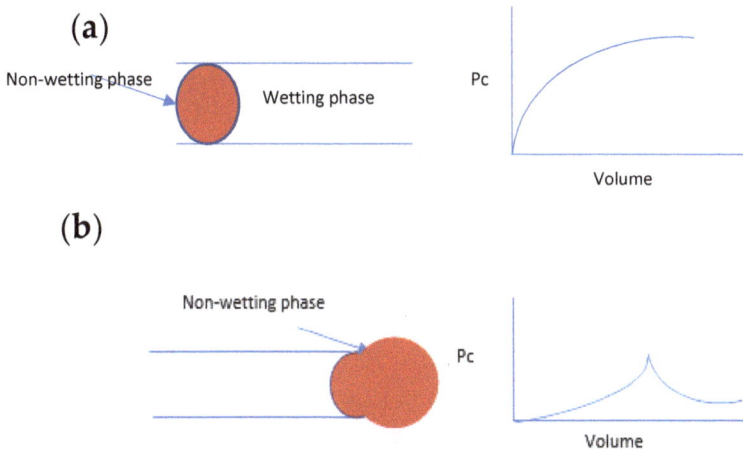

Figure 15. (**a**) Conceptual model of the movement of a non-wetting phase from the bulk reservoir to a narrowing pore throat, displacing a wetting phase. The capillary pressure increases up to the maximum value *Pc* given by Young–Laplace equation. (**b**) Movement of non-wetting phase from a channel to the bulk of the reservoir and the corresponding pressure profile.

4.1. Capillary Pressure Effects

The contact angle between two immiscible fluids and a wall depends on interfacial tension of the fluids and the nature of their contact surface. For example, Figure 16 shows the systematic reduction of interfacial tension by the addition of various surfactants (Surf 1–4) to the water phase, which reduced the contact angle with Eagle Ford wall rock. The experiment of Figure 16 primarily shows that the injection of chemical surfactants (in a laboratory test) can be effective in freeing the oil from the rock by reducing interfacial tension. The contact angles and the interfacial tension, as shown in Figure 16, were highly dependent on the fluids and the contact surface. For example, the surface tension and contact angle for mercury and air are 480 mN/m and 140°, respectively [75]. The experiment of Figure 16 is under static conditions, but a similar reduction in contact angle will occur when a ganglion moves through a rock-pore network flow path.

Figure 16. Contact angles for oil drop, adhering to oil–wet shale surface, varies with surface tension. Addition of various surfactants (Surf 1–4) to the pure interstitial water (left) leads to an increasing degree of the surface repelling the oil drop. The most effective surfactant (Surf 4) causes the most separation from the surface. Source: Dr. David Schechter, Texas A&M University.

We now seek to explore further the effects of multi-phase flow. When there is interfacial tension between two phases, then a pressure difference will occur, inside and outside an active ganglion in the confined pore space, equal to the capillary pressure. The mathematical model needs to be modified to represent the effects of the capillary pressure on the movement of a ganglion.

4.2. Movement of an Active Ganglion through an Elementary Pore-Network

The various modes of movement of a non-wetting, active ganglion (with surface tension) through an elementary pore flow model are schematically shown in Figure 17a,b. The movement of a ganglion through the pore models of Figures 3–6 was modeled neglecting any capillary pressure. With surface tension effects, the fluid blob would resist deformation and such an active ganglion may snap off in the middle as shown by Figure 17b. Below, an attempt is made to adapt our pore network flow model to account for certain surface tension effects.

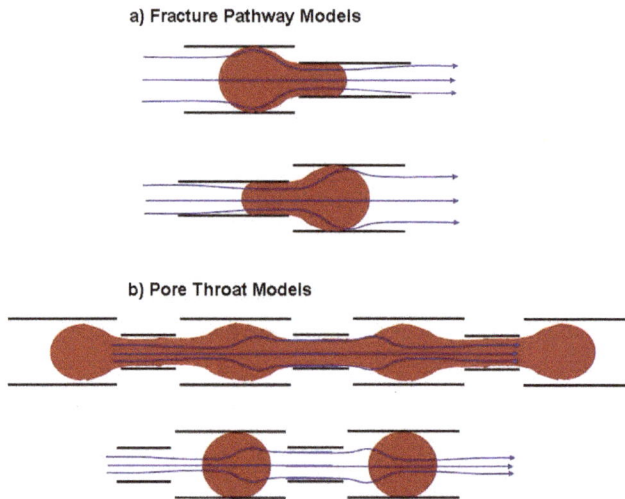

a) Fracture Pathway Models

b) Pore Throat Models

Figure 17. Principal sketches of non-wetting phase (e.g., oil, red) in interfacial contact with wetting phase (water or gas). (**a**) Fracture model shows a fracture space widening downstream (top) and narrowing (bottom). (**b**) Pore throat model showing stretching of oil ganglion during flow (top) and snapping into oil drops due to capillary forces when flow slows (bottom).

Figure 18a–d shows the progressive movement of an active ganglion through an elementary pore model representing the reservoir space. Interfacial tension between two phases during a multi-phase flow, affecting the movement of the active ganglion with a surface tension, is represented by a dipole (which can be scaled as a surface tension effect) that resisted deformation during the movement through the pores (without the snapping effect, which requires implementation of Rankine flow elements). The radius of the point dipole is inversely proportional to the square root of the far-field flow velocity. The velocity of the far-field flow is accelerated in the micro-channel, which is why the radius of the active ganglion varied when it moved through the pore network. Variations in the radius of the ganglion may be mitigated by splitting the original point dipole into a line dipole, allowing the dipole to stretch and morph into a Rankine body formed by a spaced point source and a point sink. The procedure proposed for implementing this solution is included in the final section of Appendix B, which is a potential topic of future study.

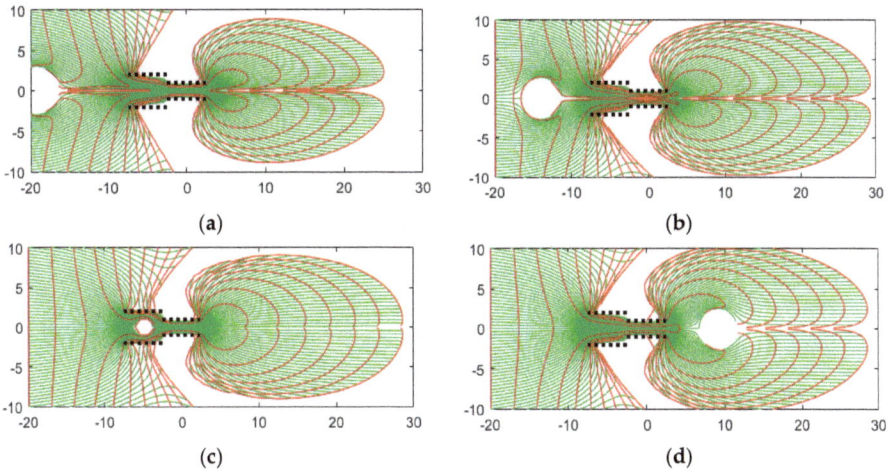

Figure 18. Movement of an active ganglion through a micro-channel. From left to right, top to bottom: (a) The initial/reference position of the ganglion. (b–d) Position and shapes of the ganglion at 60, 180, and 300 s, respectively. The radius of the starting ganglion is 4×10^{-6} m. The dimensions of the pores, strength, and duration of simulation are the same as in Figures 4 and 5. The strength of the dipole is 10 times the strength of the far-field flow. Length units are scaled in μm (both x-axis and y-axis).

4.3. Bubble Point Delay Due to Capillarity

The co-production of gas by the well (even when the reservoir is still above the bubble point pressure), occurs by dissolution of gas molecules from the single-phase oil when the produced fluid moves up the production tube and the pressure drops below bubble point at a certain depth (Figure 19). For such cases, the gas–oil ratio in early well life is mainly due to the dissolution of associated gas in the wellbore, which then provides a natural gas lift to the produced fluid. The total volume of the gas bubbles rising in the production tube can be history-matched using the production data and gas to oil ratio at any one time.

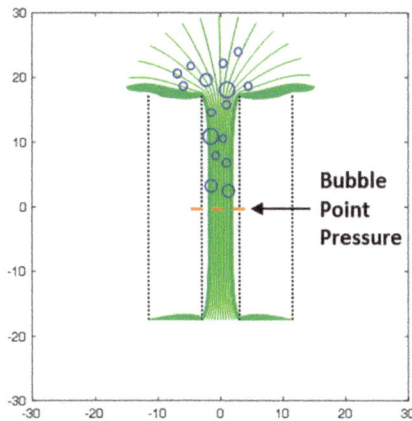

Figure 19. Dissolution of the gas molecules from the single oil phase occurring when the produced fluid moves up in the production tube and the pressure drops below bubble point at a certain depth. Length units are scaled in cm (both x-axis and y-axis).

Although an increase in the gas oil ratio increases fluid buoyancy and lift rates in the wellbore, the flow of gas–oil emulsions toward the well via the reservoir pore space may be retarded by a higher degree of multi-phase flow when the gas–oil ratio increases. Changes in the gas–oil ratio of a fluid at reservoir PVT conditions are partly controlled by pressure depletion due to continued fluid extraction. An additional factor related to the advent of bubble point pressure conditions occurs in nanoDarcy rocks due to capillary effects, which may delay an increase in the gas oil ratios by lowering of the bubble point pressure due to the occurrence of elevated capillary forces in the nano-pores [22].

When the reservoir pressure finally drops below the bubble point, gas dissolution occurs not only for fluid in the wellbore, but also in the reservoir. Multi-phase flow then occurs in the reservoir and interfacial tension effects will intensify due to the generation of expanding gas bubbles in the reservoir before the fluid reaches the well. Two extreme scenarios can be considered: (1) Large pores dominat the flow in the reservoir, which is when capillarity will not contribute to the production process of the well, and (2) the pore-space includes nm-scale pores with significant capillarity. For the latter case, capillarity may delay the bubble point pressure in the reservoir, and by preventing an early rise in the gas–oil ratio, the effectwill suppress the oil EUR of the well. The history-matched EUR based on early production data (above bubble point pressure) needs to be corrected for the capillary effect to arrive at a more accurate EUR forecast.

The capillary effect may delay the actual bubble point pressure as the fluid nears the theoretical bubble point pressure for the fluid composition of the reservoir. Calculations of the equation of state and vapor–liquid equilibrium can be linked with the velocity and pressure field solutions of our simplified model to investigate the fundamental nature of the interaction between the bubble point pressure and the capillary pressure due to interfacial tension and pore geometry. Once the EUR correction due to the delay in the bubble point pressure is quantified using the fluid composition of the reservoir, history-matching of the predicted gas–oil ratio in late well life against the actual gas ratio will tell us which of the two scenarios considered is closest to reality. We expect the capillary effect to be likely to remain limited because most reservoir fluid will reach the well via the larger pores and micro-fractures, in which case, the bubble point pressure delay due to capillarity will not occur for the bulk of the fluid produced. There would be no delay in the increase of the gas–oil ratio of the well and the bubble point pressure is reached in a growing portion of the reservoir space, unaffected by capillarity.

5. Discussion

Several fundamental aspects of flow in a porous medium with single-phase and multi-phase flow effects have been considered in the elementary CAM pore-network model. The models of Section 2 show that even a passive fluid ganglion will deform in a pore network in a way different from an up-scaled continuum model (compare Figure 6c,d). Subsequently, the BHP was modeled as a result of the transition of fluid flow from the porous reservoir to the open wellbore (Section 3). The models of dynamic bottomhole pressure presented are complementary to nodal analysis production models [76] which account for the global decline in reservoir pressure and its impact on the flow rate of the well (and vice versa). Finally, capillarity effects were considered in Section 4, including the movement of an active ganglion. Changes in the gas–oil ratio were also discussed, associated with changes in the bubble point pressure. Some additional thoughts are shared in the discussion items below.

5.1. Scaling BHP Models and Flow Velocities

The models in Section 4 of the flow transition between the pore space and the wellbore are synthetic models, not scaled for any particular prototype well. In case the model is applied to real well data, scaling of all input parameters is essential to obtain quantitative results with predictive value on well performance. Such scaling must follow common rules of dimensional analysis, with particular attention to rheological similarity [77], a dynamic aspect of flow scaling that is often overlooked. The flow velocities for the streamline models of Figures 8a, 9a, 11a and 12a are captured everywhere in the system by the set of superposed complex potentials used to construct the flow model. The magnitudes of the velocities are contoured in Figure 20a–d (comparing cases of Figures 8a, 9a, 11a and 12a). The models highlight that the fastest flow rates occur in the production tube of the well. However, flow rates in the pore throats of the pore network model are also quite high and, in the case of the underbalanced drilling model (Figure 20c), are generally higher than in the production tube. The high flow rates in narrow pores in the case of a production well (Figure 20a) will lead to sand production when the formation is poorly consolidated (which is rarely the case in shale).

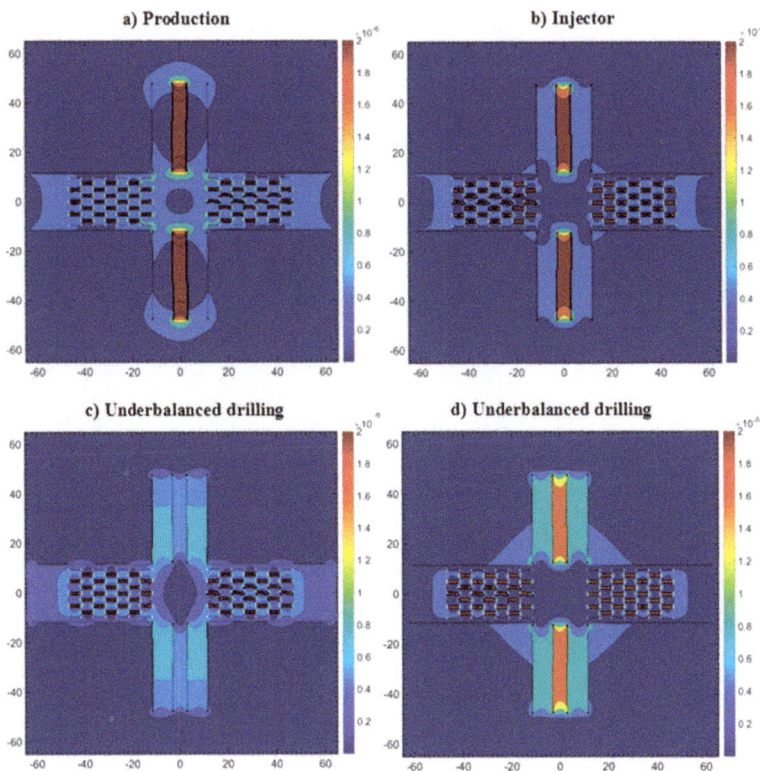

Figure 20. Absolute velocity magnitude (scaled in cm s^{-1}) for flow fields in Figures 8–11. Largest velocities occur in pore throats and in wellbore section. Length units are scaled in cm (both x-axis and y-axis).

The models of Figure 20a–d do not take into account the effects of multi-phase flow. However, the high velocity locations in the production tube and pore throats are precisely the locations where steep pressure gradients occur and also where interfacial contact lines between multi-component and multi-phase fluids will start to affect the flow dynamics. Such processes are rarely considered in wellbore stability models when planning and during drilling operations, but the gas kick will be controlled by such multi-phase flow dynamics.

For the production phase, EOR interventions involving huff-n-puff injection of fluids (liquids and gases), the detailed flow paths during injection and soaking time will be controlled by multi-phase flow effects, minimum miscibility pressure, bubble point pressure, capillarity, absorption, dissolution, etc. The injection path will be controlled by fluid PVT properties, but also by pore space geometry (Section 4.3) and fracture networks (both hydraulic and natural). In particular, the fluid-storage capacity and flow-channeling in natural fractures will play an important role in both single-phase and multi-phase flow through the reservoir space [67,78].

5.2. Upscaling of Capillarity Effects

The model of an active ganglion in Section 4.2 shows that the CAM flow model can account for certain multi-phase flow effects. However, such model adaptations are elaborate. An alternative approach is the introduction of domain functions to up-scale the local multi-phase flow effects. For example, active ganglions with high surface tension will lead to unsteady flow velocities at the pore scale, which we suggest can be represented by an oscillatory velocity flux profile (Figure 21a–c). An active ganglion will slow down at the entrance of a narrow pore throat (Figure 21a) and speed up again when exiting the throat (Figure 21b). In a representative elementary volume (REV) with a periodic pore structure, the velocity of an active ganglion can be modeled by an oscillation function (Figure 21c). The up-scaled continuum model would not "see" the local waxing and waning of fluid velocity due to interaction of the droplet interfacial tension with pore space. Nonetheless, the domainal oscillation functions can provide the required average velocity in the pore-network model for use in a continuum model. The oscillation function would need to be scaled on the basis of micro-fluidic pore network models or other physical models of the natural system.

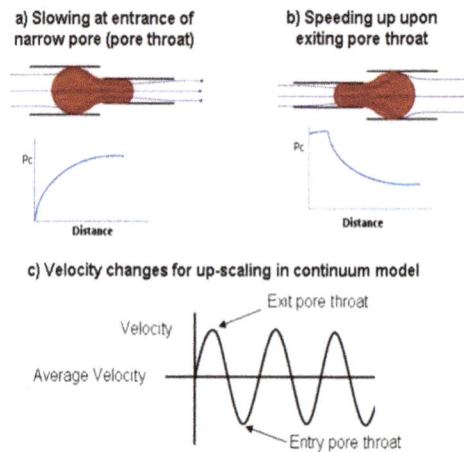

Figure 21. Conceptual model of active ganglion motion (**a**) slowing down when reaching the pore throat entrance and (**b**) speeding up when exiting pore throat. (**c**) Oscillation function capturing flow variability for specific domain in flow space.

Some fundamental insights can be gleaned from the simple pore-network models presented in this study. For single-phase flow of a given fluid, the permeability distribution alone will fix the streamlines [13]. The path of the streamlines is entirely controlled by the permeability (resistance of the medium to fluid flow) and the rate of flow along the streamlines is determined by both the permeability and porosity (fluid storage capacity of the medium) and the viscosity of the fluid (viscous retardation due to fluid only). For multi-phase flow, the permeability alone does not fix the streamlines. A combination of interactions of the *active* fluids with the pore space (capillarity, wettability) will determine the flow path. We suggest here that up-scaling may be possible using domain functions that locally scale an oscillatory flux (Figure 21), based on multi-phase flow behavior for a specific fluid composition and a specific pore network. The CAM model can accommodate such local flux control points by using so-called "areal doublets" and "areal dipole elements" [66–68].

5.3. Multi-Phase Flow and Geometry of Pore Space

We emphasize here (again) that, for multi-phase flow, the permeability alone with fluid viscosity and connected porosity are no longer sufficient to describe the flow, as would be the case for single-phase Darcy flow where streamlines are fixed by the permeability structure, and time-of-flight is affected by local storage effects related to porosity variability [13]. In multi-phase flow, the flow path will be affected by fluid interfacial tension effects and capillarity. The intensity of the interfacial tension and capillarity effects will be influenced by the geometry and ratio of pore space surface area to fluid volume, which is different for different pore shapes. Figure 22 shows that the ratio of pore space surface area to fluid volume decreases for pore space closer to spherical shapes. The graph shows how the surface area of the pore in contact with a fluid increases when pore network spaces change from spherical, via polygonal approximations of the sphere, to angular shapes confined by fewer surfaces (octahedron, pentagon, and finally a pyramid). The surface area of the pyramid (a very angular pore space with many acute inner angles) for the same fluid volume is substantially larger than for the spherical pore shapes. Consequently, reservoir fluid moving through pore shapes being angular (pyramidal) will have a much larger interfacial contact line with the pore space than for spherical pores. One may conclude that multi-phase flow effects will be more pronounced in angular pore spaces as compared to cylindrical and spherical pores.

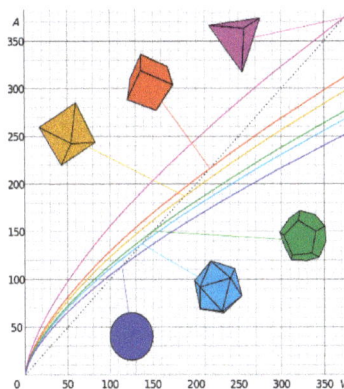

Figure 22. Curves showing that the ratio of surface area (vertical scale) to volume (horizontal scale) is smallest for pore spaces closer to spherical shapes and largest for pyramidal pore shapes. For any particular finite volume of fluid, the surface contact area with the pore wall varies with pore shape and in a non-linear fashion when pore volume increases.

Figure 23 shows two extreme cases of rock pore shapes. The angular pore space of Figure 23a will have an oscillation function showing greater variation in the velocity of the multi-phase fluid passage than would be the case for the cylindrical pore space of Figure 23b. Further work is needed to develop the envisioned oscillatory domain functions for up-scaling of multi-phase flow based on the preliminary insights reported here.

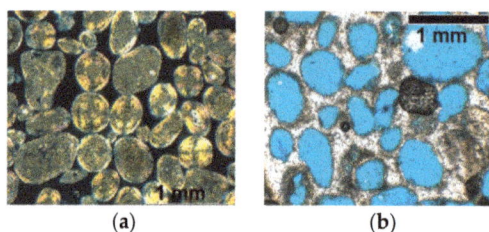

Figure 23. (**a**) Angular pores (close to pyramidal shape in Figure 22) and (**b**) spherical–cylindrical pores (close to spherical shapes in Figure 22) [79].

5.4. Future of Pore Network Models

Pore network models are needed to establish the dynamic relationship between pore morphology and connectivity to compute the up-scaled permeability tensor, with anisotropy controlled by the local porosity structure. When pores are stable and we neglect molecular sieving and interaction of molecules with pore walls and the parameters k and n are up-scaled for the continuum scale, then the flow in the porous structure can be modeled using the spatial gradients of k and n and we can assume flow is governed by Darcy's law. The permeability is the flow path controller as it determines the spatial resistance to flow and fluid composition can be taken into account by using relative permeabilities. The porosity is the scalar of the storage capacity of the porous medium and solely acts as time-of-flight controller [13].

When flow processes are more complex and variability in pore structure affects multi-phase flow behavior (immiscible), involving changes in interfacial energy due to phase changes and capillarity, pore network models can provide fundamental insight of the non-linear and time-dependent thermodynamics that control the flow processes. Direct pore scale simulators take real-pore-scale topology to model multi-phase flow and account for relative permeability, wettability, and capillarity for the specific system modeled. We may still use indirect or generic pore network models to improve our understanding of the impact of key processes on the flow behavior of multi-phase fluids in the porous reservoir. Poro-elasticity and pore failure, when local pressure breaks pore structures, may complicate continuum flow models due to time-dependency and micro-mechanical interaction with flow involving pressure differentials in the pore space and induced stress gradients in the elastic pore material.

Economic incentives to better understand, and thereby engineer or control flow in multi-phase fluids in porous media, are particularly strong when EOR interventions need to be scaled based on models of the detailed EOR process. For such applications, understanding highly heterogeneous, anisotropic, and fractured porous media remains an important and a challenging research task, computationally intensive and thus costly and time-consuming. Calculation of the equation of state, bubble point pressures, and vapor–liquid equilibrium as reservoir pressures decline over time must be linked with local velocity and pressure field solutions, capillarity and gas adsorption, slippage, Knudsen diffusion, and Klinkenberg corrections. The dynamic nature of these processes, all non-continuum effects, makes it difficult to work with simple up-scaled Darcy flow (k, n) parameters.

Pore space and throat morphology and angularity, tortuosity, and connectivity will affect immiscible multi-phase fluid flow, with a dynamic process such as adsorption, dissolution and wormhole formation, and salt/calcite precipitation with variable saturation concentrations, biomass growth, micro-fracturing, solute transport, and spatial wettability changes. Pore-scale models can provide physics-based constitutive relationships to support empirical relationships of permeability and flow in continuum models. Inevitably, simple continuum models need be modified to account for dynamic pore modifications due to micro-fracturing, dissolution, precipitation, and biomass growth, which are likely to occur over the longer time-scales involved in hydrocarbon field production.

Capillary pressure due to the co-existence of multiple phases is usually calculated by experimental means, such as the mercury injection, porous diagram method, centrifugal method, and dynamic capillary method [80]. These techniques require significant time and resources, which are not available for every project. The vast majority of EOR interventions depend on accurate representation of the complexity and heterogeneity of the reservoir. Numerical compositional reservoir simulation is required to capture the exact mechanisms, such as mass transfer between the several fluid phases [53–56]. Recently, several authors have used mathematical models to study various aspects in EOR applications, such as an optimal water alternating gas (WAG) ratio to prevent viscous fingering, partially miscible 2-phase/3-phase flow, and minimum miscibility pressure [81–85]. The present work aims to open up a new avenue to develop mathematical tools for fast reservoir models to help improve predictions of well performance, including multi-phase flow and EOR methods.

6. Conclusions

The purpose of this study is to apply recently developed algorithms based on CAM to model the particle paths of fluid flowing through a network of pores. The algorithms are also implemented to model the flow from a porous reservoir section into the wellbore. Although complex analysis methods (CAM) are limited to 2D flows, several advantages include infinite resolution and computational efficiency. A simple pore network model based on CAM resolves for the local increase and decrease of displacement rates as reservoir fluid moves in and out of narrow pore throats. Tracing passive ganglions in upscaled flow models would not deform (Figure 6d), but the same ganglions will be subjected to periodic stretching when moving through a regularly varying pore structure (Figure 6c). Previous studies have shown [13] that, for a flow in a porous medium, the streamlines are not affected by any spatial porosity changes and only vary based on spatial permeability changes, assuming single-phase flow and all other conditions are constant. However, this is not true for reservoirs with multi-phase flow, where interfacial, capillary, and inertial forces interact with the porosity structure of the formation, which all affect the fluid flow path. A preliminary model where the capillary pressure is simulated by a dipole analytical element (Figure 18) shows that the streamline patterns are deviated and different from the case without capillary pressure (Figure 5). Fluid-phase ratios and vapor–liquid equilibrium in the pore space may both be affected by the local pressure changes and interact with the capillary pressure, which will be investigated in future studies. What is unique about the CAM-based pore models presented in our study is the capability to exploit the high spatial resolution and the potential to link the model to phase-change computations.

Phase changes leading to gas bubble formation can be modeled in the pore network model by the insertion of minute dipoles in the flow description, triggered by local pressure changes. The nucleation of gas bubbles in nano-pores can be modeled using dipole nuclei, which nucleate into gas bubbles when the local pressure in a pore drops below the bubble point pressure. The gas oil ratio in well models accounting for capillary pressure in calculations of the vapor–liquid equilibrium will rise slower than without consideration of the capillary forces. Quantifying such delays in the onset of bubble point pressure due to capillarity is important because a delay in the intensification of two-phase flow may increase the volume of estimated ultimate resources (EUR), based on valid physical principles. Capillary pressure calculated from the pore-models may be used to predict the gas oil ratio (GOR) of fluids in liquid-rich shale reservoirs. This may have a significant impact on accurate estimation of bubble point pressure suppression and onset of the multi-phase flow, leading to more reliable reserve estimations.

With the basic pore network model in place, the pressure field at the transition zone between the porous reservoir and the open wellbore was modeled. In particular, the dynamic BHP profile across the wellbore is quantified for several fundamental cases. The flowing BHP profile can be estimated for underbalanced and overbalanced well sections at the level of the reservoir, after applying so-called branch-cut corrections. Several practical insights on BHP development for a flowing well can be formulated:

1. For a reservoir pressure that is underbalanced by the mud weight in the well, estimation of lateral pressure gradients in the BHP may help the operators to adjust any deficit in the density of the drilling fluids to prevent reservoir fluid reaching the surface via the annulus. The pressure of the fluid flowing in a wellbore can be effectively modeled by using a combination of a pore network model with a wellbore flow model. If the well models are properly scaled, the time for reservoir fluid to reach the surface can be predicted using time-of-flight contours. Ultra-fast rise of reservoir fluid is accompanied by a pressure kick, which may lead to the loss of well control (and is termed a blow-out when catastrophic).

2. For reservoir pressure that is overbalanced by the mud weight in the well, the estimation of lateral pressure gradients in the BHP may assist the operators to select the appropriate combination of mud weight and circulation rate that will prevent an unwarranted invasion of the reservoir space by drilling fluid, which may lead to the loss of costly drilling mud. The mud not only provides pressure balance in the well, but is also a lubricant for the cutter which may wear, break, and get stuck when lost circulation occurs.

3. Traditional wellbore stability models focus on the prevention of failure of the wellbore rock using geo-mechanical properties (elastic moduli and brittle failure criteria for certain stress concentrations). The simple models presented here show that mud circulation during drilling and pressure gradients at the transition of the reservoir to the wellbore may cause fluid flow which poses a drilling hazard if incompletely captured in concurrent geo-mechanical wellbore stability models, which focus on the elastic limit and brittle failure of the wellbore.

Author Contributions: Conceptualization, R.W.; Methodology, R.W.; Software, A.K. and R.W.; Formal Analysis, R.W. and A.K.; Investigation, R.W. and A.K.; Resources, R.W.; Data Curation, R.W.; Writing—Original Draft Preparation, R.W.; Writing—Review & Editing, R.W. and A.K.; Visualization, R.W. and A.K.; Supervision, R.W.; Project Administration, R.W.; Funding Acquisition, R.W.

Funding: Funds for this study were provided by the Texas A&M Engineering Station (TEES) and the Crisman-Berg Hughes Consortium.

Conflicts of Interest: The authors declare no conflict of interest.

Appendix A. Branch Cuts in Pressure Plot

For wellbore and pore models in this study, simply taking the real part of the complex potential may lead to inconveniently placed branch cuts, which cause discontinuity in the pressure plots. Thus, the real and complex potential need to be manually separated in order to facilitate choices about branch cut placement. Examples for branch cut solutions are presented below.

Appendix A.1. Background

The real part of the complex potential yields the potential function, which can be used to calculate the pressure field in a reservoir (as detailed in Section 4.1). Several of our prior studies include pressure field solutions for several analytical elements (such as point sources and sinks representing vertical injectors and producers, as well as line sinks, representing hydraulic fractures) and obtained excellent matches with independent pressure field solutions generated using a numerical reservoir simulator [63].

The superposed complex potential ($\Omega_{\text{sup}}(z)$) for any two analytical elements $\Omega_{FF}(z)$ and $\Omega_{AD}(z)$, is:

$$\Omega_{\text{sup}}(z) = \Omega_{AD}(z) + \Omega_{FF}(z) \ [\text{m}^2 \cdot \text{s}^{-1}]. \tag{A1}$$

The pressure change at any point z in the complex plane is calculated by selecting the real part of the superposed complex potential, $\Re\Omega_{\text{sup}}(z)$, and subsequently applying Equation (A3) below. The complex potential for a pore element is given in Equation (A4).

Appendix A.2. Demonstration of Branch Cut for a Simple Case

A brief discussion of branch points and branch cuts is warranted at this point. Assume a function $w = f(z)$, which maps points from the z plane to another domain in the w plane. If the function $f(z)$ is single-valued, for each value of z we can obtain single value of w. However, if the function $f(z)$ is multi-valued, a value of z results in several values of w. For example, consider a circular path, represented by the equation $z = z_0 + re^{i\theta}$, around a point z_0, where r is a constant greater than zero and θ varies in a counterclockwise direction around z_0. The function $f(z)$ results in different values for w as we move around the circle. Here, z_0 is defined as the branch point of the function $f(z)$ and multiple w are different branches of the function. There are two branch points for a function, one at infinity and the other at z, such that $f(z) = 0$. Finally, a branch cut is a line connecting two branch points at $f(z) = 0$ and infinity, which separates the discontinuity present in a complex plane.

Next, we calculate the pressure field for a synthetic reservoir with properties as summarized in Table A1 to show the branch cuts for a pore element. We assume initial pressure (P_0) is zero, which is an acceptable assumption for a simple synthetic case. The corresponding pressure changes (delta P) can be solved using Equation (A2) and Equation (A3):

$$\phi_{sup}(z) = \Re\Omega_{sup}(z) \; [\text{m}^2 \cdot \text{s}^{-1}], \tag{A2}$$

$$\Delta P(z) = -\frac{\phi(z)\mu}{k} \; [\text{Pa}]. \tag{A3}$$

Figure A1 shows the pressure profile for a steady-state case, where a horizontal areal doublet with finite width (represented by black dashed lines) is superposed by a far-field flow (from left to right). The fracture affects the pressure field in its vicinity, as can be inferred from the deflected isobars close to the fracture. However, an undesirable computational effect occurs beyond the right-end termination of the fracture (represented by solid white lines), where the pressure jump continues for an infinite distance in the horizontal direction toward the right. This effect is due to the occurrence of so-called branch cuts in the solution of the potential function (Equation (A2)), which becomes undefined at the vertices of the fracture. The simple model in Figure A1 has four branch points at the vertices of the fracture, which renders the complex potential undefined at those points (Equation (A4)). This results in the pressure profile becoming discontinuous at the branch cuts and the pressure change inside and outside of the fracture shows a big jump.

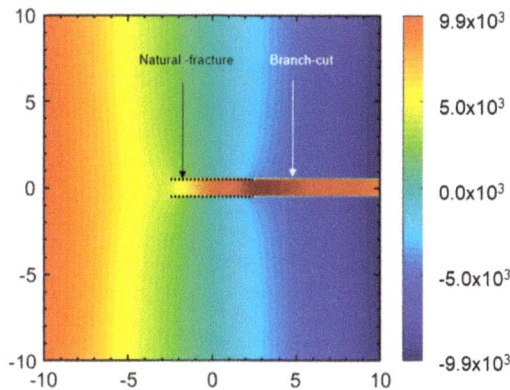

Figure A1. Pressure field for a far-field flow near a fracture. Length scale in m and pressure (rainbow colors) in Pa.

The branch in Figure A1 can be better understood if we consider a function $log(z)$, mapped in a complex plane as shown in Figure A2. The value of $log(z)$ at A, which is infinitesimally close to and above the positive x-axis, differs from that at B, which is infinitesimally close to A but is below the positive x-axis. The function $log(z)$ is discontinuous across the branch cut represented by line connecting zero to infinity. A Riemann surface can be used to represent the multiple values of ω by splitting the z plane into n parallel planes.

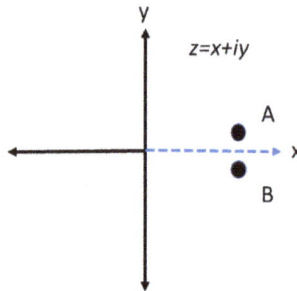

Figure A2. Representation of branch cut at positive x-axis showing discontinuity across two points A and B.

The discontinuity in Figure A1 can be analyzed by taking two arbitrary points $5 + 10i$ and $5 - 10i$ and plotting the pressure profile across the vertical cross-section between those points. This is the region with the branch cuts which extends to infinity from each vertex. The cross-section of Figure A3a demonstrates the acute jump of pressure across the branch cut and, as we move infinitesimally close to the branch cut, either from above or below it ($-0.5i$ or $0.5i$), as shown in Figure A2, the pressure gradient (ΔP) jumps from -0.5×10^4 Pa to 1.25×10^4 Pa. In effect, each pressure plane lies in two different planes of a Riemann surface as mentioned before. However, Figure A3b shows a pressure profile along $-5 + 10i$ and $-5 - 10i$ (Figure A1); the change in pressure is smooth and no discontinuity occurs as the branch cuts do not extend to negative infinity.

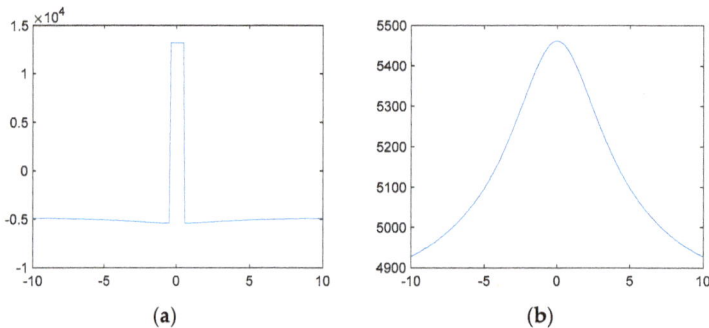

Figure A3. (a) Pressure profile along $5 + 5i$ and $5 - 5i$ in Figure A1. Horizontal scale shows distance in meters and the vertical scale is pressure in Pa. (b) Pressure profile along $-5 + 5i$ and $-5 - 5i$ in Figure A1. Horizontal scale shows distance in meters and the vertical scale is pressure in Pa.

Appendix A.3. Proposed Solution to the Branch Cut Placement

The method adopted here to overcome branch cut effects is to separate the real and imaginary parts manually and calculate the phase angles by using tangent function as shown in Equation (A6) below for all the logarithmic terms in Equation (A4). This results in a smooth pressure profile without any unrealistic pressure jumps as shown in Figure A3a.

The following template is used to manually separate the real and imaginary terms in Equation (A4) and generate the potential plot solution of Figure 15b:

$$\Omega(z) = \frac{-i \cdot v(t)}{2\pi hn \cdot L \cdot W} \cdot e^{-i\gamma} \cdot [(z + z_{a2})(\log(-e^{-i\gamma}(z - z_{a2})) - (z + z_{a1})(\log(-e^{-i\gamma}(z - z_{a1})) + (z + z_{b1})(\log(-e^{-i\gamma}(z - z_{b1})) - (z + z_{b2})(\log(-e^{-i\gamma}(z - z_{b2}))] \quad [\text{m}^2 \cdot \text{s}^{-1}], \quad (A4)$$

$$\Omega(z,t) = \frac{v(t)}{2\pi hn \cdot L \cdot W} [\Re[A - B + C - D] + i \cdot \Im[A - B + C - D]] \, [\text{m}^2 \cdot \text{s}^{-1}]. \quad (A5)$$

Due to the symmetrical nature of the Equation (A4), only one of the four vertices (among, z_{a1}, z_{a2}, z_{b1}, and z_{b2}) comprised in the terms A, B, C, and D (Equation (A5)) needs to be simplified, for which we use:

$$A = -i \cdot e^{-i\gamma}[(z + z_{a2})(\log(-e^{-i\gamma}(z - z_{a2}))]. \quad (A6)$$

Term A defined in Equation (A6) can be expanded as follows:

$$A = i \cdot \log R_1 \cdot \cos\gamma \cdot (x - x_{a2}) - \cos\gamma \cdot (x - x_{a2}) \cdot \arctan\frac{b_1}{a_1} - \log R_1 \cdot \cos\gamma \cdot (y - y_{a2}) - i \cdot \cos\gamma \cdot (y - y_{a2}) \cdot \arctan\frac{b_1}{a_1} + \log R_1 \cdot \sin\gamma \cdot (x - x_{a2}) + i \cdot \sin\gamma \cdot (x - x_{a2}) \cdot \arctan\frac{b_1}{a_1} + i \cdot \log R_1 \cdot \sin\gamma \cdot (y - y_{a2}) - \sin\gamma \cdot (y - y_{a2}) \cdot \arctan\frac{b_1}{a_1} \quad (A7)$$

The real and imaginary parts of Equation (A7) can be separated as follows:

$$\Re[A] = \log R_1 \cdot [\sin\gamma \cdot (x - x_{a2}) - \cos\gamma \cdot (y - y_{a2})] - \arctan\frac{b_1}{a_1}[\sin\gamma \cdot (y - y_{a2}) + \cos\gamma \cdot (x - x_{a2})], \quad (A8)$$

$$\Im[A] = \log R_1 \cdot [\cos\gamma \cdot (x - x_{a2}) - \sin\gamma \cdot (y - y_{a2})] - \arctan\frac{b_1}{a_1}[\cos\gamma \cdot (y - y_{a2}) - \sin\gamma \cdot (x - x_{a2})], \quad (A9)$$

where,

$$\begin{aligned}
z &= x + i \cdot y \\
z_{a2} &= x_{a2} + i \cdot y_{a2} \\
a_1 &= -\sin\gamma \cdot (y - y_{a2}) - \cos\gamma \cdot (x - x_{a2}) \\
b_1 &= +\sin\gamma \cdot (x - x_{a2}) - \cos\gamma \cdot (y - y_{a2}) \\
R_1 &= \sqrt{a_1^2 + b_1^2}
\end{aligned} \quad (A10)$$

Other termss in Equation (A5) related to vertices z_{a1} (B), z_{b1} (C), and z_{b2} (D) can be formulated similar to Equation (A6) and Equation (A7) to separate the real and imaginary terms as follows.

$$\Re[B] = \log R_2 \cdot [\sin\gamma \cdot (x - x_{a1}) - \cos\gamma \cdot (y - y_{a1})] - \arctan\frac{b_2}{a_2}[\sin\gamma \cdot (y - y_{a1}) + \cos\gamma \cdot (x - x_{a1})] \quad (A11)$$

$$\Im[B] = \log R_2 \cdot [\cos\gamma \cdot (x - x_{a1}) - \sin\gamma \cdot (y - y_{a1})] - \arctan\frac{b_2}{a_2}[\cos\gamma \cdot (y - y_{a1}) - \sin\gamma \cdot (x - x_{a1})] \quad (A12)$$

$$\Re[C] = \log R_3 \cdot [\sin\gamma \cdot (x - x_{b1}) - \cos\gamma \cdot (y - y_{b1})] - \arctan\frac{b_3}{a_3}[\sin\gamma \cdot (y - y_{b1}) + \cos\gamma \cdot (x - x_{b1})] \quad (A13)$$

$$\Im[C] = \log R_3 \cdot [\cos\gamma \cdot (x - x_{b1}) - \sin\gamma \cdot (y - y_{b1})] - \arctan\frac{b_3}{a_3}[\cos\gamma \cdot (y - y_{b1}) - \sin\gamma \cdot (x - x_{b1})] \quad (A14)$$

$$\Re[D] = \log R_4 \cdot [\sin\gamma \cdot (x - x_{b2}) - \cos\gamma \cdot (y - y_{b2})] - \arctan\frac{b_4}{a_4}[\sin\gamma \cdot (y - y_{b2}) + \cos\gamma \cdot (x - x_{b2})] \quad (A15)$$

$$\Im[D] = \log R_4 \cdot [\cos\gamma \cdot (x - x_{b2}) - \sin\gamma \cdot (y - y_{b2})] - \arctan\frac{b_4}{a_4}[\cos\gamma \cdot (y - y_{b2}) - \sin\gamma \cdot (x - x_{b2})] \quad (A16)$$

Equation (A5) with terms A, B, C, and D, as defined in Equations (A8)–(A16), results in a continuous potential function plots (isobars, Figure A4) for the reservoir defined in Table A1.

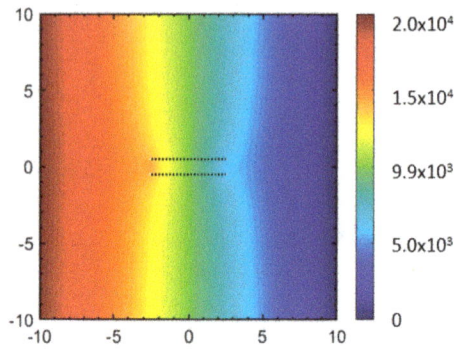

Figure A4. Pressure profile for fracture superposed with far field flow by separating the real and imaginary terms individually. The branch cut seen in Figure A1 has been removed by the applied procedure. Length scale in meters and pressure (rainbow colors) in Pa.

Table A1. Properties for pressure field generation.

Physical Quantity	Symbol	Value	Units
Depth	h	1	m
Porosity	n	20	%
Permeability	k	9.87×10^{-16}	m^2
Viscosity	μ	0.01	Pa·s
Far-field velocity	u_x	9.5×10^{-9}	$m \cdot s^{-1}$
Angle of far-field flow	α	0	°
Fracture center	z_c	0	m
Fracture length	L	5	m
Fracture width	W	1	m
Angle	γ	0	°
Angle between the corner points	β	90	°
Strength of fracture	ν	9.5×10^{-9}	$m^4 \cdot s^{-1}$

The velocity component obtained by differentiating the expression in Equation (A4) is used to generate the streamlines and TOFC for the pore models in this study. Equations (A10)–(A15), detailed later, obtained by splitting Equation (A4), is used to calculate the pressure plots along with Equation (A3).

Appendix B. Dipole Strength and Relationship Radius with Far-Field Flow Rate

Appendix B.1. Velocity Potential

Consider a local dynamic system occupying a certain bounded domain, $\Re$, represented by a singularity doublet (or dipole) (Figure A5) of a certain strength $D_m(t)$ that may or may not vary with time, t, and has dimension [m^3/s].

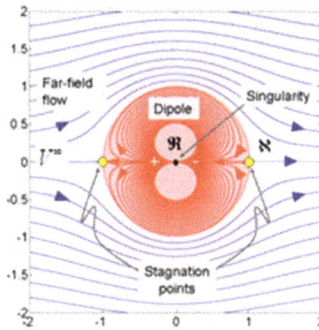

Figure A5. Flow paths tracked for far-field fluid (blue particles) and injection fluid (red particles). The doublet singularity has non-dimensional strength $D_m = 1$ and the far-field flow-rate $V^\infty = 1$. Upstream and downstream stagnation points are marked (yellow dots). Local domain radius is controlled by the relative strength of the dipole and far-field flow rate in the external domain [61].

The velocity potential for a single singularity doublet of strength, D_m, with dimension $[m^3/s]$, and in antipolar alignment with a superposed, uniform far-field flow of velocity, V^∞, with dimension $[m/s]$, is [61]:

$$V(z) = V^\infty e^{-\alpha i} + \left[\frac{-D_m e^{+\theta i}}{(z - z_d)^2}\right] \, [m/s]. \tag{A17}$$

The angle α gives the counter-clockwise inclination of the far-field flow V^∞ with respect to the x-axis; angle θ is the clockwise inclination of the polarity axis of the doublet, also with respect to the x-axis. In case $\alpha = \theta = 0$, and the doublet is located in the origin, as in Figure A5, Equation (A17) simplifies to:

$$V(z) = V^\infty + \left(\frac{-D_m}{z^2}\right) \, [m/s]. \tag{A18}$$

Appendix B.2. Dipole Radius and Revolution Time

The singularity doublet ensures a circular cylindrical space is maintained (Figure A5). Outside the boundary of $\mathfrak{R}$ occurs an external flow domain, $\aleph$, itself a dynamic system whose physical manifestation of existence is comprised of a uni-directional far field flow $V^\infty(t)$ with dimension $[m/s]$. The external flow and the local dipole are oriented anti-polar (Figure A5). Due to the interaction between the local and external dynamic system, the radius, $a(t)$, of the circular cylindrical space occupied by the dipole at any time, t, depends on the relative strength of the dipole and the far-field flow:

$$a(t) = \sqrt{\frac{D_m(t)}{V^\infty(t)}} \, [m]. \tag{A19}$$

Equation (A19) becomes non-dimensional when all lengths are normalized by a typical length scale, in our case such that the initial radius $a_i = 1$, irrespective of the dimension of unit time $t_i = 1$ [s].

Any initial unit radius and the subsequent decrease in the radius of domain $\mathfrak{R}$ at three other times in that same time dimension [s] depend solely on the relative strength of the dipole and the far-field flow at each time instant (Equation (A20)) and are shown in Figure A6a–c. We have thus defined the nature of the local dynamic system and its length scale is set by the circular boundary with the external domain space for each moment in time.

If the far-field flow increases while the dipole strength remains unchanged, the radius of the local domain will shrink over time (Figure A6). The local revolution time of the dipole is:

$$t_\omega = \pi a^3(t)/D_m(t) \, [s]. \tag{A20}$$

The smaller the dipole radius, the faster the local clock. Reversely, expansion of the dipole slows down revolution time.

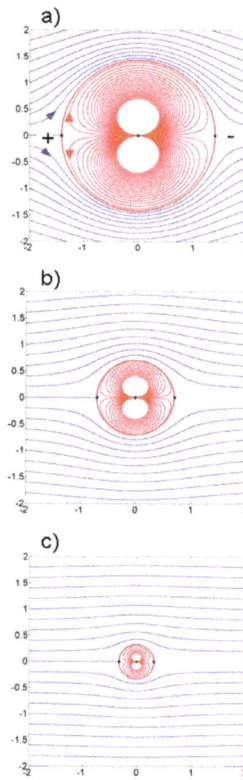

Figure A6. Local revolution time shortens as the radius of the local domain shrinks. The far-field velocity increases as (**a**) $U_x{}^* = 1$; (**b**) $U_x{}^* = 2$; (**c**) $U_x{}^* = 10$ [61].

Appendix B.3. Complex Potential and Velocities in Polar Coordinates

Now we can adopt a complex potential for the singularity doublet and a superposed uniform far-field flow (Figure 1) and drop the time dependency:

$$W(z) = V^\infty \left(z + \frac{a^2}{z} \right) \; [\text{m}^2/\text{s}]. \tag{A21}$$

Converting to polar coordinates, using $W(z) = \phi + i\psi$ and $z = x + iy = r\cos\theta + ir\sin\theta = re^{-i\theta}$, gives:

$$\psi(r,\theta) = V^\infty r \left(1 - \frac{a^2}{r^2} \right) \sin\theta \; [\text{m}^2/\text{s}], \tag{A22}$$

$$\phi(r,\theta) = -V^\infty r \left(1 + \frac{a^2}{r^2} \right) \cos\theta \; [\text{m}^2/\text{s}]. \tag{A23}$$

Recall the radial and tangential velocity components that can be obtained from:

$$v_r = \frac{1}{r} \left(\frac{\partial \psi}{\partial \theta} \right) = \frac{\partial \phi}{\partial r} \; [\text{m/s}], \tag{A24}$$

$$v_\theta = -\frac{\partial \psi}{\partial r} = \frac{1}{r}\left(\frac{\partial \phi}{\partial \theta}\right) \text{ [m/s]}. \tag{A25}$$

Applying the differentiation of Equations (A24) and (A25) to Equations (A22) and (A23) respectively and reintroducing time dependency gives:

$$v_r(t) = V^\infty(t)\left[1 - \left(\frac{a^2(t)}{r}\right)\right]\cos\theta \text{ [m/s]}, \tag{A26}$$

$$v_\theta(t) = V^\infty(t)\left[1 + \left(\frac{a^2(t)}{r}\right)\right]\sin\theta \text{ [m/s]}. \tag{A27}$$

For a certain adopted time unit, the radial and tangential velocity vectors anywhere in the external (r,θ) space of the local dynamic system, made up of a dipole oriented anti-polar to a far-field flow, can be found from Equations (A26) and (A27).

Appendix B.4. Velocities of Internal and External Domains

What is relevant for the present discourse is how velocities in the local (internal) and external domains are constrained and connected. The boundary between $\Re$ and $\aleph$ comprises two points where the flow velocity remains invariant and zero at all times, the so-called stagnation points (Figure A5). The velocities of fluid particles in the local system are, at any time, mostly faster than those of the external dynamic system. There is one streamline that physically limits the velocities in each domain, namely the travel path of particles moving along the periphery of the dipole with velocity $v_\theta(t) = 2V^\infty(t)\sin\theta$. For a given radius and certain dipole strength, the corresponding far-field flow is known and fixed:

$$V^\infty(t) = D_m(t)/a^2(t) \text{ [m/s]}. \tag{A28}$$

Velocities in the external domain are nowhere faster than near the apex points of the dipole rim $2V^\infty$. Particles in the local dynamic system reach higher, infinitely fast velocities at the center of the dipole; elsewhere, in the local domain, particles are fastest when they cross the imaginary vertical line $x = 0$.

Appendix B.5. Volume Conservation in Local Domain

If the local domain, $\Re$, is occupied by an incompressible fluid volume, a change in volume (or radius) occurring when the relative strength of the dipole and far-field flow changes (Equation (A19); Figure A6) is mathematically plausible but physically unrealistic. The change in volume of the dipole area can be mitigated by allowing the singularity doublet to split up into a spaced doublet. A spaced doublet that stays aligned with the far-field flow ensures a Rankine flow space is maintained (Figure A7). The Rankine body is actually made up of a point source and a point sink, the source positioned upstream and the sink downstream. Superposing far-field flow with velocity V_x^∞ and angle α onto the vector field for a point source and point sink with strengths m_1 and m_2, respectively, yields the following vector field:

$$V(z) = V_x e^{-\alpha i} + \frac{m_1}{z - z_1} + \frac{m_2}{z - z_2} \text{ [m/s]}. \tag{A29}$$

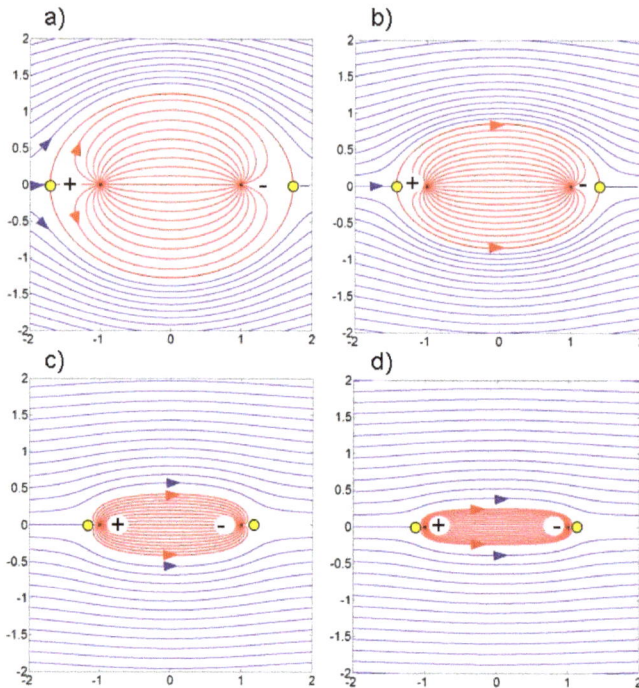

Figure A7. Rankine body outlines effective sweep region for injection fluid (red curves). Well strengths of the doublet are constant and equal for all cases ($m^*_{injector}$ = +1 and $m^*_{producer}$ = −1). Rankine body region is flattened by faster far-field flow-rates: (**a**) U_x^* = 1; (**b**) U_x^* = 2; (**c**) U_x^* = 5, and (**d**) U_x^* = 10 [61].

Before deriving the algorithm to transform a singularity doublet to a Rankine object, an assumption is made such that the new radius of the point doublet, which is calculated from the square root of the ratio of strength of the dipole and local far-field velocity at a certain time, is taken as the half width (h) of the evolved Rankine object. The height (radius) of the point dipole at initial time (t_0), represented by h_0 [m] is a constant and is calculated from:

$$h_0 = \sqrt{\frac{D_m}{V_0^\infty}} \; (\text{constant}) \; [\text{m}], \tag{A30}$$

where D_m [m^3/s] is the strength of the dipole and V_0^∞ [m/s] is the initial local far-field flow velocity. The area occupied by the point dipole at time (t_0) is a constant calculated from the following Equation:

$$A_{ref} = \pi h_0^2 \; (\text{constant}) \; [\text{m}^2]. \tag{A31}$$

We assume that the center of the active object is moving under the influence of the pores and the far-field flow. Based on the particle tracking procedure, Z_a [m] is calculated for the center of the dipole at a certain time t_j from initial time t_0:

$$z_a(t_j) = z_{j-1}(t_{j-1}) + v(z_{j-1}(t_{j-1}))\Delta t \; [\text{m}]. \tag{A32}$$

The local far-field velocity at time t is calculated by superposing the far-field velocity with the velocity potential of singularity doublet (Equation (A18)). The condensed superposed velocity is:

$$V^\infty(z,t) = v_x - iv_y \; [\text{m/s}]. \tag{A33}$$

At a certain time Δt from initial time t_0, the height (radius) of the dipole shrinks as the magnitude of the far-field velocity increases. The new height at time 't' is calculated as follows.

$$h(t) = \sqrt{\frac{D_m}{V^\infty(t)}} \tag{A34}$$

We assume that the $h(t)$ at time t calculated from Equation (A34) becomes the height $h(t)$ of the Rankine oval at a certain time. If we next assume the Rankine oval approximates an ellipse shape, its surface area can be approximated by $\pi h(t) Z_{sp}$, where Z_{sp} is the line segment between the center of the Rankine oval (Z_a) and the stagnation points. The figure of reference Rankine oval is given in Figure A8. Based on the earlier constraint, this area of the ellipse (A) should be equal to the area of the dipole calculated initially:

$$A = \pi h(t) Z_{sp}, A_{ref} = \pi h_0^2 = \pi h(t) Z_{sp}, Z_{sp}(t) = \frac{h_0^2}{h(t)}. \tag{A35}$$

If we assume two-point objects, a source and a sink, each with a strength m_1 [m^2/s] and $-m_1$ [m^2/s] at points z_1 [m] and z_2 [m], respectively, and a center at z_a [m]. If we assume the object is in a complex plane, $b = z_a - z_1$, then $z_2 = 2z_a - z_1$. The stagnation point can also be calculated from the equation given by Weijermars and Van Harmelen [61], in their appendix B3):

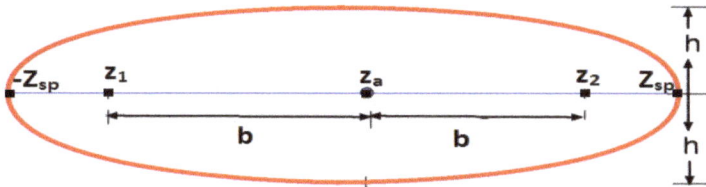

Figure A8. Rankine oval and definition of half width h and source-sink spacing 2b. [86].

$$Z_{sp}(t) = z_a \pm \sqrt{\left(z_1 - z_a - \frac{m_1}{V_x}\right)^2 + \frac{m_1^2}{(V_x)^2}} \; [\text{m}], \tag{A36}$$

$$\frac{h_0^2}{h(t)} = z_a \pm \sqrt{\left(z_1 - z_a - \frac{m_1}{V_x}\right)^2 + \frac{m_1^2}{(V_x)^2}}. \tag{A37}$$

Finally, the half width, $h(t)$, of the Rankine oval is given by the following Equation [87]:

$$h(t) = (z_a - z_1)\cot\left(\frac{V^\infty h}{m_1}\right) \; [\text{m}], z_1 = \frac{h}{\cot\left(\frac{V^\infty h}{m_1}\right)} - z_a. \tag{A38}$$

The cotangent term in Equation (A38) includes strength m (here scaled with 2π included), which is why 2π does not appear in the expression. Expression (A38) can be used to dynamically scale the strengths of the doublet source/sink pair of the Rankine oval, such that its area $\pi h(t)(Z_{sp}(t))$ stays the same as that of the initial dipole, which was scaled by Equation (A30). The calculation uses an iterative process as shown in Figure A9.

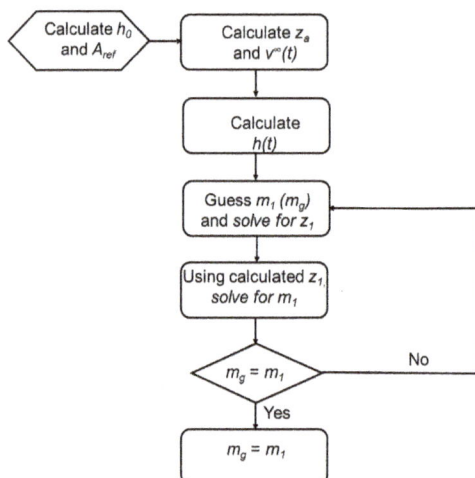

Figure A9. Flow chart for evaluation of m_1 and z_1 by iterative process.

References

1. Loucks, R.G.; Reed, R.M.; Ruppel, S.C.; Jarvie, D.M. Morphology, genesis, and distribution of nanometer-scale pores in siliceous mudstones of the Mississippian Barnett Shale. *J. Sedimen. Res.* **2009**, *79*, 848–861. [CrossRef]
2. Loucks, R.G.; Reed, R.M.; Ruppel, S.C.; Hammes, U. Preliminary classification of matrix pores in mudrocks. *Gulf Coast Assoc. Geol. Soc. Trans.* **2010**, *60*, 435–441.
3. Loucks, R.G.; Reed, R.M.; Ruppel, S.C.; Hammes, U. Spectrum of pore types and networks in mudrocks and a descriptive classification for matrix-related mudrock pores. *AAPG Bulletin* **2012**, *96*, 1071–1098. [CrossRef]
4. Pommer, M.; Milliken, K. Pore types and pore-size distributions across thermal maturity, Eagle Ford Formation, Southern Texas. In Proceedings of the 2014 AAPG Annual Convention and Exhibition, Houston, TX, USA, 6–9 April 2014.
5. Alfi, M.; Chai, Z.; Yan, B.; Stimpson, B.C.; Barrufet, M.A.; Killough, J. Advances in simulation of hydrocarbon production from shale reservoirs. In Proceedings of the 2017 Unconventional Resources Technology Conference, Austin, TX, USA, 24–26 July 2017. [CrossRef]
6. Alfi, M.; Banerjee, D.; Nasrabadi, H. Effect of confinement on the dynamic contact angle of hydrocarbons. *Energy Fuels* **2016**, *30*, 8962–8967. [CrossRef]
7. Alfi, M.; Nasrabadi, H.; Banerjee, D. Experimental investigation of confinement effect on phase behavior of hexane, heptane and octane using lab-on-a-chip technology. *Fluid Phase Equilibr.* **2016**, *423*, 25–33. [CrossRef]
8. Weijermars, R.; Alves, I.N. High-resolution visualization of flow velocities near frac-tips and flow interference of multi-fracked Eagle Ford wells, Brazos County, Texas. *J. Petrol. Sci. Eng.* **2018**, *165*, 946–961. [CrossRef]
9. Khanal, A.; Weijermars, R. Pressure depletion and drained rock volume near hydraulically fractured parent and child wells. *J. Pet. Sci. Eng.* **2019**, *172*, 607–626. [CrossRef]
10. Curtis, M.E.; Cardott, B.J.; Sondergeld, C.H.; Rai, C.S. The development of organic porosity in the woodford shale as a function of thermal maturity. In Proceedings of the 2012 SPE Annual Technical Conference and Exhibition, San Antonio, TX, USA, 8–10 October 2012. [CrossRef]
11. Erdman, N.; Drenzek, N. Integrated preparation and imaging techniques for the microstructural and geochemical characterization of shale by scanning electron microscopy. *AAPG Memoir* **2013**, *102*, 7–14. [CrossRef]
12. Bahadur, J.; Melnichenko, Y.B.; Mastalerz, M.; Furmann, A.; Clarkson, C.R. Hierarchical pore morphology of cretaceous shale: A small-angle neutron scattering and ultra-small-angle neutron scattering study. *Energy Fuel* **2014**, *28*, 6336–6344. [CrossRef]
13. Zuo, L.; Weijermars, R. Rules for flight paths and time of flight for flows in porous media with heterogeneous permeability and porosity. *Geofluids* **2018**, *3*, 1–18. [CrossRef]

14. Teklu, T.W.; Alharthy, N.; Kazemi, H.; Yin, X.; Graves, R.M. Hydrocarbon and non-hydrocarbon gas miscibility with light oil in shale reservoirs. In Proceedings of the 2014 SPE Improved Oil Recovery Symposium, Tulsa, OK, USA, 12–16 April 2014. [CrossRef]

15. Li, J.; Yu, T.; Liang, X.; Zhang, P.; Chen, C.; Zhang, J. Insights on the gas permeability change in porous shale. *Adv. Geo Energy Res.* **2017**, *1*, 63–67. [CrossRef]

16. Jin, L.; Ma, Y.; Jamili, A. Investigating The effect of pore proximity on phase behavior and fluid properties in shale formations. In Proceedings of the 2013 SPE Annual Technical Conference and Exhibition, New Orleans, LA, USA, 30 September–2 October 2013. [CrossRef]

17. Alharthy, N.S.; Nguyen, T.; Kazemi, H.; Teklu, T.; Graves, R. Multiphase compositional modeling in small-scale pores of unconventional shale reservoirs. In Proceedings of the 2013 SPE Annual Technical Conference and Exhibition, New Orleans, LA, USA, 30 September–2 October 2013. [CrossRef]

18. Nojabaei, B.; Johns, R.T.; Chu, L. Effect of capillary pressure on phase behavior in tight rocks and shales. *SPE Reserv. Eval. Eng.* **2013**, *16*, 281–289. [CrossRef]

19. Richardson, Y.; Drobek, M.; Julbe, A.; Blin, J.; Pinta, F. Biomass gasification to produce syngas. *Recent Adv. Thermo Chem. Convers. Biomass* **2015**, *2015*, 213–250.

20. Civan, F.; Rai, C.S.; Sondergeld, C.H. Shale-gas permeability and diffusivity inferred by improved formulation of relevant retention and transport mechanisms. *Transp. Porous Med.* **2011**, *86*, 925. [CrossRef]

21. Fairbrass, M.; Tanguy, D. *The Scale of Things: Mind-Blowing Proportions, Remarkable Ratios, Extraordinary Facts*; Quadrille Publishing: London, UK, 2017.

22. Stimpson, B.C.; Barrufet, M.A. Constructing oil-gas capillary pressure and relative permeability curves from a distribution of pores in shale reservoirs. In Proceedings of the 2017 Unconventional Resources Technology Conference, Austin, TX, USA, 24–26 July 2017. [CrossRef]

23. Stimpson, B.C.; Barrufet, M.A. Effects of confined space on production from tight reservoirs. In Proceedings of the 2016 SPE Annual Technical Conference and Exhibition, Dubai, UAE, 26–28 September 2016. [CrossRef]

24. Khoshghadam, M.; Khanal, A.; Lee, W.J. Numerical study of impact of nano-pores on gas-oil ratio and production mechanisms in liquid-rich shale oil reservoirs. In Proceedings of the 2015 Unconventional Resources Technology Conference, San Antonio, TX, USA, 20–22 July 2015. [CrossRef]

25. Khoshghadam, M.; Khanal, A.; Yu, C.; Rabinejadganji, N.; Lee, W.J. Producing gas-oil ratio behavior of unconventional volatile-oil reservoirs, and its application in production diagnostics and decline curve analysis. In Proceedings of the 2017 Unconventional Resources Technology Conference, Austin, TX, USA, 24–26 July 2017. [CrossRef]

26. Berg, S.; Ott, H.; Klapp, S.A.; Schwing, A.; Neiteler, R.; Brussee, N.; Makurat, A.; Leu, L.; Enzmann, F.; Schwarz, J.-O.; et al. Real-time 3D imaging of Haines jumps in porous media flow. *Proc. Natl. Acad. Sci. USA* **2013**, *110*, 3755–3759. [CrossRef]

27. Berkowitz, B. Characterizing flow and transport in fractured geological media: A review. *Adv. Water Resour.* **2002**, *25*, 861–884. [CrossRef]

28. Neuman, S.P. Trends, prospects and challenges in quantifying flow and transport through fractured rocks. *Hydrogeol. J.* **2005**, *13*, 124–147. [CrossRef]

29. Geiger, S.; Dentz, M.; Neuweiler, I. A novel multi-rate dual-porosity model for improved simulation of fractured and multi-porosity reservoirs. In Proceedings of the 2011 SPE Reservoir Characterisation and Simulation Conference and Exhibition, Abu Dhabi, UAE, 9–11 October 2011. [CrossRef]

30. Flemisch, B.; Berre, I.; Boon, W.; Fumagalli, A.; Schwenck, N.; Scotti, A.; Tatomir, A. Benchmarks for single-phase flow in fractured porous media. *Adv. Water Resour.* **2018**, *111*, 239–258. [CrossRef]

31. Warren, J.E.; Root, P.J. The behavior of naturally fractured reservoirs. *Soc. Petrol. Eng. J.* **1963**, *3*, 1–11. [CrossRef]

32. Lu, H.; Donato, G.D.; Blunt, M.J. General transfer functions for multiphase flow in fractured reservoirs. *SPE J.* **2008**, *13*, 289–297. [CrossRef]

33. Al-Kobaisi, M.; Ozkan, E.; Kazemi, H. A hybrid numerical-analytical model of finite-conductivity vertical fractures intercepted by a horizontal well. *SPE Res. Eval. Eng.* **2006**, *9*, 345–355. [CrossRef]

34. Mason, G.; Fischer, H.; Morrow, N.; Ruth, D. Correlation for the effect of fluid viscosities on counter-current spontaneous imbibition. *J. Petrol. Sci. Eng.* **2010**, *72*, 195–205. [CrossRef]

35. Wtherspoon, P.; Wang, J.; Iwai, K.; Gale, J. Validity of cubic law for fluid flow in a deformable rock fracture. *Tech. Inf. Rep. No. 23* **1980**. [CrossRef]

36. Zimmerman, R.W.; Chen, D.; Cook, N.G. The effect of contact area on the permeability of fractures. *J. Hydrol.* **1992**, *139*, 79–96. [CrossRef]

37. Zimmerman, R.; Kumar, S.; Bodvarsson, G. Lubrication theory analysis of the permeability of rough-walled fractures. *Int. J. Rock Mech. Mining Sci. Geomech. Abstr.* **1991**, *28*, 325–331. [CrossRef]

38. Pyrak-Nolte, L.; Morris, J. Single fractures under normal stress: The relation between fracture specific stiffness and fluid flow. *Int. J. Rock Mech. Mining Sci.* **2000**, *37*, 245–262. [CrossRef]

39. Sisavath, S.; Al-Yaarubi, A.; Pain, C.C.; Zimmerman, R.W. A simple model for deviations from the cubic law for a fracture undergoing dilation or closure. In *Thermo-Hydro-Mechanical Coupling in Fractured Rock*; Kümpel, H.-J., Ed.; Springer: Basel, Switzerland, 2003; pp. 1009–1022. [CrossRef]

40. Zimmerman, R.W. A simple model for coupling between the normal stiffness and the hydraulic transmissivity of a fracture. In Proceedings of the 42nd US Rock Mechanics, 2nd US–Canada Rock Mechanics Symposium, San Francisco, CA, USA, 29 June–2 July 2008.

41. Zimmerman, R.W.; Al-Yaarubi, A.H.; Pain, C.C.; Grattoni, C.A. Nonlinear regimes of fluid flow in rock fractures. *Int. Rock Mech.* **2004**, *41*, 384. [CrossRef]

42. Vilarrasa, V.; Koyama, T.; Neretnieks, I.; Jing, L. Shear-induced flow channels in a single rock fracture and their effect on solute transport. *Transp. Porous Media* **2010**, *87*, 503–523. [CrossRef]

43. Chen, Z.; Qian, J.; Qin, H. Experimental study of the non-darcy flow and solute transport in a channeled single fracture. *J. Hydrodyn. Ser. B* **2011**, *23*, 745–751. [CrossRef]

44. Yasuhara, H.; Polak, A.; Mitani, Y.; Grader, A.; Halleck, P.; Elsworth, D. Evolution of fracture permeability through fluid–rock reaction under hydrothermal conditions. *Earth Plan. Sci. Lett.* **2006**, *244*, 186–200. [CrossRef]

45. Chaudhuri, A.; Rajaram, H.; Viswanathan, H.; Zyvoloski, G.; Stauffer, P. Buoyant convection resulting from dissolution and permeability growth in vertical limestone fractures. *Geophys. Res. Lett.* **2009**, *36*, L03401. [CrossRef]

46. Lake, L.W.; Johns, R.T.; Rossen, W.R.; Pope, G.A. *Fundamentals of Enhanced Oil Recovery*; Society of Petroleum Engineers: Richardson, TX, USA, 2014.

47. Blunt, M.J. *Multiphase Flow in Permeable Media: A Pore-Scale Perspective*; Cambridge University Press: Cambridge, UK, 2017.

48. Melrose, J.C. Thermodynamic aspects of capillarity. *Ind. Eng. Chem.* **1968**, *60*, 53–70. [CrossRef]

49. Melrose, J. Role of capillary forces in determining microscopic displacement efficiency for oil recovery by waterflooding. *J. Can. Petrol. Technol.* **1974**, *13*, 1–13. [CrossRef]

50. Chatzis, I.; Morrow, N.R.; Lim, H.T. Magnitude and detailed structure of residual oil saturation. *Soc. Petrol. Eng. J.* **1983**, *23*, 311–326. [CrossRef]

51. Mandal, A.; Achinta, B. Modeling of flow of oil-in-water emulsions through porous media. *Petrol. Sci.* **2015**, *12*, 273–281. [CrossRef]

52. Thiele, M.R.; Batycky, R.P.; Blunt, M.J. A streamline-based 3d field-scale compositional reservoir simulator. In Proceedings of the 1997 SPE Annual Technical Conference and Exhibition, San Antonio, TX, USA, 5–8 October 1997. [CrossRef]

53. Alpak, F.O.; Riviere, B.; Frank, F. A phase-field method for the direct simulation of two-phase flows in pore-scale media using a non-equilibrium wetting boundary condition. *Comput. Geosci.* **2016**, *20*, 881–908. [CrossRef]

54. Frank, F.; Liu, C.; Alpak, F.O.; Riviere, B. A finite volume/discontinuous Galerkin method for the advective Cahn–Hilliard equation with degenerate mobility on porous domains stemming from micro-CT imaging. *Comput. Geosci.* **2018**, *22*, 543–563. [CrossRef]

55. Frank, F.; Liu, C.; Alpak, F.O.; Berg, S.; Riviere, B. Direct numerical simulation of flow on pore-scale images using the phase-field method. *SPE J.* **2018**, *23*, 1833–1850. [CrossRef]

56. Frank, F.; Liu, C.; Scanziani, A.; Alpak, F.O.; Riviere, B. An energy-based equilibrium contact angle boundary condition on jagged surfaces for phase-field methods. *J. Colloid Interface Sci.* **2018**, *523*, 282–291. [CrossRef] [PubMed]

57. Riewchotisakul, S.; Akkutlu, I.Y. Adsorption enhanced transport of hydrocarbons in organic nanopores. *SPE J.* **2015**, *21*, 1–10. [CrossRef]

58. Blunt, M.J.; Jackson, M.D.; Piri, M.; Valvatne, P.H. Detailed physics, predictive capabilities and macroscopic consequences for pore-network models of multiphase flow. *Adv. Water Resour.* **2002**, *25*, 1069–1089. [CrossRef]

59. Xiong, Q.; Baychev, T.G.; Jivkov, A.P. Review of pore network modelling of porous media: Experimental characterisations, network constructions and applications to reactive transport. *J. Contam. Hydrol.* **2016**, *192*, 101–117. [CrossRef] [PubMed]

60. Weijermars, R.; Van Harmelen, A.; Zuo, L.H. Controlling flood displacement fronts using a parallel analytical streamline simulator. *J. Petrol. Sci. Eng.* **2016**, *139*, 23–42. [CrossRef]

61. Weijermars, R.; van Harmelen, A. Breakdown of doublet re-circulation and direct line drives by far-field flow: Implications for geothermal and hydrocarbon well placement. *Geophys. J. Int.* **2016**, *206*, 19–47. [CrossRef]

62. Weijermars, R.; Van Harmelen, A. Advancement of sweep zones in waterflooding: Conceptual insight and flow visualizations of oil-withdrawal contours and waterflood time-of-flight contours using complex potentials. *J. Petrol. Explor. Prod. Technol.* **2017**, *7*, 785–812. [CrossRef]

63. Weijermars, R.; van Harmelen, A.; Zuo, L.; Nascentes Alves, I.; Yu, W. High resolution visualization of flow interference between frac clusters (Part 1): Model verification and basic cases. In Proceedings of the 2017 SPE/AAPG/SEG Unconventional Resources Technology Conference, Austin, TX, USA, 24–26 July 2017. [CrossRef]

64. Weijermars, R.; van Harmelen, A.; Zuo, L. Flow interference between frac clusters (Part 2): Field example from the midland basin (wolfcamp formation, spraberry trend field) with implications for hydraulic fracture design. In Proceedings of the 2017 SPE/AAPG/SEG Unconventional Resources Technology Conference, Austin, TX, USA, 24–26 July 2017. [CrossRef]

65. Weijermars, R.; Harmelen, A.V.; Zuo, L.; Alves, I.N.; Yu, W. Flow interference between hydraulic fractures. *SPE Reserv. Eval. Eng.* **2018**, *21*, 942–960. [CrossRef]

66. Van Harmelen, A.; Weijermars, R. Complex analytical solutions for flow in hydraulically fractured hydrocarbon reservoirs with and without natural fractures. *Appl. Math. Model.* **2018**, *56*, 137–157. [CrossRef]

67. Khanal, A.; Weijermars, R. Modeling flow and pressure fields in porous media with discrete fractures and smart placement of branch cuts for variant and invariant complex potentials. *Appl. Math. Model.* **2019**. In Review.

68. Khanal, A.; Weijermars, R. Visualization of drained rock volume (DRV) in hydraulically fractured reservoirs with and without natural fractures using complex analysis methods (CAM). *Petrol. Sci.* **2019**. [CrossRef]

69. Parsegov, S.G.; Nandlal, K.; Schechter, D.S.; Weijermars, R. Physics-driven optimization of drained rock volume for multistage fracturing: Field example from the Wolfcamp Formation, Midland Basin. In Proceedings of the 2018 Unconventional Resources Technology Conference, Houston, TX, USA, 23–25 July 2018. [CrossRef]

70. Weijermars, R.; Van Harmelen, A. Shale Reservoir Drainage Visualized for a Wolfcamp Well (Midland Basin, West Texas, USA). *Energies* **2018**, *11*, 1665. [CrossRef]

71. Kahan, W. Branch Cuts for Complex Elementary Functions, or Much Ado about Nothing'S Sign Bit. In *The State of the Art in Numerical Analysis*; Iserles, A., Powell, M.J.D., Eds.; Clarendon Press: Oxford, UK, 1987.

72. Holzbecher, L. Streamline visualization of potential flow with branch cuts, with applications to groundwater. *J. Flow Vis. Image Process.* **2018**, *25*, 119–144. [CrossRef]

73. Wang, J.; Weijermars, R. New interface for assessing potential instability at Critical Wellbore Pressure with deviatoric stress distributions and various failure criteria. *Rock Mech. Rock Eng.* **2019**. In Review.

74. Weijermars, R.; Zhang, X.; Schultz-Ela, D. Geomechanics of fracture caging in wellbores. *Geophys. J. Int.* **2013**, *193*, 1119–1132. [CrossRef]

75. Wang, S.; Javadpour, F.; Feng, Q. Confinement correction to mercury intrusion capillary pressure of shale nanopores. *Sci. Rep.* **2016**, *6*, 20160. [CrossRef]

76. Jansen, J.D. *Nodal Analysis of Oil and Gas Production Systems*; Society of Petroleum Engineers: Richardson, TX, USA, 2017.

77. Weijermars, R.; Schmeling, H. Scaling of Newtonian and non-Newtonian fluid dynamics without inertia for quantitative modelling of rock flow due to gravity (including the concept of rheological similarity). *Phys. Earth Planet. Inter.* **1986**, *43*, 316–330. [CrossRef]

78. Nandlal, K.; Weijermars, R. Drained rock volume around hydraulic fractures in porous media: planar fractures versus fractal networks. *Petrol. Sci.* **2019**. In Press.

79. Eberli, G.P.; Baechle, G.T.; Anselmetti, F.S.; Incze, M.L. Factors controlling elastic properties in carbonate sediments and rocks. *The Lead. Edge* **2003**, *22*, 654–660. [CrossRef]
80. Goudarzi, A.; Delshad, M.; Sepehrnoori, K. A chemical EOR benchmark study of different reservoir simulators. *Comput. Geosci.* **2016**, *94*, 96–109. [CrossRef]
81. Juanes, R.; Lie, K. A front-tracking method for efficient simulation of miscible gas injection processes. In Proceedings of the SPE Reservoir Simulation Symposium, The Woodlands, TX, USA, 31 January–2 February 2005. [CrossRef]
82. Juanes, R.; Blunt, M.J. Impact of viscous fingering on the prediction of optimum WAG ratio. In Proceedings of the SPE/DOE Symposium on Improved Oil Recovery, Tulsa, OK, USA, 22–26 April 2006. [CrossRef]
83. LaForce, T.C.; Jessen, K. Analytical and numerical investigation of multicomponent multiphase WAG displacements. In Proceedings of the SPE Annual Technical Conference and Exhibition, Anaheim, CA, USA, 11–14 November 2007. [CrossRef]
84. Seto, C.J.; Orr, F.M. Analytical solutions for multicomponent, two-phase flow in porous media with double contact discontinuities. *Transp. Porous Med.* **2009**, *78*, 161–183. [CrossRef]
85. Pires, A.P.; Bedrikovetsky, P.G.; Shapiro, A.A. A splitting technique for analytical modelling of two-phase multicomponent flow in porous media. *J. Pet. Sci. Eng.* **2006**, *51*, 54–67. [CrossRef]
86. Rankine Oval. 2005. Available online: http://www-mdp.eng.cam.ac.uk/web/library/enginfo/aerothermal_dvd_only/aero/fprops/poten/node36.html (accessed on 26 January 2019).
87. Kundu, P.K.; Cohen, I.M. *Fluid Mechanics*; Academic Press: San Diego, CA, USA, 2002.

energies

MDPI

Article

Investigation of the Effect of Natural Fractures on Multiple Shale-Gas Well Performance Using Non-Intrusive EDFM Technology

Wei Yu [1,2,*], Xiaohu Hu [1,3], Malin Liu [4,*] and Weihong Wang [1,3]

[1] State Key Laboratory of Shale Oil and Gas Enrichment Mechanisms and Effective Development, Beijing 100083, China; huxhu.syky@sinopec.com (X.H.); wangwh.syky@sinopec.com (W.W.)
[2] Hildebrand Department of Petroleum and Geosystems Engineering, The University of Texas at Austin, Austin, TX 78712, USA
[3] Key Laboratory of Shale Oil/Gas Exploration & Production, SINOPEC, Beijing 100083, China
[4] Institute of Nuclear and New Energy Technology, Tsinghua University, Beijing 100084, China
* Correspondence: yuwei127@gmail.com (W.Y.); liumalin@mail.tsinghua.edu.cn (M.L.); Tel.: +1-512-574-0080 (W.Y.); +86-10-8979-6092 (M.L.)

Received: 4 January 2019; Accepted: 6 March 2019; Published: 10 March 2019

Abstract: The influence of complex natural fractures on multiple shale-gas well performance with varying well spacing is poorly understood. It is difficult to apply the traditional local grid refinement with structured or unstructured gridding techniques to accurately and efficiently handle complex natural fractures. In this study, we introduced a powerful non-intrusive embedded discrete fracture model (EDFM) technology to overcome the limitations of exiting methods. Through this unique technology, complex fracture configurations can be easily and explicitly embedded into structured matrix blocks. We set up a field-scale two-phase reservoir model to history match field production data and predict long-term recovery from Marcellus. The effective fracture properties were determined thorough history matching. In addition, we extended the single-well model to include two horizontal wells with and without including natural fractures. The effects of different numbers of natural fractures on two-well performance with varying well spacing of 200 m, 300 m, and 400 m were examined. The simulation results illustrate that gas productivity almost linearly increases with the number of two-set natural fractures. Furthermore, the difference of well performance between different well spacing increases with an increase in natural fracture density. A larger well spacing is preferred for economically developing the shale-gas reservoirs with a larger natural fracture density. The findings of this study provide key insights into understanding the effect of natural fractures on well performance and well spacing optimization.

Keywords: well spacing; shale gas; natural fractures; embedded discrete fracture model; well interference

1. Introduction

Typical shale gas reservoirs consist of free gas and adsorbed gas. Economic shale gas production has been enabled by the advanced technologies of multiple horizontal wells and multi-stage hydraulic fracturing. The U.S. Energy Information Administration (EIA) estimates that dry shale gas production (0.48 trillion cubic meters) accounts for around 62% of the total U.S. dry natural gas production in 2017 [1]. The Marcellus shale formation is the most productive gas field in the United States. It has been reported that natural fractures are commonly observed in many shale gas formations based on outcrops, cores and image logs [2]. Engelder et al. [3] observed that there are two sets of natural fractures or two regional joint sets (J_1 and J_2 sets) in the Marcellus shale formation, which significantly affect the successful hydraulic fracturing treatment. A better understanding of natural fracture effects on well performance plays an important role in well spacing optimization of shale gas reservoirs.

The reactivation of pre-existing natural fractures during the hydraulic fracturing process is an important factor to create complex fracture networks, which have been observed and predicted by many examples of microseismic event patterns [4,5] and complex fracture propagation models [6–11]. Complex fracture networks will create a large productive fracture surface area, which is necessary to maximize shale-gas production. Based on the history matching results with an actual shale-gas well, Yu et al. [12] compared the long-term well productivity between simple and complex fracture geometries and found that complex fracture networks can produce 36.4% more gas recovery after 30 years than the simple fractures. Yu et al. [13] built a synthetic single shale-gas well model including 11 planar hydraulic fractures and 200 natural fractures. The authors found that the well performance of 200 two-set natural fractures is much better than that of 200 one-set natural fractures due to the formation of a much more complex connected fracture network. An increase of gas recovery of about 23.2% was achieved by the 200 two-set natural fractures when compared to the base model without natural fractures. Although there are many existing reservoir simulation studies for well spacing optimization in shale gas reservoirs [14–19], the influence of natural fractures on multiple shale-gas well performance with varying well spacing has not been well examined and understood.

Although there are many analytical or semi-analytical models to simulate shale-gas reservoirs [20–23], numerical reservoir simulation is needed to accurately model multiple shale-gas well production due to complexity of natural and hydraulic fracture configurations and complex two-phase flow physics [24–27]. The traditional local grid refinement (LGR) method with structured grids is difficult to model complex fractures explicitly [28,29]. Although the LGR method with unstructured grids has the capability to handle complex fractures, the computational efficiency is a big issue. In addition, advanced parallel computing power is generally needed. Furthermore, when performing sensitivity studies and history matching with varying fracture geometries such as fracture number, length, and height, re-gridding of matrix blocks containing fractures is required. In our previous work, we developed an innovative non-intrusive embedded discrete fracture model (EDFM) technology in conjunction with any third-party reservoir simulators with structured grids to accurately and efficiently handle any complex hydraulic and natural fractures [30,31], which provides a unique solution to overcome the above limitations of existing methods. The non-intrusive feature means that we do not need to get access to the source code of commercial reservoir simulators and just modifying their input files.

In this study, we introduced the non-intrusive EDFM technology in conjunction with a commercial reservoir simulator of CMG-GEM [32] to simulate shale gas production considering complex natural fractures and two-phase flow (gas and water). Based on an actual shale-gas well with available gas and water flow rates from the Marcellus shale formation, we build a field-scale reservoir model to perform history matching with gas and water flow rates under the flowing bottomhole pressure (BHP) constraint. After history matching, we predict a 30-year production forecasting. Subsequently, we extend the history-matched reservoir model with effective fracture properties to include two horizontal wells with three different well spacing values of 200, 300 and 400 m. In addition, the impacts of different numbers of natural fractures such as 100, 1000, 5000, and 100,000 on multiple shale-gas well performance are investigated. The effect of natural fracture density on well spacing optimization and pressure distribution is discussed.

2. Methodology

The non-intrusive EDFM technology in conjunction with the commercial reservoir simulator was originally developed by Xu et al. [30,31]. Here, we only briefly introduced this powerful technology. Based on the non-intrusive EDFM technology, both 3D complex hydraulic and natural fractures in the physical domain can be directly and explicitly embedded into the simple structured matrix grids, as demonstrated in Figure 1a. Based on the intersections between complex fractures and matrix grids, a number of extra fracture grids will be generated, as shown in Figure 1b [30]. In Figure 1, two 3D fractures are explicitly embedded into three matrix grids in the physical domain. Correspondingly, an additional four fracture grids are created in the computational domain. Multi-phase flow such as gas and water between matrix and fracture grids can be conveniently simulated through calculating transmissibility between these non-neighboring connections (NNCs) grids, as shown in different colors of arrows of Figure 1b, which is a general feature for commercial reservoir simulators to handle faults before. A non-intrusive EDFM preprocessor has been developed to automatically check the complex intersections of fractures and matrix grids and calculate the transmissibility and other physical properties of the additional fracture grids such as porosity and depth, which are needed to input for the commercial reservoir simulator.

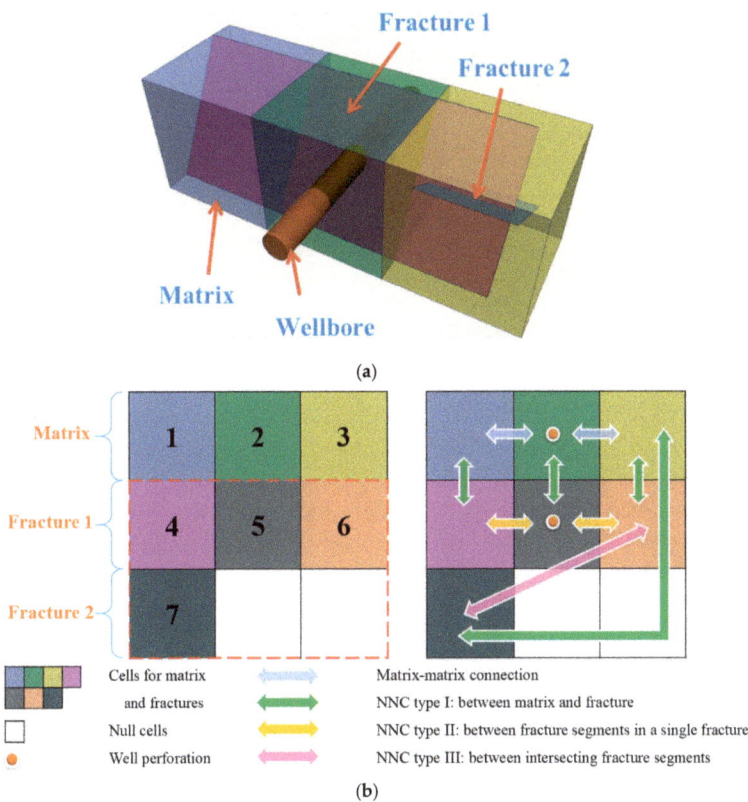

(a)

(b)

Figure 1. The non-intrusive EDFM technology in conjunction with commercial reservoir simulator to efficiently model 3D complex fractures: (**a**) Physical domain; (**b**) Computational domain [30].

There are three types of NNCs including NNC type I, which is the connection between matrix and fracture, NNC type II, which is the connection between fracture segments in a single fracture, and NNC type III, which is the connection between intersecting fracture segments [30]. The flow rate (q_l, m^3/s) of phase l between NNC grids is calculated by the following equation:

$$q_l = \lambda_l T_{NNC} \Delta p_l \tag{1}$$

where λ_l represents the relative mobility of phase l (cp^{-1}), Δp_l represents the pressure difference between NNC grids (Pa), T_{NNC} represents the transmissibility factor of NNC grids (mD-m), which can be calculated by:

$$T_{NNC} = \frac{k_{NNC}A_{NNC}}{d_{NNC}} \tag{2}$$

where k_{NNC} represents the matrix grid permeability for NNC type I and average fracture permeability for NNC types II and III (mD); A_{NNC} represents the contact area between NNC grids (m^2); d_{NNC} represents the connection distance between different NNC grids (m). It should be mentioned that the non-intrusive EDFM technology only deals with the transmissibility factor and does not involve relative phase mobility calculation. Hence, it can be applied in both black oil simulation and compositional simulation.

For the connection between fractures and wellbore, the following effective wellbore index (*WI*, mD-m) will be calculated [30]:

$$WI = \frac{2\pi k_f w_f}{\ln(0.14\sqrt{L^2 + W^2}/r_w)} \tag{3}$$

where k_f is the fracture permeability (mD), w_f is fracture width (m), L is fracture segment length (m), W is fracture segment height (m), and r_w is the wellbore radius (m).

The validation of this methodology against the traditional LGR method can be found in our previous work [30,31]. The efficiency of the EDFM method is much higher than the LGR method, especially when dealing with multiple wells with a large number of fractures [33]. It has been widely applied to model well interference due to complex fracture hits [34,35], automatic history matching for shale reservoirs [36–38], gas injection for enhanced unconventional oil recovery [39–41], and naturally fractured reservoir simulation [42].

3. Field Case Study of Single Shale-Gas Well

3.1. History Matching

An actual Marcellus shale-gas well with available 486-day gas and water production data was selected to perform history matching. The well has 11 perforations stages with a cluster spacing of 15.24 m and was hydraulically fractured using 16,770 m^3 water and 2,895,956 kg proppant. The lateral length of horizontal wellbore is about 1382 m. There are 10 stages with nine clusters per stage and one stage with seven clusters, so 97 effective hydraulic fractures were assumed in the following simulation study.

We set up a field-scale shale-gas reservoir model with two-phase flow (gas and water) using a compositional numerical reservoir simulator [32], as shown in Figure 2. The gas type is methane. The dimension of the reservoir model without embedding hydraulic fractures is 1585 m × 622 m × 27.6 m and it is assumed that the reservoir thickness was fully penetrated by hydraulic fractures. Hence, the effective hydraulic fracture height is 27.6 m. The matrix grid size is 6 m × 12 m × 27.6 m in x, y, and z directions.

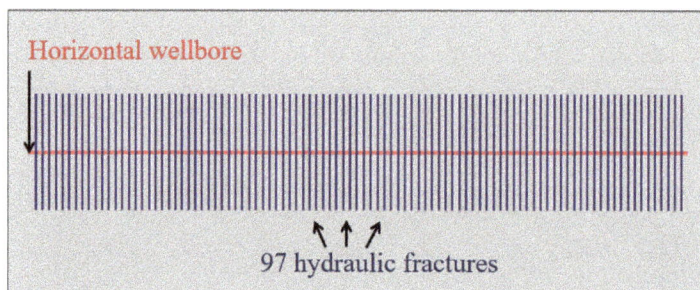

Figure 2. The field-scale shale-gas reservoir model with two-phase flow including one horizontal well and 97 hydraulic fractures.

The number of matrix grids is 13,260. Hydraulic fractures were modeled using the non-intrusive EDFM method. After embedding 97 hydraulic fractures with fracture half-length of 127 m into matrix grids, an extra number of fracture grids of 2040 was generated along the x direction. The fracture grid size remains the same as the matrix grid size. The model dimension with hydraulic fractures becomes 1828.8 m × 622 m × 27.6 m. It is assumed that both matrix and fracture grids have the same gas-water relative permeability curve, which is from the work by Yu et al. [38]. In addition, the experimental measurements of gas desorption are considered in the simulation model [38]. The other reservoir and fracture parameters are listed in Table 1.

Table 1. Reservoir and fracture properties used in the field-scale shale-gas model.

Properties	Value	Unit
Initial reservoir pressure	27.44	MPa
Reservoir temperature	54.5	°C
Reservoir depth	1889	m
Residual water saturation	20%	-
Porosity	12.44%	-
Total compressibility	3×10^{-7}	kPa^{-1}
Reference pressure for compressibility	27.44	MPa
Fracture aperture	0.003	m

During the history-matching process, we applied the measured flowing BHP for the reservoir simulation constraint, as shown in Figure 3a. In order to achieve good match results with measured gas and water flow rates, we mainly tune matrix permeability, fracture half-length, fracture conductivity, and fracture water saturation because these parameters are very sensitive for tuning and performing history matching.

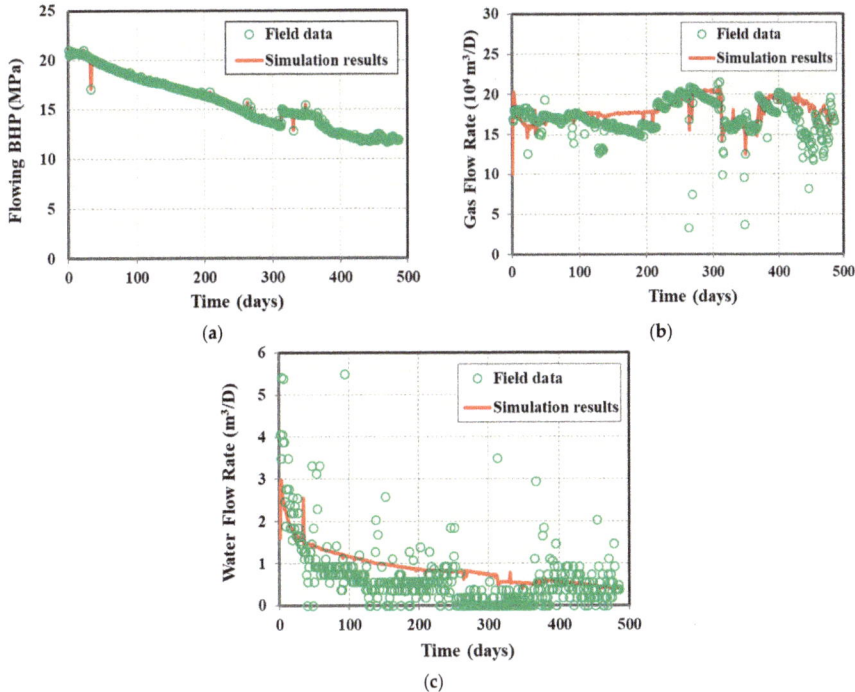

Figure 3. Comparison of flowing BHP and gas and water flow rates between filed data and simulation results: (**a**) Flowing BHP; (**b**) Gas flow rate; (**c**) Water flow rate.

It should be mentioned that the fracture water saturation is higher than the matrix water saturation because water is more difficult to transport into the matrix than into the fracture due to lower matrix permeability. Figure 3b,c present the comparison of gas and water flow rates between filed data and simulation results. The relative error of history-matching results for gas flow rate is about 10%. It can be illustrated that reasonable agreements were obtained. The final history-matching tuning parameters were determined as matrix permeability of 0.000525 mD, fracture half-length of 127 m, fracture conductivity of 0.73 mD-m, and fracture water saturation of 37.4%. It should be noted that fracture half-length has an important impact on multiple shale-gas well placement. In general, longer fracture half-length will result in larger well spacing.

3.2. Production Forecasting

After the history-matching process, we performed long-term production forecasting for 30 years. Since there are no available actual well control information for prediction, a constant flowing BHP of 3.45 MPa after the history-matching period was assumed for the simulation constraint, as illustrated in Figure 4a. Figure 4b,c show the prediction of gas flow rate and cumulative gas production incorporating a short-term field data, respectively. As shown, the estimated ultimate recovery (EUR) after 30 years for this shale-gas well is about 413×10^6 m^3. Figure 4d,e show the prediction of water flow rate and cumulative water production incorporating the short-term field data, respectively. It can be seen that the water flow rate becomes negligible after around three years. The cumulative water production after 30 years for this shale-gas well is about 747 m^3. Figure 5 compares the pressure distribution after the history-matching period and 30 years of production, clearly illustrating different drainage area and production interference intensity between hydraulic fractures. It can be observed that a stronger production interference between hydraulic fractures occures after 30 years of production when compared to that after the history-matching period. In addition, a larger effective drainage area can be generated after 30 years of production.

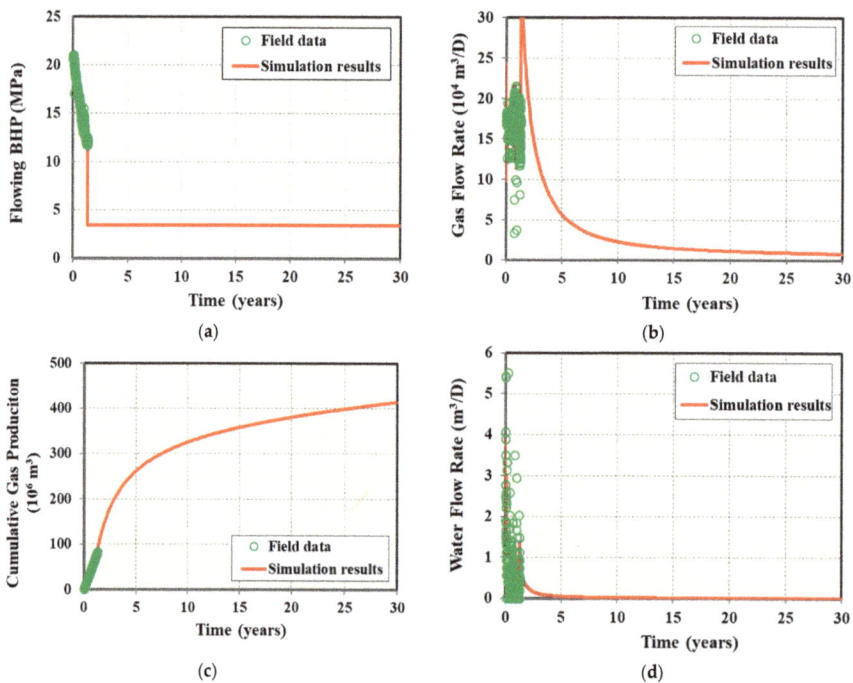

Figure 4. *Cont.*

(e)

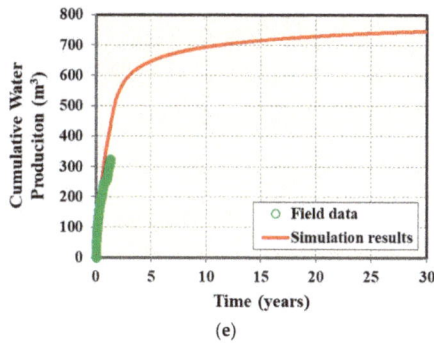

Figure 4. Production forecasting for 30 years incorporating the short-term filed production data: (**a**) Flowing BHP; (**b**) Gas flow rate; (**c**) Cumulative gas production; (**d**) Water flow rate; (**e**) Cumulative water production.

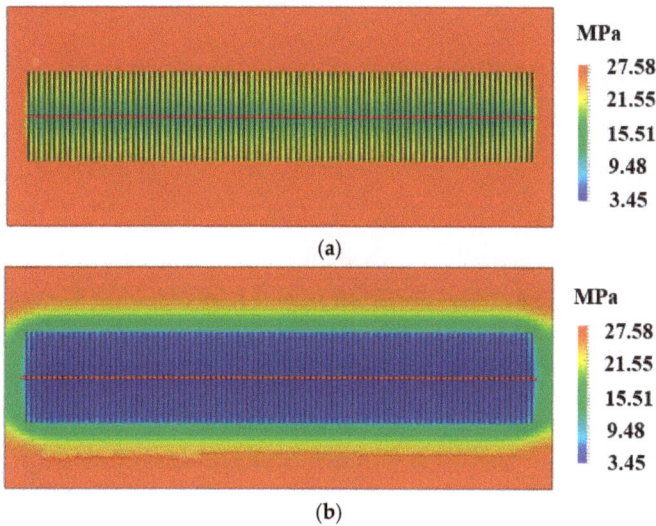

(a)

(b)

Figure 5. Comparison of pressure distribution at different times: (**a**) After the history-matching period; (**b**) After 30 years of production.

4. Basic Reservoir Model of Two Shale-Gas Wells

4.1. Varying Well Spacing without Natural Fractures

Based on the history-matching results from the field-scale shale-gas model, we extended the reservoir model to include two horizontal shale-gas wells with varying well spacing and each well has 97 hydraulic fractures. The effect of natural fractures was not considered in this basic model. The model dimension without embedding hydraulic fractures is 1585 m × 975 m × 27.6 m and fractures fully penetrate the reservoir thickness. The matrix grid size is 6 m × 12 m × 27.6 m in x, y, and z directions. The number of matrix grids is 20,800. Three different well spacing values such as 200 m, 300 m, and 400 m were investigated, as depicted in Figure 6. The placement of hydraulic fractures from two horizontal wells has a staggered pattern. As shown in Figure 6a, there is about 21% overlap of fracture half-length. The distance between neighboring fracture tips of two wells is 47.5 m and 148 m for well spacing of 300 m and 400 m, respectively, as shown in Figure 6b,c. After embedding 194 hydraulic fractures into matrix grids, an extra number of fracture grids of 4320 were generated along the x direction. The grid size of fracture grids is the same as the matrix grid size. The final model dimension becomes 1914 m × 975 m × 27.6 m. All fracture and reservoir properties remain the same as those of field case study with good history matching results. A constant flowing BHP of 3.45 MPa was applied in the 30-year simulation constraint.

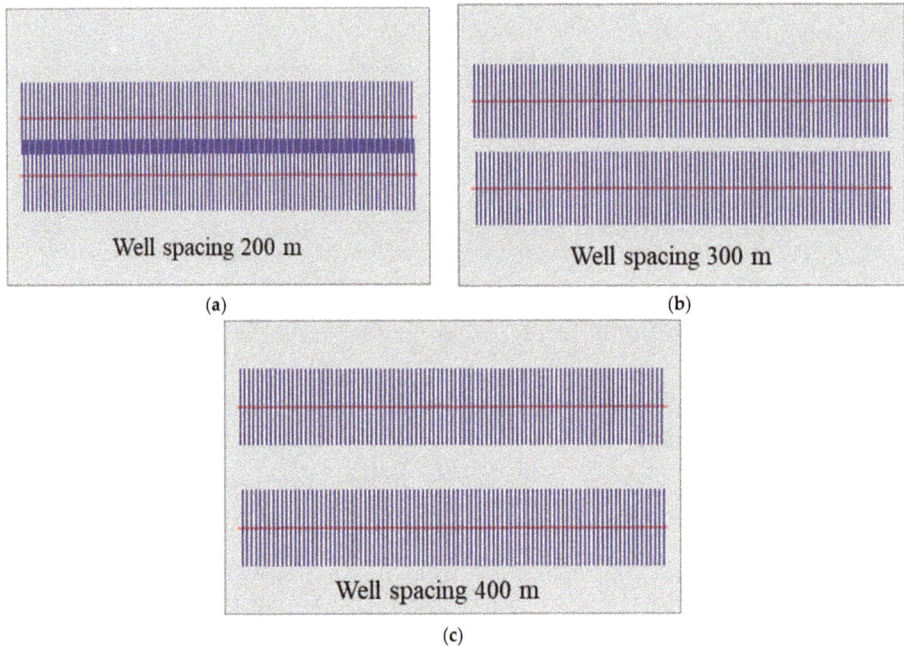

(a)

(b)

(c)

Figure 6. The basic reservoir simulation model including two horizontal wells and 97 hydraulic fractures for each well: (**a**) Well spacing of 200 m; (**b**) Well spacing of 300 m; (**c**) Well spacing of 400 m.

4.2. Varying Well Spacing with Natural Fractures

Next, the effect of natural fractures with different numbers was considered in the basic reservoir model with varying well spacing. Four different numbers of natural fractures such as 100, 1000, 5000, and 10,000 were investigated in this study. It should be mentioned that hydraulic fractures are non-planar if considering the interaction between pre-existing natural fracture with hydraulic fracture using fracture propagation models [6–8]. However, hydraulic fractures are assumed to be planar in this study. We did not apply the fracture propagation model to predict complex fracture network. Figure 7 displays the natural fracture distribution of four different cases using an example of the well spacing of 300 m. We applied a statistical method to generate the locations of natural fractures, which are normally distributed in the reservoir model. For each case, there are two sets of natural fractures and each set has the same number of natural fractures. One set of natural fractures has an orientation of 45 degrees with respect to the x axis. Another set has an orientation of 135 degrees along the x axis. The natural fractures have a range of length from 30 m to 91 m. The natural fracture conductivity for each case is assumed to be 0.03 mD-m with natural fracture aperture of 0.003 m. The natural fractures fully penetrate the reservoir thickness. It should be mentioned that the impacts of natural fracture conductivity, length, and orientation are not examined in this study. After embedding natural and hydraulic fractures together into matrix grids with well spacing of 300 m, an extra number of fracture grids of 5520, 15,840, 62,000, and 119,440 were generated for the number of natural fractures of 100, 1000, 5000, 10,000, respectively, along the x direction. As can be seen that, the extra number of fractures grids is larger than the number of matrix grids of 20,800 for the 5000 and 10,000 natural fractures.

With the increasing number of natural fractures, there are more interactions between natural and hydraulic fractures, resulting in more well interference due to complex fracture connections.

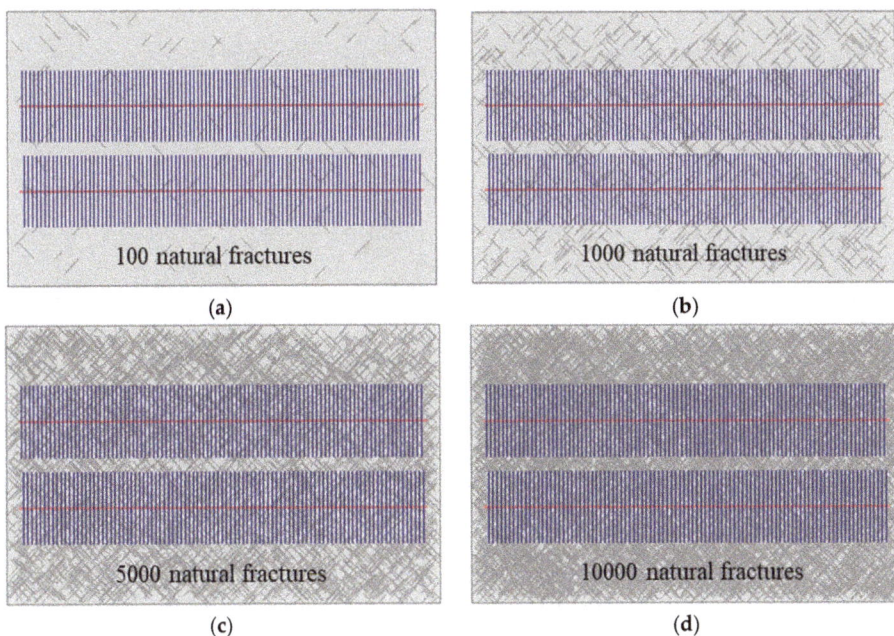

Figure 7. The basic reservoir simulation model including two horizontal wells with well spacing of 300 m and four different numbers of natural fractures: (**a**) 100 natural fractures; (**b**) 1000 natural fractures; (**c**) 5000 natural fractures; (**d**) 10,000 natural fractures.

5. Results and Discussion

5.1. Effect of Natural Fractures on Well Performance

The effects of different numbers of natural fractures on cumulative gas production are compared in Figures 8–10 for well spacings of 200, 300 and 400 m, respectively. It can be clearly observed that a larger number of natural fractures contributes to more production. The incremental EUR after 30 years compared to the case with well spacing of 200 m without natural fractures is about 1.5%, 13.8%, 69%, and 130.7% for the fracture number of 100, 1000, 5000, and 10,000, respectively. When compared to the case with well spacing of 300 m, the incremental EUR after 30 years of production is about 1.4%, 13.9%, 66.7%, and 124.3% for the fracture number of 100, 1000, 5000, and 10,000, respectively. When compared to the case with well spacing of 400 m, the incremental EUR after 30 years of production is about 1.6%, 15.3%, 72.6%, and 131% for the fracture number of 100, 1000, 5000, and 10,000, respectively. For 100 natural fractures, there is only a smaller number of natural fractures connecting with hydraulic fractures, leading to a slight increase of cumulative gas production. However, for 10,000 natural fractures, there is a much more complex connected fracture network between natural and hydraulic fractures, resulting in a big increase of contact area between fracture and matrix and a large increase of cumulative gas production. The CPU time for the model with well spacing of 300 m is about 5, 7 min 17 and 30 min, corresponding to the number of natural fractures of 100, 1000, 5000, and 10,000, respectively. Hence, the natural fractures play an important role in long-term multiple shale-gas well performance, which can be easily and efficiently handled by the non-intrusive EDFM technology.

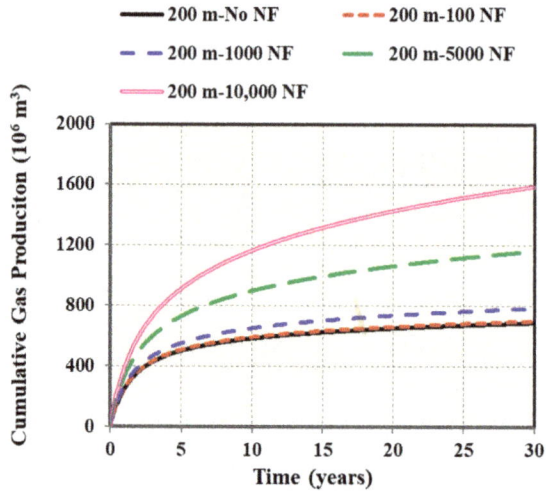

Figure 8. Effect of different numbers of natural fractures on cumulative gas production for the reservoir model with well spacing of 200 m (NF in the legend represents natural fractures).

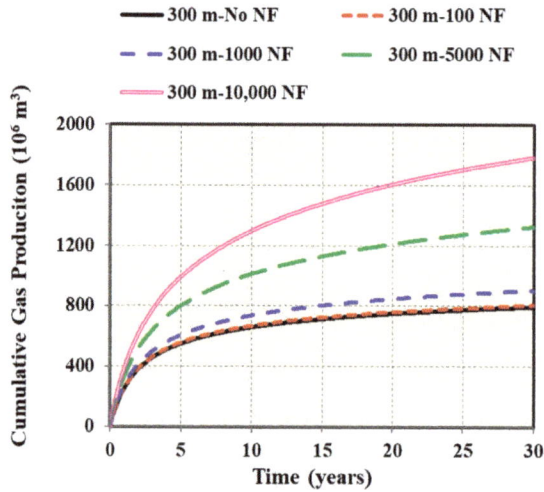

Figure 9. Effect of different numbers of natural fractures on cumulative gas production for the reservoir model with well spacing of 300 m (NF in the legend represents natural fractures).

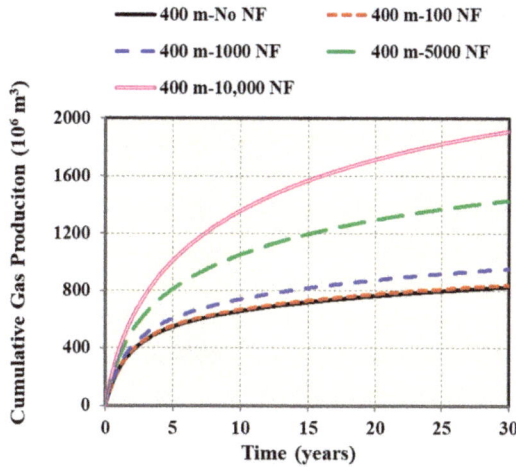

Figure 10. Effect of different numbers of natural fractures on cumulative gas production for the reservoir model with well spacing of 400 m (NF in the legend represents natural fractures).

5.2. Effect of Natural Fractures on Well Spacing Optimization

The effect of natural fracture density on the incremental gas production after 30 years under different well spacing is presented in Figure 11. Here, the natural fracture density refers to the number of natural fractures per 929 m^2, which is about 0.06, 0.6, 3, and 6 for the total number of natural fractures of 100, 1000, 5000, and 10,000, respectively. The incremental gas production after 30 years means that the difference of cumulative gas production with and without natural fractures under a given well spacing. It can be clearly seen that the incremental gas production of different well spacing with and without natural fractures almost linearly increases with the increasing natural fracture density. In addition, the difference of incremental gas production between different well spacing increases with an increase in natural fracture density. The difference of incremental gas production between well spacing of 200 m and 400 m under the natural fracture density of 0.06, 0.6, 3 and 6 per 929 m^2 is about 3.2 × 10^6 m^3, 31.2 × 10^6 m^3, 125.6 × 10^6 m^3, and 184.7 × 10^6 m^3, respectively. Consequently, it can be implied that a larger well spacing is preferred for the shale-gas reservoir with a larger natural fracture density.

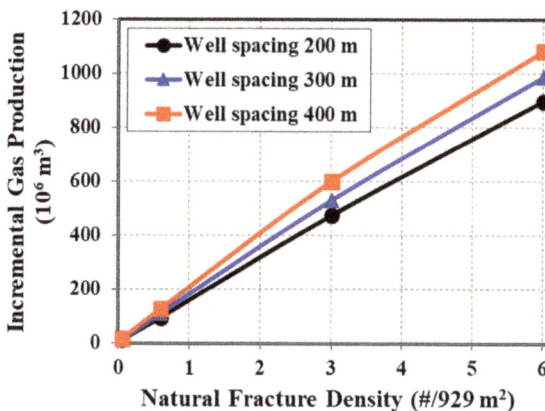

Figure 11. Effect of natural fracture density on the incremental gas production after 30 years under different well spacing.

5.3. Effect of Natural Fractures on Pressure Distribution

Figures 12 and 13 compare pressure distribution after one and 30 years of production under different well spacing with and without natural fractures. The extreme case with 10,000 natural fractures was plotted for comparison. As shown, different drainage area at early and later time of production can be clearly observed. In addition, a stronger well interference between two wells after 30 years under a smaller well spacing occurs. Especially, the drainage area expands more when considering a bigger number of natural fractures.

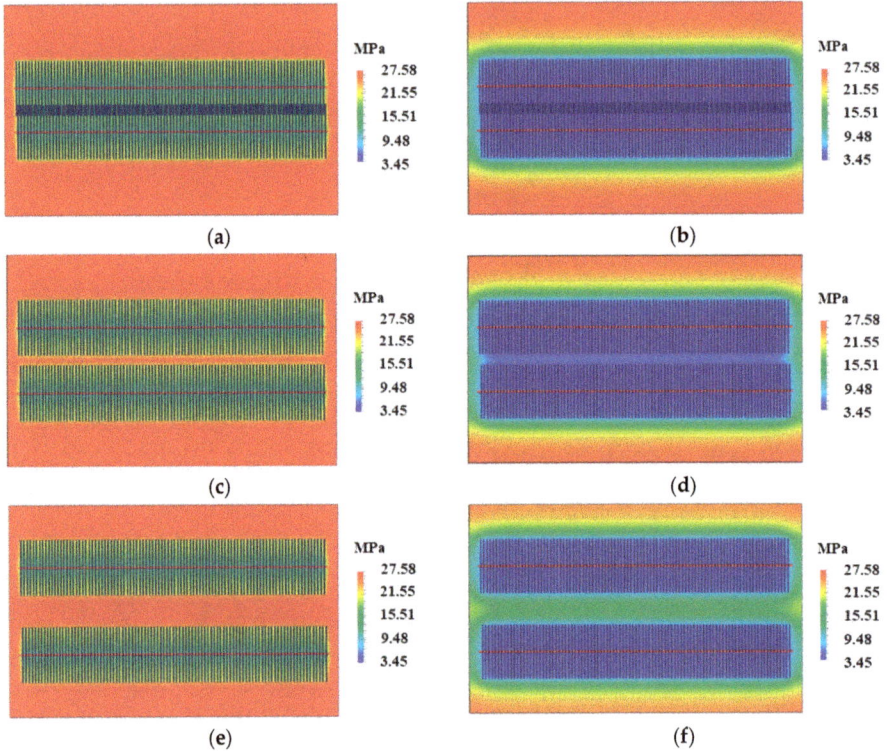

Figure 12. Comparison of pressure distribution after 1 and 30 years of production under different well spacing without natural fractures: (**a**) After 1 year of production with well spacing 200 m; (**b**) After 30 years of production with well spacing 200 m; (**c**) After 1 year of production with well spacing 300 m; (**d**) After 30 years of production with well spacing 300 m; (**e**) After 1 year of production with well spacing 400 m; (**f**) After 30 years of production with well spacing 400 m.

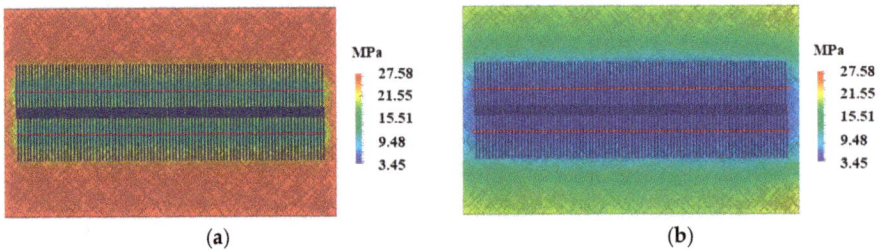

Figure 13. *Cont.*

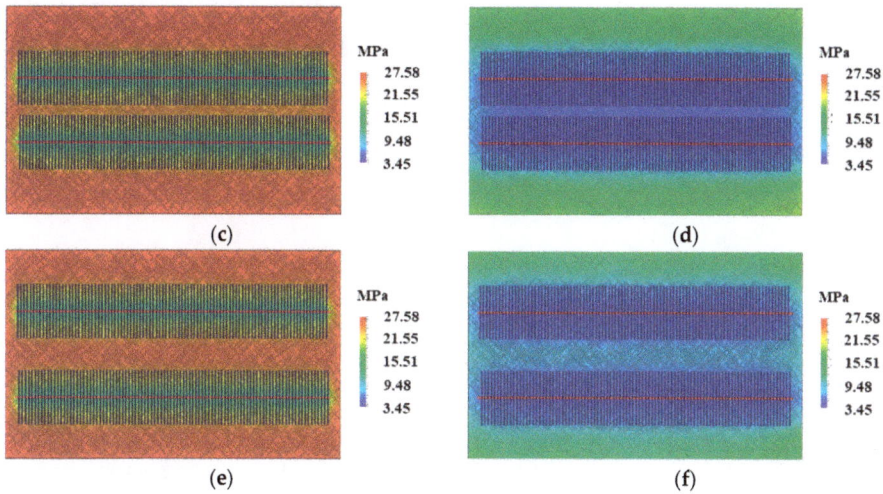

Figure 13. Comparison of pressure distribution after one and 30 years of production under different well spacing with 10,000 natural fractures: (**a**) After one year of production with well spacing 200 m; (**b**) After 30 years of production with well spacing 200 m; (**c**) After one year of production with well spacing 300 m; (**d**) After 30 years of production with well spacing 300 m; (**e**) After one year of production with well spacing 400 m; (**f**) After 30 years of production with well spacing 400 m.

6. Conclusions

We applied the non-intrusive EDFM technology as well as a compositional reservoir simulator to perform shale-gas two-phase flow simulations. The impact of natural fractures on multiple shale-gas well performance with varying well spacing was investigated based on history matching results of an actual shale-gas well from Marcellus shale formation. The following conclusions can be drawn from this study:

(1) The effective matrix and fracture properties were obtained based on good history matching results such as matrix permeability of 0.000525 mD, fracture half-length of 127 m, fracture conductivity of 0.73 mD-m, and fracture water saturation of 37.4%.

(2) The EUR and cumulative water production after 30 years of the actual shale-gas well were determined as 413×10^6 m^3 and 747 m^3, respectively.

(3) The effect of natural fractures on two shale-gas well performance almost linearly increases with the increasing number of natural fractures. For example, the natural fractures contribute to the incremental EUR after 30 years of 1.4%, 13.9%, 66.7%, and 124.3% for the well spacing of 300 m with fracture number of 100, 1000, 5000, and 10,000, respectively.

(4) The CPU time for the model with well spacing of 300 m is about 30 min when dealing with 10,000 natural fractures based on the non-intrusive EDFM technology.

(5) A larger number of natural fractures is easier to form a more complex connected fracture network with hydraulic fractures, resulting in a higher well productivity.

(6) The difference of well performance between different well spacing increases with the increasing natural fracture density. For example, the difference of well performance between well spacing of 200 m and 400 m is about 3.2×10^6 m^3, 31.2×10^6 m^3, 125.6×10^6 m^3, and 184.7×10^6 m^3 corresponding to the natural fracture density of 0.06, 0.6, 3 and 6 per 929 m^2, respectively.

(7) This study provides a better understanding of the impact of complex natural fractures on multiple shale-gas well performance with varying well spacing. A larger well spacing is suggested when the shale-gas reservoir has a larger natural fracture density.

Author Contributions: Conceptualization, W.Y. and M.L.; methodology and investigation W.Y. and M.L.; writing—original draft preparation, W.Y.; writing—review and editing, X.H.; project administration, W.W.; funding acquisition, W.Y., M.L. and W.W.

Funding: This research was funded by "National Science and Technology Major Project of China, grant number 2016ZX05061003-003", "National Natural Science Foundation of China, grant number 51728401".

Acknowledgments: We would like to acknowledge Computer Modelling Group Ltd. and SimTech LLC for providing the CMG-GEM and EDFM software for this investigation.

Conflicts of Interest: The authors declare no conflict of interest.

References

1. U.S. Energy Information Administration. Available online: https://www.eia.gov/tools/faqs/faq.php?id= 907&t=8 (accessed on 3 October 2018).
2. Gale, J.F.W.; Laubach, S.E.; Olson, J.E.; Eichhubl, P.; Fall, A. Natural fractures in shale: A review and new observations. *AAPG Bull.* **2014**, *98*, 2165–2216. [CrossRef]
3. Engelder, T.; Lash, G.G.; Uzcategui, R.S. Joint sets that enhance production from Middle and Upper Devonian gas shales of the Appalachian Basin. *AAPG Bull.* **2009**, *93*, 857–889. [CrossRef]
4. Fisher, M.K.; Wright, C.A.; Davidson, B.M.; Steinsberger, N.P.; Buckler, W.S.; Goodwin, A.; Fielder, E.O. Integrating fracture mapping technologies to improve stimulations in the Barnett shale. *SPE Prod. Facil.* **2005**, *20*, 85–93. [CrossRef]
5. Fisher, M.K.; Warpinski, N.R. Hydraulic-fracture-height growth: Real data. *SPE Prod. Oper.* **2012**, *27*, 8–19. [CrossRef]
6. Wu, R.; Kresse, O.; Weng, X.; Cohen, C.; Gu, H. Modeling of interaction of hydraulic fractures in complex fracture networks. In Proceedings of the SPE Hydraulic Fracturing Technology Conference, The Woodlands, TX, USA, 6–8 February 2012.
7. Wu, K.; Olson, J.E. Investigation of the impact of fracture spacing and fluid properties for interfering simultaneously or sequentially generated hydraulic fractures. *SPE Prod. Oper.* **2013**, *28*, 427–436. [CrossRef]
8. Wu, K.; Olson, J.E. Simultaneous multifracture treatments: Fully coupled fluid flow and fracture mechanics for horizontal wells. *SPE J.* **2015**, *20*, 337–346. [CrossRef]
9. Yue, K.; Olson, J.E.; Schultz, R.A. Layered modulus effect on fracture modeling and height containment. In Proceedings of the SPE/AAPG/SEG Unconventional Resources Technology Conference, Houston, TX, USA, 23–25 July 2018.
10. Tang, J.; Wu, K.; Zeng, B.; Huang, H.; Hu, X.; Guo, X.; Zuo, L. Investigate effects of weak bedding interfaces on fracture geometry in unconventional reservoirs. *J. Pet. Sci. Eng.* **2018**, *165*, 992–1009. [CrossRef]
11. Xie, J.; Huang, H.; Ma, H.; Zeng, B.; Tang, J.; Yu, W.; Wu, K. Numerical investigation of effect of natural fractures on hydraulic-fracture propagation in unconventional reservoirs. *J. Nat. Gas Sci. Eng.* **2018**, *54*, 143–153. [CrossRef]
12. Yu, W.; Wu, K.; Liu, M.; Sepehrnoori, K.; Miao, J. Production forecasting for shale gas reservoirs with nanopores and complex fracture geometries using an innovative non-intrusive EDFM method. In Proceedings of the SPE Annual Technical Conference and Exhibition, Dallas, TX, USA, 24–26 September 2018.
13. Yu, W.; Xu, Y.; Liu, M.; Wu, K.; Sepehrnoori, K. Simulation of shale gas transport and production with complex fractures using embedded discrete fracture model. *AIChE J.* **2018**, *64*, 2251–2264. [CrossRef]
14. Díaz de Souza, O.C.; Sharp, A.J.; Martinez, R.C.; Foster, R.A.; Simpson, M.R.; Piekenbrock, E.J.; Abou-Sayed, I. Integrated unconventional shale gas reservoir modeling: A worked example from the Haynesville Shale, De Soto Parish, North Lousiana. In Proceedings of the SPE Americas Unconventional Resources Conference, Pittsburgh, PA, USA, 5–7 June 2012.
15. Yu, W.; Sepehrnoori, K. Optimization of multiple hydraulically fractured horizontal wells in unconventional gas reservoirs. *J. Pet. Eng.* **2013**, *2013*. [CrossRef]
16. Tung, Y.; Virues, C.; Cumming, J.; Gringarten, A. Multiwell deconvolution for shale gas. In Proceedings of the SPE Europec Featured at 78th EAGE Conference and Exhibition, Vienna, Austria, 30 May–2 June 2016.
17. Manestar, G.J.; Thompson, A. Dual porosity modeling for shale gas wells in the Vaca Muerta formation. In Proceedings of the SPE Latin America and Caribbean Petroleum Engineering Conference, Buenos Aires, Argentina, 17–19 May 2017.

18. Mehranfar, R.; Marquez, L.; Altman, R.; Kolivand, H.; Orantes, R.; Espinola, O. Optimization under uncertainty for reliable unconventional play evaluation. A case study in Vaca Muerta shale gas blocks, Argentina. In Proceedings of the SPE Trinidad and Tobago Section Energy Resources Conference, Port of Spain, Trinidad and Tobago, 25–26 June 2018.

19. Shahkarami, A.; Wang, G.; Rohland, Z. Impacts of field depletion on future infill drilling plans in the Marcellus shale. In Proceedings of the SPE/AAPG Eastern Regional Meeting, Pittsburgh, CA, USA, 7–11 October 2018.

20. Zhou, W.; Banerjee, R.; Poe, B.D.; Spath, J.; Thambynayagam, M. Semi-analytical production simulation of complex hydraulic-fracture networks. *SPE J.* **2014**, *19*, 6–18. [CrossRef]

21. Yu, W.; Huang, S.; Wu, K.; Sepehrnoori, K.; Zhou, W. Development of a semi-analytical model for simulation of gas production in shale gas reservoirs. In Proceedings of the SPE/AAPG/SEG Unconventional Resources Technology Conference, Denver, FL, USA, 25–27 August 2014.

22. Yang, R.; Huang, Z.; Yu, W.; Li, G.; Ren, W.; Zuo, L.; Tan, X.; Sepehrnoori, K.; Tian, S.; Sheng, M. A comprehensive model for real gas transport in shale formations with complex non-planar fracture networks. *Sci. Rep.* **2016**, *6*, 36673. [CrossRef] [PubMed]

23. Chen, Z.; Liao, X.; Zhao, X.; Lyu, S.; Zhu, L. A comprehensive productivity equation for multiple fractured vertical wells with non-linear effects under steady-state flow. *J. Pet. Sci. Eng.* **2017**, *149*, 9–24. [CrossRef]

24. Yu, W.; Sepehrnoori, K. Simulation of gas desorption and geomechanics effects for unconventional gas reservoirs. *Fuel* **2014**, *116*, 455–464. [CrossRef]

25. Liang, Y.; Sheng, J.; Hildebrand, J. Dynamic permeability models in dual-porosity system for unconventional reservoirs: Case studies and sensitivity analysis. In Proceedings of the SPE Reservoir Characterization and Simulation Conference and Exhibition, Abu Dhabi, UAE, 8–10 May 2017.

26. Tang, H.; Hasan, R.; Killough, J. Development and application of a fully implicitly coupled wellbore/reservoir simulator to characterize the transient liquid loading in horizontal gas wells. *SPE J.* **2018**, *23*, 1615–1629. [CrossRef]

27. Yu, W.; Sepehrnoori, K. *Shale Gas and Tight Oil Reservoir Simulation*, 1st ed.; Elsevier: Cambridge, MA, USA, 2018; ISBN 978-0-12-813868-7.

28. AITwaijri, M.; Xia, Z.; Yu, W.; Qu, L.; Hu, Y.; Xu, Y.; Sepehrnoori, K. Numerical study of complex fracture geometry effect on two-phase performance of shale-gas wells using the fast EDFM method. *J. Pet. Sci. Eng.* **2018**, *164*, 603–622. [CrossRef]

29. Miao, J.; Yu, W.; Xia, Z.; Zhao, W.; Xu, Y.; Sepehrnoori, K. A simple and fast EDFM method for production simulation in shale reservoirs with complex fracture geometry. In Proceedings of the 2nd International Discrete Fracture Network Engineering Conference, Seattle, WA, USA, 20–22 June 2018.

30. Xu, Y.; Cavalcante Filho, J.S.A.; Yu, W.; Sepehrnoori, K. Discrete-fracture modeling of complex hydraulic-fracture geometries in reservoir simulators. *SPE Reserv. Eval. Eng.* **2017**, *20*, 403–422. [CrossRef]

31. Xu, Y.; Yu, W.; Sepehrnoori, K. Modeling dynamic behaviors of complex fractures in conventional reservoir simulators. In Proceedings of the SPE/AAPG/SEG Unconventional Resources Technology Conference, Austin, TX, USA, 24–26 July 2017.

32. Computer Modeling Group (CMG). *GEM-GEM User's Guide*; Computer Modeling Group Ltd.: Calgary, AB, Canada, 2017.

33. Xu, Y.; Yu, W.; Li, N.; Lolon, E.; Sepehrnoori, K. Modeling well performance in Piceance Basin Niobrara formation using embedded discrete fracture model. In Proceedings of the SPE/AAPG/SEG Unconventional Resources Technology Conference, Houston, TX, USA, 23–25 July 2018.

34. Yu, W.; Miao, J.; Sepehrnoori, K. Well interference in shale reservoirs with complex fracture geometry. In Proceedings of the 2nd International Discrete Fracture Network Engineering Conference, Seattle, WA, USA, 20–22 June 2018.

35. Yu, W.; Xu, Y.; Weijermars, R.; Wu, K.; Sepehrnoori, K. A numerical model for simulating pressure response of well interference and well performance in tight oil reservoirs with complex-fracture geometries using the fast embedded-discrete-fracture-model method. *SPE Reserv. Eval. Eng.* **2018**, *21*, 489–502. [CrossRef]

36. Dachanuwattana, S.; Jin, J.; Zuloaga-Molero, P.; Li, X.; Xu, Y.; Sepehrnoori, K.; Yu, W.; Miao, J. Application of proxy-based MCMC and EDFM to history match a Vaca Muerta shale oil well. *Fuel* **2018**, *220*, 490–502. [CrossRef]

37. Dachanuwattana, S.; Xia, Z.; Yu, W.; Qu, L.; Wang, P.; Liu, W.; Miao, J.; Sepehrnoori, K. Application of proxy-based MCMC and EDFM to history match a shale gas condensate well. *J. Pet. Sci. Eng.* **2018**, *167*, 486–497. [CrossRef]

38. Yu, W.; Tripoppoom, S.; Sepehrnoori, K.; Miao, J. An automatic history-matching workflow for unconventional reservoirs coupling MCMC and non-intrusive EDFM methods. In Proceedings of the SPE Annual Technical Conference and Exhibition, Dallas, TX, USA, 24–26 September 2018.

39. Zhang, Y.; Di, Y.; Yu, W.; Sepehrnoori, K. A comprehensive model for investigation of CO2-EOR with nanopore confinement in the Bakken tight oil reservoir. In Proceedings of the SPE Annual Technical Conference and Exhibition, San Antonio, TX, USA, 9–11 October 2017.

40. Yu, W.; Zhang, Y.; Varavei, A.; Sepehrnoori, K.; Zhang, T.; Wu, K.; Miao, J. Compositional simulation of CO2 huff-n-puff in Eagle Ford tight oil reservoirs with CO2 molecular diffusion, nanopore confinement and complex natural fractures. In Proceedings of the SPE Improved Oil Recovery Conference, Tulsa, OK, USA, 14–18 April 2018.

41. Sun, R.; Yu, W.; Xu, F.; Pu, H.; Miao, J. Compositional simulation of CO2 huff-n-puff process in Middle Bakken tight oil reservoirs with hydraulic fractures. *Fuel* **2019**, *236*, 1446–1457. [CrossRef]

42. Xu, F.; Yu, W.; Li, X.; Miao, J.; Zhao, G.; Sepehrnoori, K.; Li, X.; Jin, J.; Wen, G. A fast EDFM method for production simulation of complex fractures in naturally fractured reservoirs. In Proceedings of the SPE/AAPG Eastern Regional Meeting, Pittsburgh, PA, USA, 7–11 October 2018.

energies

MDPI

Article

Fracture Detection and Numerical Modeling for Fractured Reservoirs

Lihua Zuo [1], Xiaosi Tan [2,*], Wei Yu [3] and Xiaodong Hu [4]

[1] College of Business, Engineering and Technology, Texas A&M University-Texarkana, Texarkana, TX 75503, USA; lihuazuo@gmail.com
[2] National Mobile Communication Research Laboratory, Southeast University, Nanjing 211100, China
[3] Department of Petroleum and Geosystems Engineering, University of Texas at Austin, Austin, TX 78712, USA; yuwei127@gmail.com
[4] Department of Petroleum Engineering, China University of Petroleum, Beijing 102202, China; XiaodongHU-CUPB@hotmail.com
* Correspondence: tanxiaosi@seu.edu.cn; Tel.: +86-18650786651

Received: 17 January 2019; Accepted: 25 January 2019; Published: 26 January 2019

Abstract: The subsurface fractures could impact the fluid mechanisms dramatically, which makes the modeling of the hydraulic and natural fractures an essential step for fractured reservoirs simulations. However, because of the complexities of fracture patterns and distributions, it is difficult to detect and quantify the fracture networks. In this study, line detection techniques are designed and applied to quantify the fracture segments from fracture figures. Using this fracture detection algorithm, the fracture segments could be located by detecting the endpoints and the intersections of fractures, thus that the fracture patterns could be accurately captured and characterized. The proposed method is applied to two previous well-known field cases and the pressure distribution results are consistent with the micro-seismic data profiles. These two field cases are simulated and computed by using a semianalytical model and Embedded Discrete Fracture Model (EDFM) respectively. The third case is constructed by the fracture outcrop figure and simulated by a numerical simulator with EDFM implemented. The simulation results are accurate and clearly illustrate the important role fractures play in unconventional reservoirs. The technology proposed in this study could be used to quantify the fracture input data for reservoir simulations and be easily expanded for fracture detection and characterization problems in other fields.

Keywords: fractured reservoir; line detection; semi-analytical model; EDFM; fracture modeling

1. Introduction

Due to geomechanical imbalances, various fractures such as natural fractures and hydraulic fractures are generated [1]. The presence of these fractures in the rocks will dramatically change the mechanical and transport properties of the subsurface area [2]. Fracture quantification and fracture modeling are very important in many disciplines of earth sciences, such as geophysics, petroleum and rock mechanics. Different combinations of fracture properties and rock matrix properties will generate different flow storage and transport mechanisms in the fractured reservoirs [3–8]. For example, in the case of high matrix porosity and low matrix permeability, the majority of hydrocarbon will be stored in the matrix, but fractures will act as the transport channel for producing wells. While in the case of low matrix porosity and permeability, the fractures will be in charge of the majority of storage and transport channels. Due to the importance of the fractures, it is very important to quantify the fractures both physically and geometrically, thus that the exact role of the fractures can be analyzed efficiently.

While fracture detection and characterization are inevitable procedures in practical applications, the complexity of fracture physical properties and fracture networks make the quantification and modeling of fractures difficult. Fracture lengths vary from 10^{-2} m to 10^3 m in order of magnitude [9] and fracture orientations are complicated due to the different local subsurface stresses, and fracture aperture distributions and fracture spatial density also have a large range of variations.

To better quantify fracture properties and locations, many algorithms have been developed to detect fracture lines and curves from fracture outcrop figures. The classical work of detecting lines was done by Hough [10], in which the author proposed a method for machine recognition of complex lines in photographs or other pictorial representations. Hough designed a so-called Hough transform to convert the lines to a slope and intersects, thus that the fracture lines could be grouped and analyzed systematically. The Hough transform was then expanded by Duda and Hart [11] to detect both straight lines and curves, by replacing the use of the slope-intercept parameters with angle-radius parameters, thus that it could be applied to more general curve detections. The transform becomes popular in the computer vision field due to the work of Ballard [12], in which the author generalized the Hough transform to the detection of some analytic curves in grey level images, specifically lines, circles, and parabolas. Since then, there have been many applications about the Hough transform. For example, line and curve detection have been used for object recognition [13,14], transportation monitoring and management [15,16], and medical imaging [17]. Another trend of the Hough transform application is applying machine learning techniques such as clustering methods combined with the Hough transform to detect lines and edges. For example, Achtert et al. [18] proposed a method for finding arbitrarily oriented subspace clusters based on the Hough transform, in which the clustering algorithm was used to find the clusters that are lying in a very noisy environment. For more applications with the Hough transform, please refer to the reviews [19–21].

For most cases in the petroleum industry, the only data available are the fracture figures from the surface or outcrops, and sometimes the fracture network is so complicated that manually setting up the fracture network is not practical. That is why an automatic fracture detection algorithm is needed to efficiently detect and quantify the fracture networks. The quantification results are generated to be compatible with the reservoir simulation tools for reservoir simulations. Based on different algorithms, several software packages have been developed. For example, DigiFract [22] is a software package written in Python designed to work directly with fracture data digitized from outcrops and it is based on a geographical information system core that is applied for mapping real fracture data sets and studying the impact of fractures geometries on flow [23]. An integrated workflow for stress and flow modelling using outcrop-derived discrete fracture networks was designed after the fracture image was obtained with Unmanned Aerial Vehicles, in which DigiFract is used to digitize the fracture information [24]. Another software Engauge Digitizer [25] has the capacity of removing the axis and interpolating them with certain regression functions (such as trigonometric functions or polynomial functions) after the users choose certain points to be interpolcated. However, this software could not read fracture lines directly without the user filtering some points manually. FracPaQ [26] is an open-access software package, designed to analyze and quantify fracture patterns in two dimensions from digital data. The Hough transform is used when the input is JPG/JPEG image format. However, there are two disadvantages for FracPaQ. First, because the Hough transform is applied to detect lines when analyzing the original image input, it takes a lot of computation time. Second, FracPaQ focuses on statistical quantities such as fracture length, segment length, segment orientation, intensity, density, and connectivity but it could not calculate the nodes coordinates, the intersections of fracture segments. And when the input is in an image format, it cannot even return nodes locations, which are important information for reservoir simulations. The existence and continuous development of these software packages justify the necessity of a fracture detection and characterization workflow. However, most of the above software packages concentrate more on analyzing the statistical information of the fracture networks and only accept input data of coordinates of fracture endpoints, otherwise, the users need to manually set up points as input for these software packages. They are capable of analyzing a large amount of fractures, but the computation efficiency is low due to the complexity related to the Hough transform or other similar algorithms. For petroleum engineers, what is really needed is a workflow that could detect and quantify the fracture nodes and the line segments to be used for further simulation by using various reservoir simulation techniques such as analytical methods, semi-analytical methods, and numerical methods. That is exactly the purpose of this study.

In this study, we concentrate on fracture detection and quantification applications for some common cases in the petroleum industry. In most unconventional reservoir simulation cases, the number of fractures is less than 100 and all fractures are planar. The fracture geometry is relatively simple, thus that there is no need to use the Hough transform. Two-dimensional fracture figures are used as inputs. The fracture lines and their intersection points are detected and calculated. After detecting fractures lines and providing input data for two reservoir simulation methods, the semianalytical model [8,27] and numerical methods with the Embedded Discrete Fracture Model (EDFM) [28] implemented, the pressure profile of the reservoir is computed.

The rest of this study is organized as follows. First, the methodology of point detection and line detection are explained. Second, the workflow proposed in this study is used to detect the fracture lines in three field cases, with two well-known fracture networks and the third is built from the fracture outcrop figure. The fracture quantification results are then imported to two reservoir simulation methods, a semianalytical model and a numerical method with EDFM implemented to simulate the flow transport. In the end limitations and future work of our methods are discussed, before the conclusions are drawn.

2. Methodology

Three main steps are used in our study to detect and quantify the fracture networks. First, using fracture figures as input, the fracture endpoints and intersections are detected by point detection algorithms. Second, using these points and the original figure, fracture lines are determined and quantified. Third, the fracture length and flow orientation are computed using fracture endpoints. In this section, the methodologies of these three steps are explained one by one.

2.1. Pre-Processing

The point detection algorithms designed in this study concentrates on 1-pixel figures. In real applications, the input figures could be multi-pixels, in which case the well-established clustering algorithms and sharpening/thinning algorithms are used first to convert the original figures to a 1-pixel format. Robust clustering and sharpening algorithms are provided in many existing software such as Matlab (R2017b) for convenient use. In this study, for the simplicity of our discussion, we will assume the input figures are already in a 1-pixel format.

2.2. Points Detection

After the figures are converted to a 1-pixel format, all endpoints and intersection points need to be detected. In the 1-pixel format, each node is a cell with the 1-pixel side length and has at most 8 neighbor cells. To easily distinguish cells, the neighborhood relation for each cell is converted to a neighborhood index by assuming a binary format neighborhood relation. The bit locations used in this study are illustrated in Figure 1a, with the target cell locating at the center. Based on this definition, the neighborhood index for each cell could only have 256 possible values, from 0–255. In fact, the neighborhood connection type is calculated using the following equation:

$$N = \sum_{i=1}^{8} 2^{i-1}, \tag{1}$$

Figure 1b,c show two examples for this conversion. In Figure 1b, the neighborhood index will be calculated as $N = 2^3 + 2^4 + 2^6 = 88$, because it has neighbor cells on bit locations 4, 5, and 7. Similarly, the connection type in Figure 1c is calculated as $N = 1 + 2^2 + 2^3 + 2^4 + 2^6 = 93$ since it has neighbor cells on bit locations 1, 3, 4, 5, and 7. After this conversion, all dots in the figure are converted to integer values within the range of 1 to 256 and each number is distinguishable for different neighborhood relations. With this conversion, the end points and intersection points could easily be detected out. For example, the endpoints will be dots with only one neighbor, thus there are only 8 possible values if one point is the endpoint, that is, $N = 2^{i-1}$, $i = 1, 2, \ldots, 8$. When the corresponding neighborhood index equals to one of these 8 values, it will be marked as endpoints. And similarly, the intersection points are points with more than 3 neighbor dots, such as the center cell in Figure 1b.

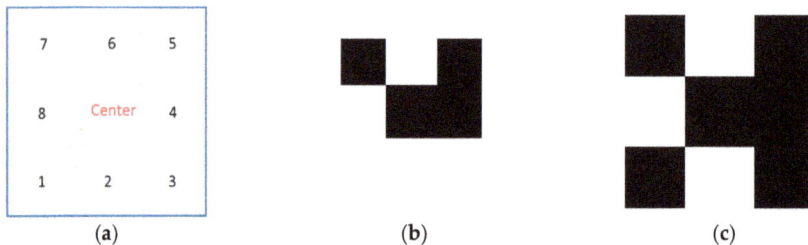

7	6	5
8	Center	4
1	2	3

(a) (b) (c)

Figure 1. Illustrations of pixel binary format neighborhood connections: (**a**) The ordering of 8 bit; (**b**) the center cell with $N = 88$ neighborhood index; (**c**) the center cell with $N = 93$ neighborhood index.

Using this conversion technique, all special points (endpoints and intersection points) will be detected and stored for further use in the next step.

2.3. Line Segments Detection

After locating all the special points, we proceed to the detection of the line segments. There are many line segments detecting software based on the classic method of Hough transform to recognize line segments patterns from complicated colored images. These techniques are also utilized in software like FraqPac [26] to decide the geometry of fractures from geology photos. However, in this study, we focus on the detection of fracture segments from the input of 1-pixle format images, which are based on the endpoints and intersections detected as described in the previous section. We aim at getting an output file including the description of each fracture segment with coordinates of its start and end points, length, slope, mid points, and the connection relationship and relative connection relationships with the well, which is used as an input for the semianalytical model and numerical methods with EDFM implemented. Hence, instead of the Hough transform based methods, we propose a more intuitive but efficient way to detect the fracture segments from 1-pixel images.

2.3.1. Detection of Line Segments with Known Endpoints: Basic Idea

First, we describe the procedure of testing whether there is a line segment existing between two known points. With the coordinates of two points, an equation of a straight line that connects them can be directly derived. Coordinates of all the possible points on this straight line in the binary image can also be computed. As shown in Figure 2a, due to 1-pixel image assumption, sometimes the lines on the binary image may not be perfectly straight (black connected cells), thus the point detection algorithm might lead to a straight line not perfectly passing the black cells. Hence, besides the computed points on the 1-pixel line, we should also include neighborhood points (the region within the dashed blue lines) in Figure 2b.

Define the set of all the black points on the binary image as P and the set of the computed possible points as P_c and suppose they contain N and N_c elements, respectively. The points in P_c on the binary 1-pixel image are determined by considering the intersection $P \cap P_c$ and suppose it contains N_i elements. Consider the ratio between two sets, $\epsilon = N_i/N_c$, if ϵ is larger than a chosen criterion (90% for the cases in this study), we mark it as a line segment. Using this method, all existing line segments could be detected. The 90% value is found by trial and errors among all the values ranging from 0 to 1. For different cases, this value might be different and should be modified accordingly to achieve the best detection results. The larger this value is, the less line segments will be detected and vice versa.

For example, in the extreme case of being 100%, only the lines with all the pixels in this range will be detected, which is in some cases hard to achieve.

(a)   (b)

Figure 2. Illustrations of line detection technique: (**a**) Example of a not-perfectly-straight line after 1-pixel pre-processing; (**b**) neighborhood points considered in the line detection process.

A disadvantage of this method should be pointed out. When there are two very close parallel fractures, it is hard to distinguish between these two tight parallel fractures using this method. At that time, the detection accuracy would decrease significantly.

2.3.2. Detection of the Lines with Endpoints and Intersections

As explained in Section 2.2, all the end points on the fracture image can be detected first. Intuitively, with the method introduced in Section 2.3.1, we can iterate through all the end points and decide the existence of line segments between any pair of the end points. By comparing slope and length (equivalently radius and angles) of lines starting from the same point, we can simply exclude all the repeated lines. A large number of micro-seismic data at one location is usually related to the larger aperture of that fracture or a large number of fractures in that location. In that case, a large number of fracture segments could be detected and quantified. In this study, one fracture segment is simply used to represent all possible fractures and more sophisticated cases will be studied in the future.

However, it should be pointed out that a large error exists when an image with a complex fracture structure is processed with this method. The preprocessing error of the image (especially the "sharp/thin" operation) may bring distortion to the straight lines, which will affect the detection of the intersection points. However, it is noticed from our numerical experiments that the endpoints can usually be detected correctly. Hence when the original image has a complicated fracture formation, the line segments can be detected using another method. First, the endpoints are detected, which only has negligible errors. Then we detect the line segments between these endpoints, using the technique introduced in Section 2.3.1. In fact, the result of this step already has enough for the simulation with the EDFM method. While semianalytical methods require more information for the fracture lines, for example, the intersections of these line segments, which constitute all the points needed for the semianalytical method. In this way, the complex images such as Case 2 analyzed in Section 3 can be processed accurately. The intersection points may not be marked in the exact position but we can get a good approximation to simulate fracture structure from a complicated image, with acceptable errors.

After line detection is finished, the angles and lengths of the fractures could be calculated using the fracture lines with the endpoints and stored for the use of the semianalytical method.

2.4. Flow Direction Determination for the Semianalytical Model

For the semianalytical method, extra fracture analysis work should be done. As shown later in the application, the semianalytical method needs the flow direction between two end points of the fractures. Our computation for flow directions is based on the relative connective relationship of each point to the wellbore.

To achieve this purpose, the shortest path problem needs to be solved, defined as finding the shortest path from each node to the wellbore. The fracture structure can be viewed as an undirected graph [29], by regarding all the points as the nodes, and all the fracture segments as the edges. We assign each edge with the same unit weight, then by solving the shortest path problem from each point to the wellbore, we can find the distances from the points to its closest connected wellbore. The flow direction between the two points is defined by these distances. Specifically, in the models investigated in this study, because the same fracture aperture is utilized for all the fractures, for a pair of connected nodes, our assumption is that flow goes from the point with a larger distance to the well to the point of smaller distance. For the output file, all the points can also be sorted according to the distance. When fractures have different apertures, this assumption will not hold any more [30].

In summary, the workflow for our method described in Section 2 is illustrated in Figure 3. Next, our technique will be applied for three unconventional reservoir cases in Section 3.

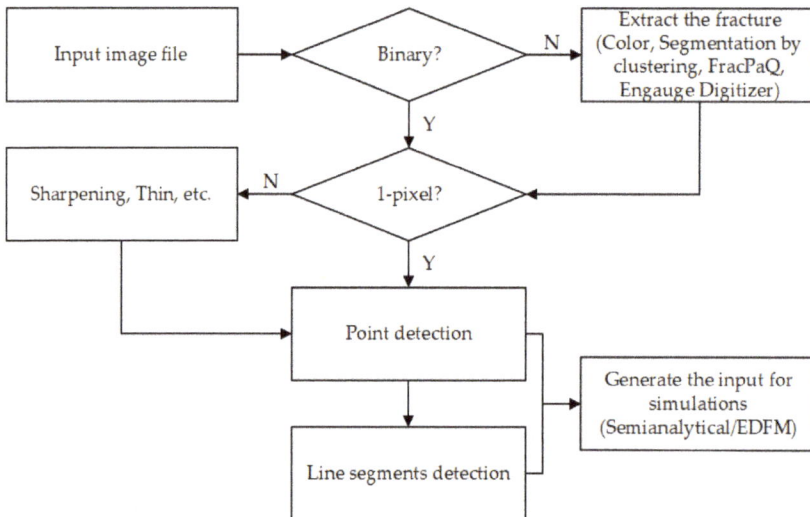

Figure 3. A flow diagram of the fracture detection algorithm. With the input of an image file with the traced fracture network, a text file will be the output, which can be used as the input file for the semianalytical/EDFM simulation.

3. Application

With the help of horizontal wells drilling and hydraulic fracturing techniques, unconventional plays have become a revolution for the energy industry since the late 1980s. In unconventional reservoirs, fractures modeling is an inevitable procedure, in which the fracture quantification needs to be done beforehand in order to simulate the fractures. The input data for fracture quantification might be micro-seismic data, fracture outcrop figures, or other information. Applying the fracture detection and quantification proposed above, the fracture information can be extracted for different reservoir simulation methods.

Efficient simulation methods of modeling fractured reservoirs include analytical methods, semianalytical methods [6,8,27], and numerical methods [31,32]. The dual-porosity/dual-permeability approach is also often applied to model complex fractures and simulate the effect of fracture networks. However, the resolution of these approaches is not enough to capture the details of fractures.

In this section, a semianalytical method and a numerical method with EDFM implemented will be used to model the shale gas productions in the fractured reservoirs. The technique proposed in this study will be applied in three cases to show its effectiveness. The first two cases use micro-seismic data as input and the third case uses the natural fracture outcrop. After the fracture image is processed and fracture data is quantified, two reservoir simulation methods will be used to simulate the reservoir. One is a semianalytical model and the other uses a numerical method with EDFM implemented. More details of these two methods will be explained next.

3.1. Semianalytical Model

Assume the reservoir matrix is homogeneous, there are many analytical and semianalytical models dedicated to efficiently simulating the gas transport in shales. Gringarten et al. [33] used source and Green's functions to solve unsteady flow problems for a wide variety of reservoir flow problems. Cinco-Lay and Samaniego [34] presented a new technique for analyzing pressure transient data for wells intercepted by a finite conductivity vertical fracture. Wan and Aziz [35] proposed a new semianalytical solution for horizontal wells with multiple fractures with different strike angles and partially penetrating the formation in the vertical direction. Then the authors calculated the well index combined with a numerically computed gridblock pressure. Lin and Zhu [36] studied the well performance for fractured horizontal wells in an infinite slab reservoir, by using the instantaneous solution of plane sources, in which the fractures could be fully or partially penetrated. Zhou et al. [27] designed a semianalytical model to simulate the gas production in Barnett shale with fully penetrated planar fractures. Yu et al. [8] extended Zhou's model to simulate well production for reservoirs with fully a penetrated non-planar fracture as well as a planar fracture with varying fracture width and fracture permeability. Their model was verified with a numerical reservoir simulator for a single fracture case before being applied to several case studies based on Marcellus shale properties. In this section, the model built in Yu et al. [8] is used to simulate the shale gas reservoirs.

3.2. EDFM Method

As one of the conventional methods for fracture modeling, dual-porosity/dual-permeability approaches are often applied to model complex fractures and simulate the effect of fracture networks. However, the resolution of these solutions is not high enough to capture the details of fractures geometry. To solve that issue, discrete-fracture models (DFM), using finite-volume or finite-difference methods, have been developed. To be compatible with the complex geometries of fractures, such as nonplanar fractures and fractures with variable aperture, unstructured grids-based reservoir simulator are also developed [31,37–40] in order to explicitly model the fractures. However, the use of unstructured grids leads to high computational cost and they are still limited in real field studies. As a solution, the EDFM was developed [3,5,41,42] to honor the accuracy of DFMs while keeping the efficiency offered by structured gridding. In this study, a numerical method with EDFM implemented by Shakiba and Sepehrnoori [28] is used to compute the pressure distributions. Due to the very small matrix permeability, pressure diffusion is very slow in the matrix, as it is apparent in the figures below. This can imply some numerical errors in the EDFM method because of the singular behavior in $\sqrt{t}$ of pressure close to a fracture is badly captured. In these cases, the MINC (Multiple Interacting Continua) methods for capturing short time behavior can be applied [43,44].

3.3. Case Study

Case 1. Vertical Barnett Shale Well

The first example comes from Fisher et al. [45], where the micro-seismic mapping data and conjectured fracture network in this area are given, with one active vertical well. Previously, if engineers want to use this fracture pattern, a tedious and time-consuming fracture quantification process needs to be taken, during which the researchers need to manually locate the fracture lines and endpoints. With the method proposed in this study, the only work needed to do is to set up the color of the fractures in Figure 4a. By using clustering and sharpening methods provided by Matlab, we can get the fracture lines in binary 1-pixel format. The methods introduced in Section 2 could then be applied to read the nodes and detect the fracture lines. The nodes and lines results are shown in Figure 4a.

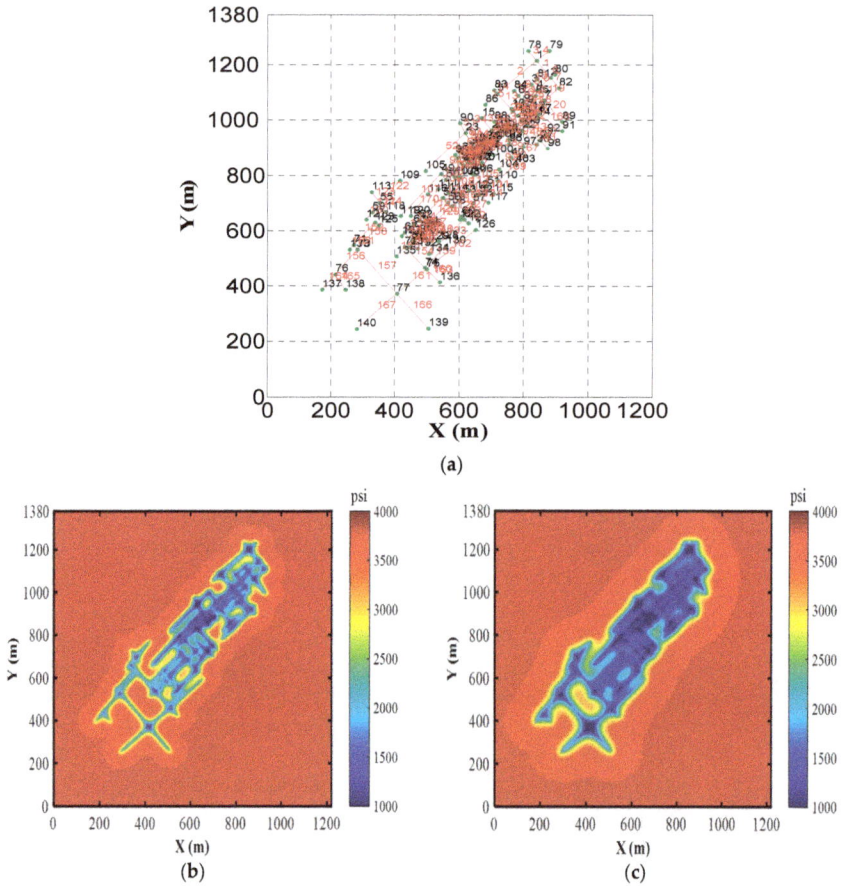

(a)

(b)

(c)

Figure 4. Shale gas model with fractures in Barnett shale with one vertical well in Case 1: (**a**) Detection results of the nodes and lines with the proposed method; (**b**) the pressure profile after 3 years production; (**c**) the pressure profile after 10 years production.

Using the nodes and line segments information provided in Figure 4a, a semianalytical method is used to calculate the pressure distributions. The reservoir information and other properties are listed in Table 1. The pressure profiles after 3 years and 10 years are plotted in Figure 4b,c respectively. The stimulated rock volume could be identified from this pressure results.

Table 1. Reservoir parameters.

Parameter	Value
Reservoir dimension, m	$1200 \times 1328 \times 20$
Depth, m	2134
Reservoir temperature, °C	82
Initial reservoir pressure, psi	3800
Wellbore pressure, psi	1000
Porosity, %	3
Matrix permeability, m^2	1×10^{-20}
Water saturation, %	30
Fracture conductivity, m^2·m	6×10^{-21}

Figure 4a illustrates accurate detection result of all the fractures. This case is a good example of the convenience of the auto detection workflow compared to manually locating the lines and nodes. In this case, there are over 170 fracture segments, it will be time-consuming to manually locate all the 340 plus fracture end nodes. With the help of our algorithm, it just takes several seconds to detect all the nodes and fracture segments.

Figure 4b,c show a high resolution pressure profile after 3 years and 10 years, respectively. In the top right corner, where the density of fractures is high, the pressure diffuses much faster than the bottom left corner, where there are only a few coarse fractures. This validates the important role of natural fractures in altering the shale gas flow transport mechanisms.

Case 2. Horizontal Barnett Shale Well

The second case is also in Barnett shale [45] but with a horizontal well. Applying the methods introduced in this study, the nodes and line segments are detected in Figure 5a. The Reservoir dimensions and other information is listed in Table 2. It is a shale gas reservoir of 1829 m × 1829 m × 91 m large, with 120 × 120 × 1 mesh dimension.

Table 2. Reservoir parameters.

Parameter	Value
Matrix permeability, mD	0.0001
Porosity, %	6
Reservoir Temperature, °C	82
Initial pressure, psi	3800
Initial saturation, %	30
Constant wellbore pressure, psi	1000
Fracture conductivity, m^2·m	5×10^{-12}
Horizontal well locations, m	(1891, 3811), (3801, 1801)

Applying the fracture detection workflow introduced in this study, the fracture segments are quantified and illustrated in Figure 5a, with nodes in black colors and line segments in red colors, the unit is still ft to make the readers easy to compare with the original fracture figure of reference [45]. Because this case has relatively simple fracture patterns, our detection workflow gets perfect detection of all fracture segments.

Using the fracture patterns detected in Figure 5a and the horizontal well location as shown in yellow in the figure, an in-house shale gas reservoir simulator with EDFM methods implemented is used to run the reservoir simulation. The parameters are listed in Table 2.

After 3 years and 30 years, the pressure profile is plotted in Figure 5b,c, respectively. Again, the pressure diffuses much faster in the high-density areas at the left part and the bottom right part. The pressure diffuses much slower in the top part because there are only four fractures.

(a)

(b)

(c)

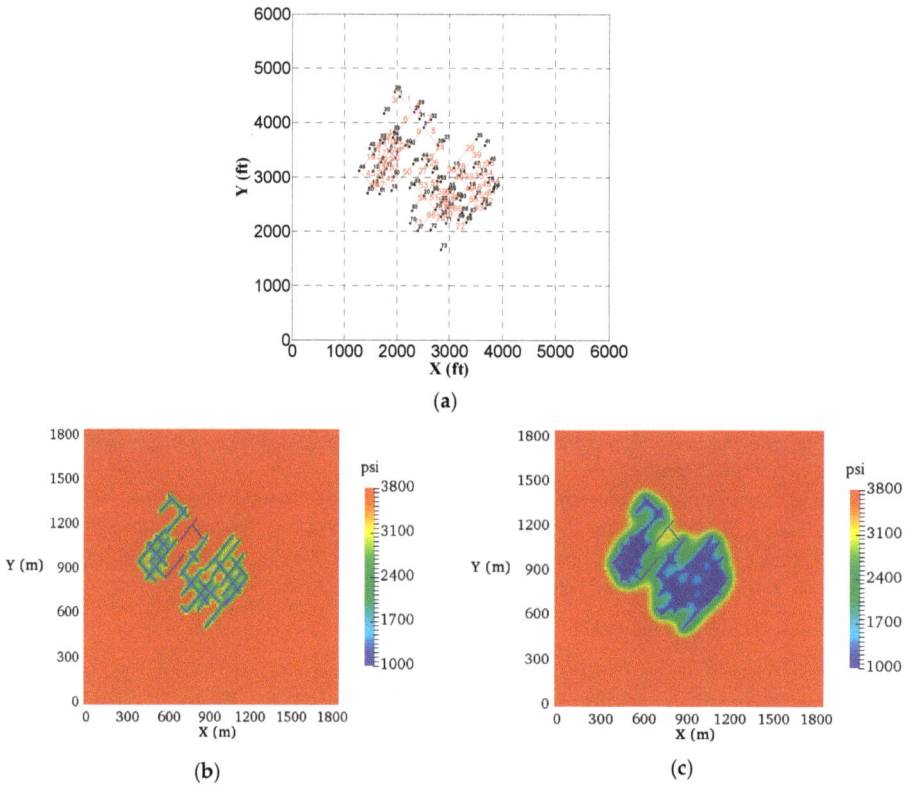

Figure 5. Shale gas model with fractures in Barnett shale with one horizontal well in Case 2: (a) Detection results of the nodes and lines with the proposed method; (b) the pressure profile after 3 years production; (c) the pressure profile after 10 years production.

Case 3. Actual Fracture Outcrop Case (EDFM Method)

To show the versatility of our method, in the last case, an original fracture outcrop figure is used to get the fracture information. The fracture lines are drawn manually as in Figure 6b, which is used as the input for our method and the nodes, line segments results are plotted in Figure 6c. Here we only detect the endpoints and line segments, which are enough for the EDFM method inputs. Using the reservoir parameters as listed in Table 3, with different dimensions, the pressure result in 3 years is shown in Figure 6d.

Table 3. Reservoir parameters.

Parameter	Value
Depth, m	2134
Matrix permeability, m^2	1×10^{-19}
Porosity, %	6
Reservoir Temperature, °C	82
Initial pressure, psi	3800
Initial saturation, %	30
Constant wellbore pressure, psi	1000
Fracture conductivity, $m^2 \cdot m$	5×10^{-12}

Due to the limitation of our clustering algorithm, the fracture lines in Figure 6a are still drawn manually but the fracture nodes and lines are detected based on Figure 6b using the workflow of this study. Figure 6c has the perfect match of all the fracture segments and nodes. Using the fracture information shown in Figure 6c, our in-house reservoir simulator with EDFM method implemented is used to run the simulation for 3 years. The pressure result (in psi) is shown in Figure 6d.

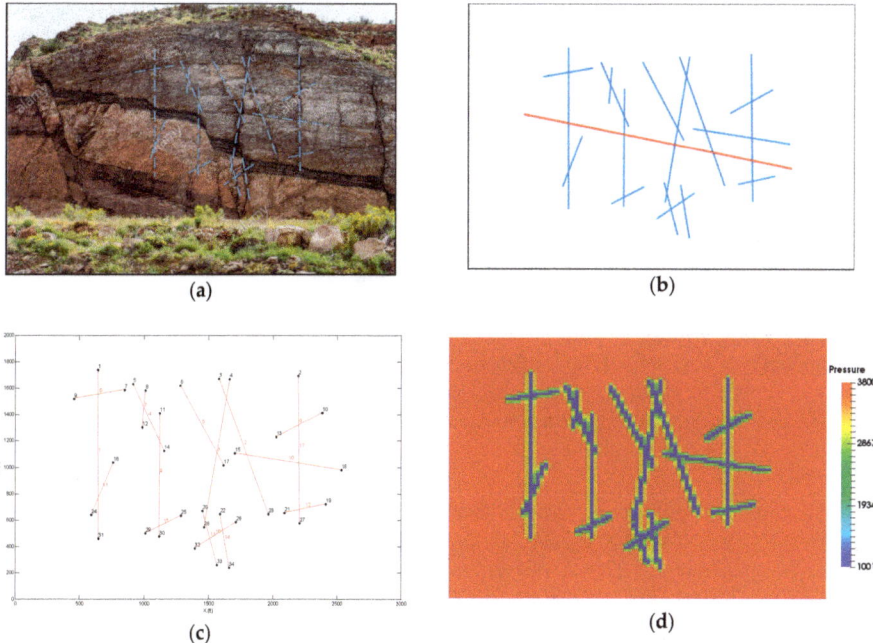

(a)

(b)

(c)

(d)

Figure 6. Field fracture outcrop case in Utah, USA. (**a**) A network of faults and fractures in a rock outcrop. Southern Utah, USA [46]; (**b**) fracture network pattern (blue lines) and well location (red line) (range length:height = 3:2); (**c**) fracture nodes and line segment detection results; (**d**) the pressure profile after 3 years production.

4. Discussion

Fracture quantification and modeling are an inevitable procedure in an unconventional reservoir simulation. Accurate fracture quantification could build a strong foundation for fracture modeling because fractures could largely impact the fluid transport mechanism in unconventional reservoirs. In this study, using two-dimensional fracture figures as input, a fracture detection, and modeling workflow is proposed to quantify the fracture networks in fractured reservoirs.

Using clustering algorithms, two-dimensional fracture figures could be converted to a binary 1-pixel format. The connection types of all the fracture nodes are converted to integers in order to detect the end points and intersection points of fracture segments. Then the fracture segments are detected by checking the overlaps between existing nodes between two points and the lines connecting two points to see whether the overlap ratio is higher than 90%, the line segment is accepted. For semianalytical methods of reservoir simulation, the extra shortest distance problem needs to be solved to get the flow direction in any fracture segment. The fracture information computed by this method provides all the necessary input data for fracture modeling in semianalytical methods and EDFM methods. To validate our model, three reservoir cases are constructed using either the micro-seismic data or the field fracture outcrop data, before the semianalytical method and a numerical method with EDFM implemented are used to simulate the reservoir model and compute the pressure profiles.

To the best of the authors' knowledge, this work is the first work of applying image processing techniques to directly quantify fracture patterns for fracture modeling in unconventional reservoirs. Due to the complexity of the fracture networks, the detection accuracy is not 100% for some complicated fracture patterns, which will be further improved by refining our clustering techniques.

On the other hand, because this study concentrates on the fracture segments detection, the input data is assumed to be the binary and 1-pixel format in the 2D plane. However, in practical cases, there are only the original JPEG format figures and they are in 3D space. In the future, with more advancement in machine learning and clustering and with the help of advanced image processing algorithms to handle the clustering work, original multi-pixel colored figures can be accepted as input. It is also possible to reconstruct some 3D fracture models having mainly observed 2D traces as well as well data, in which the hydraulic fractures are likely to be 2D planar objects that intersect natural fractures along some segments. Some techniques have been proposed to model transient fluid flows in 3D DFN by mapping the fractures to an equivalent pipe network [47,48]. The resulting problem is similar to the one got using EDFM methods or others. Meanwhile, multiphase flow can also be addressed using the mixing of dual porosity models and streamlines models [49–51]. In the future, more field cases will be studied with these advanced techniques.

Despite the above limitations, the methods proposed in this work could inspire further applications of image processing in the energy field, especially in the oil and gas industry. Image processing ability is a must capacity for future exploration robots or equipment. It is worthwhile to put more effort into this area, in order to get ready for assembling the first automated exploration subsurface robot and further propel the development of the oil and gas industry.

5. Conclusions

To facilitate the fracture modeling in reservoir simulations by semianalytical methods and numerical simulation with EDFM methods implemented, a workflow of fracture detection and quantification was designed. This workflow was applied in three cases to show the effectiveness of this method. Based on the work in this study, the following conclusions could be drawn.

1. The endpoints and intersection points of fracture segments are detected by converting the figure format connection types to integer values.

2. The fracture line segments are located by checking the overlaps between existing nodes and nodes lying on the connecting lines of two end points.

3. Solving the shortest path problems helps us to determine the flow direction within each fracture segment, to be used for semianalytical methods.

4. Three unconventional reservoir cases are constructed to validate the effectiveness of the methods proposed in this study.

Author Contributions: Conceptualization, W.Y.; Methodology, L.Z., X.T.; Software, L.Z., X.T.; Investigation, L.Z., X.T.; Resources, W.Y., X.H.; Writing-Original Draft Preparation, X.T., L.Z.; Writing-Review & Editing, L.Z.; Supervision, W.Y.

Funding: This research received no external funding.

Conflicts of Interest: The authors declare no conflict of interest.

References

1. Rutqvist, J. Fractured rock stress-permeability relationships from in situ data and effects of temperature and chemical-mechanical couplings. *Geofluids* **2015**, *15*, 48–66. [CrossRef]

2. Nick, H.M.; Paluszny, A.; Blunt, M.J.; Matthai, S.K. Role of geomechanically grown fractures on dispersive transport in heterogeneous geological formations. *Phys. Rev. E* **2011**, *84*, 056301. [CrossRef] [PubMed]

3. Lee, S.H.; Lough, M.F.; Jensen, C.L. Hierarchical modeling of flow in naturally fractured formations with multiple length scales. *Water Resour. Res.* **2001**, *37*, 443–455. [CrossRef]

4. Cipolla, C.L.; Lolon, E.P.; Erdle, J.C.; Rubin, B. Reservoir Modeling in Shale-Gas Reservoirs. *SPE Reserv. Eval. Eng.* **2010**, *13*, 638–653. [CrossRef]

5. Hajibeygi, H.; Karvounis, D.; Jenny, P. A hierarchical fracture model for the iterative multiscale finite volume method. *J. Comput. Phys.* **2011**, *230*, 8729–8743. [CrossRef]

6. Yang, R.; Huang, Z.; Yu, W.; Li, G.; Ren, W.; Zuo, L.; Tan, X.; Sepehrnoori, K.; Tian, S.; Sheng, M. A Comprehensive Model for Real Gas Transport in Shale Formations with Complex Non-planar Fracture Networks. *Sci. Rep.* **2016**, *6*, 36673. [CrossRef] [PubMed]

7. Yang, C.; Sharma, V.K.; Datta-Gupta, A.; King, M.J. A Novel Approach for Production Transient Analysis of Shale Gas/Oil Reservoirs. In Proceedings of the Unconventional Resources Technology Conference, San Antonio, TX, USA, 20–22 July 2015.

8. Yu, W.; Wu, K.; Zuo, L.; Tan, X.; Weijermars, R. Physical Models for Inter-Well Interference in Shale Reservoirs: Relative Impacts of Fracture Hits and Matrix Permeability. In Proceedings of the Unconventional Resources Technology Conference, San Antonio, TX, USA, 1–3 August 2016.

9. Gutierrez, M.; Youn, D.-J. Effects of fracture distribution and length scale on the equivalent continuum elastic compliance of fractured rock masses. *J. Rock Mech. Geotech. Eng.* **2015**, *7*, 626–637. [CrossRef]

10. Hough, P.V.C. Method and Means for Recognizing Complex Patterns. U.S. Patent No. 3069654A, 18 December 1962.

11. Duda, R.O.; Hart, P.E. Use of the Hough Transformation to Detect Lines and Curves in Pictures. *Commun. ACM* **1972**, *15*, 11–15. [CrossRef]

12. Ballard, D.H. Generalizing the Hough transform to detect arbitrary shapes. *Pattern Recognit.* **1981**, *13*, 111–122. [CrossRef]

13. Lowe, D.G. Distinctive Image Features from Scale-Invariant Keypoints. *Int. J. Comput. Vis.* **2004**, *60*, 91–110. [CrossRef]

14. Wang, M.; Wang, X.; Li, Z. Recognition of multibreak patterns by 8-neighborhood-based General Hough Transform. *Optik* **2010**, *121*, 2254–2258. [CrossRef]

15. Satzoda, R.K.; Sathyanarayana, S.; Srikanthan, T.; Sathyanarayana, S. Hierarchical Additive Hough Transform for Lane Detection. *IEEE Embed. Syst. Lett.* **2010**, *2*, 23–26. [CrossRef]

16. Huang, S.; Gao, C.; Meng, S.; Li, Q.; Chen, C.; Zhang, C. Circular road sign detection and recognition based on hough transform. In Proceedings of the 2012 5th International Congress on Image and Signal Processing, Sichuan, China, 16–18 October 2012; pp. 1214–1218.

17. De Marchi, L.; Baravelli, E.; Cera, G.; Speciale, N.; Marzani, A. Warped Wigner-Hough transform for defect reflection enhancement in ultrasonic guided wave monitoring. *Math. Probl. Eng.* **2012**, *2012*. [CrossRef]

18. Achtert, E.; Böhm, C.; David, J.; Kröger, P.; Zimek, A. Global correlation clustering based on the Hough transform. *Stat. Anal. Data Min. ASA Data Sci. J.* **2008**, *1*, 111–127. [CrossRef]

19. Illingworth, J.; Kittler, J. A survey of the hough transform. *Comput. Vis. Graph. Image Process.* **1988**, *44*, 87–116. [CrossRef]

20. Hassanein, A.S.; Mohammad, S.; Sameer, M.; Ragab, M.E. A Survey on Hough Transform, Theory, Techniques and Applications. *arXiv* **2015**, arXiv:1502.02160.

21. Kaur, G.; Kumar, D. Lane Detection Techniques: A Review. *Int. J. Comput. Appl.* **2015**, *112*, 4–8. [CrossRef]

22. Hardebol, N.J.; Bertotti, G. DigiFract: A software and data model implementation for flexible acquisition and processing of fracture data from outcrops. *Comput. Geosci.* **2013**, *54*, 326–336. [CrossRef]

23. Hardebol, N.J.; Maier, C.; Nick, H.; Geiger, S.; Bertotti, G.; Boro, H. Multiscale fracture network characterization and impact on flow: A case study on the Latemar carbonate platform. *J. Geophys. Res. Solid Earth* **2015**, *120*, 8197–8222. [CrossRef]

24. Bisdom, K.; Nick, H.M.; Bertotti, G. An integrated workflow for stress and flow modelling using outcrop-derived discrete fracture networks. *Comput. Geosci.* **2017**, *103*, 21–35. [CrossRef]

25. Mitchell, M.; Muftakhidinov, B.; Winchen, T. Engauge Digitizer Software. Version 10.11. Available online: http://markummitchell.github.io/engauge-digitizer (accessed on 27 December 2018).

26. Healy, D.; Rizzo, R.E.; Cornwell, D.G.; Farrell, N.J.C.; Watkins, H.; Timms, N.E.; Gomez-Rivas, E.; Smith, M. FracPaQ: A MATLAB™ toolbox for the quantification of fracture patterns. *J. Struct. Geol.* **2017**, *95*, 1–16. [CrossRef]

27. Zhou, W.; Banerjee, R.; Poe, B.D.; Spath, J.; Thambynayagam, M. Semianalytical Production Simulation of Complex Hydraulic-Fracture Networks. *SPE J.* **2013**, *19*, 6–18. [CrossRef]

28. Shakiba, M.; Sepehrnoori, K. Using Embedded Discrete Fracture Model (EDFM) and Micro-seismic Monitoring Data to Characterize the Complex Hydraulic Fracture Networks. In Proceedings of the Society of Petroleum Engineers, Houston, TX, USA, 28–30 September 2015.

29. Diestel, R. *Graph Theory*, 5th ed.; Springer: Heidelberg, Germany; New York, NY, USA, 2010; ISBN 978-3-642-14278-9.

30. Bisdom, K.; Bertotti, G.; Nick, H.M. The impact of in-situ stress and outcrop-based fracture geometry on hydraulic aperture and upscaled permeability in fractured reservoirs. *Tectonophysics* **2016**, *690*, 63–75. [CrossRef]

31. Hoteit, H.; Firoozabadi, A. Compositional Modeling of Discrete-Fractured Media Without Transfer Functions by the Discontinuous Galerkin and Mixed Methods. *SPE J.* **2006**, *11*, 341–352. [CrossRef]

32. Marcondes, F.; Sepehrnoori, K. An element-based finite-volume method approach for heterogeneous and anisotropic compositional reservoir simulation. *J. Pet. Sci. Eng.* **2010**, *73*, 99–106. [CrossRef]

33. Gringarten, A.C.; Ramey, H.J.J. The Use of Source and Green's Functions in Solving Unsteady-Flow Problems in Reservoirs. *Soc. Pet. Eng. J.* **1973**, *13*, 285–296. [CrossRef]

34. Cinco, L.H.; Samaniego, V.F.; Dominguez, A.N. Transient Pressure Behavior for a Well with a Finite-Conductivity Vertical Fracture. *Soc. Pet. Eng. J.* **1978**, *18*, 253–264. [CrossRef]

35. Wan, J.; Aziz, K. Multiple Hydraulic Fractures in Horizontal Wells. In Proceedings of the Society of Petroleum Engineers, Anchorage, AK, USA, 26–27 May 1999.

36. Lin, J.; Zhu, D. Modeling well performance for fractured horizontal gas wells. *J. Nat. Gas Sci. Eng.* **2014**, *18*, 180–193. [CrossRef]

37. Noorishad, J.; Mehran, M. An upstream finite element method for solution of transient transport equation in fractured porous media. *Water Resour. Res.* **1982**, *18*, 588–596. [CrossRef]

38. Matthäi, S.K.; Mezentsev, A.; Belayneh, M. Control-Volume Finite-Element Two-Phase Flow Experiments with Fractured Rock Represented by Unstructured 3D Hybrid Meshes. In Proceedings of the Society of Petroleum Engineers, The Woodlands, TX, USA, 31 January–2 February 2005.

39. Sandve, T.H.; Berre, I.; Nordbotten, J.M. An efficient multi-point flux approximation method for Discrete Fracture–Matrix simulations. *J. Comput. Phys.* **2012**, *231*, 3784–3800. [CrossRef]

40. Olorode, O.; Freeman, C.M.; Moridis, G.; Blasingame, T.A. High-Resolution Numerical Modeling of Complex and Irregular Fracture Patterns in Shale-Gas Reservoirs and Tight Gas Reservoirs. *SPE Reserv. Eval. Eng.* **2013**, *16*, 443–455. [CrossRef]

41. Li, L.; Lee, S.H. Efficient Field-Scale Simulation of Black Oil in a Naturally Fractured Reservoir Through Discrete Fracture Networks and Homogenized Media. *SPE Reserv. Eval. Eng.* **2008**, *11*, 750–758. [CrossRef]

42. Xu, Y.; Cavalcante Filho, J.S.A.; Yu, W.; Sepehrnoori, K. Discrete-Fracture Modeling of Complex Hydraulic-Fracture Geometries in Reservoir Simulators. *SPE Reserv. Eval. Eng.* **2017**, *20*, 403–422. [CrossRef]

43. Ding, D.Y.; Farah, N.; Bourbiaux, B.; Wu, Y.S.; Mestiri, I. Simulation of Matrix/Fracture Interaction in Low-Permeability Fractured Unconventional Reservoirs. *SPE J.* **2018**, *23*. [CrossRef]

44. Farah, N.; Delorme, M.; Ding, D.Y.; Wu, Y.S.; Codreanu, D.B. Flow modelling of unconventional shale reservoirs using a DFM-MINC proximity function. *J. Pet. Sci. Eng.* **2019**, *173*, 222–236. [CrossRef]

45. Fisher, M.K.; Wright, C.A.; Davidson, B.M.; Goodwin, A.K.; Fielder, E.O.; Buckler, W.S.; Steinsberger, N.P. Integrating Fracture Mapping Technologies to Optimize Stimulations in the Barnett Shale. In Proceedings of the Society of Petroleum Engineers, San Antonio, TX, USA, 29 September–2 October 2002.

46. Alamy Alamy Stock Photo. Available online: https://c8.alamy.com/comp/EA788J/a-network-of-faults-and-fractures-in-a-rock-outcrop-southern-utah-EA788J.jpg (accessed on 11 November 2018).

47. Noetinger, B.; Roubinet, D.; Russian, A.; Le Borgne, T.; Delay, F.; Dentz, M.; Gouze, P. Random walk methods for modeling hydrodynamic transport in porous and fractured media from pore to reservoir scale. *Transp. Porous Media* **2016**, *115*, 345–385. [CrossRef]

48. Noetinger, B. A quasi steady state method for solving transient Darcy flow in complex 3D fractured networks accounting for matrix to fracture flow. *J. Comput. Phys.* **2015**, *283*, 205–223. [CrossRef]

49. Mesbah, M.; Vatani, A.; Siavashi, M. Streamline simulation of water-oil displacement in a heterogeneous fractured reservoir using different transfer functions. *Oil Gas Sci. Technol.* **2018**, *73*, 14. [CrossRef]

Energies **2019**, *12*, 386

50. Bourbiaux, B. Fractured Reservoir Simulation: A Challenging and Rewarding Issue. *Oil Gas Sci. Technol.* **2010**, *65*, 227–238. [CrossRef]
51. Zuo, L.H.; Yu, W.; Miao, J.; Varavei, A.; Sepehrnoori, K. Efficient modeling of fluid transport in naturally fractured porous medium using EDFM and streamline method. *Pet. Explor. Dev.* **2019**, *46*, 1–7. [CrossRef]

energies

MDPI

Article

Cyclic CH$_4$ Injection for Enhanced Oil Recovery in the Eagle Ford Shale Reservoirs

Yuan Zhang [1,*]**, Yuan Di** [2]**, Yang Shi** [3] **and Jinghong Hu** [1]

[1] Beijing Key Laboratory of Unconventional Natural Gas Geology Evaluation and Development Engineering, China University of Geosciences (Beijing), Beijing 100083, China; hujinghong@cugb.edu.cn

[2] College of Engineering, Peking University, Beijing 100871, China; diyuan@mech.pku.edu.cn

[3] Research Institute of Petroleum Exploration and Development, PetroChina, Beijing 100083, China; shy312@petrochina.com.cn

* Correspondence: zhangyuan@cugb.edu.cn; Tel.: +86-151-2007-7038

Received: 24 October 2018; Accepted: 7 November 2018; Published: 9 November 2018

Abstract: Gas injection is one of the most effective enhanced oil recovery methods for the unconventional reservoirs. Recently, CH$_4$ has been widely used; however, few studies exist to accurately evaluate the cyclic CH$_4$ injection considering molecular diffusion and nanopore effects. Additionally, the effects of operation parameters are still not systematically understood. Therefore, the objective of this work is to build an efficient numerical model to investigate the impacts of molecular diffusion, capillary pressure, and operation parameters. The confined phase behavior was incorporated in the model considering the critical property shifts and capillary pressure. Subsequently, we built a field-scale simulation model of the Eagle Ford shale reservoir. The fluid properties under different pore sizes were evaluated. Finally, a series of studies were conducted to examine the contributions of each key parameter on the well production. Results of sensitivity analysis indicate that the effect of confinement and molecular diffusion significantly influence CH$_4$ injection effectiveness, followed by matrix permeability, injection rate, injection time, and number of cycles. Primary depletion period and soaking time are less noticeable for the well performance in the selected case. Considering the effect of confinement and molecular diffusion leads to the increase in the well performance during the CH$_4$ injection process. This work, for the first time, evaluates the nanopore effects and molecular diffusion on the CH$_4$ injection. It provides an efficient numerical method to predict the well production in the EOR process. Additionally, it presents useful insights into the prediction of cyclic CH$_4$ injection effectiveness and helps operators to optimize the EOR process in the shale reservoirs.

Keywords: Cyclic CH$_4$ injection; enhanced oil recovery; nanopore confinement; molecular diffusion; sensitivity analysis

---

1. Introduction

As reported, around 40% of the natural gas reserves are contained in the unconventional reservoirs all over the world [1]. The Eagle Ford shale is one of the productive oil shale reservoirs in the North America, which is located in the northwest of Texas. The main thickness of production varies from 50 to 300 feet [2,3]. The technologies of horizontal drilling and multistage hydraulic fracturing have attracted much attention, especially for the micro- and nano-pores in the unconventional reservoirs [4,5]. The combination of these technologies is extensively used to exploit the reserves in the tight and shale reservoirs [6,7]. However, Dejam et al. [8,9] pointed out that low permeability may increase the threshold pressure gradient, and large amount of oil still reserves in the formations, which requires gas injection for the production enhancement [10–12].

Due to the low permeability of shale rocks, waterflooding cannot perform as effective as that in the conventional resources. Hence, the attention has been attracted to gas injection in the unconventional reservoirs. Recent theoretical and experimental studies have shown that CH_4 injection is more impressive than CO_2 because it has high compressibility and the sources are rich [13,14]. Therefore, CH_4 can take the place of CO_2 in some situations. Alfarge et al. [15] pointed out that extending soaking period and increasing injection volume are benefit to improve the well production. Meng and Sheng [16] conducted the experiment of CH_4 Huff-n-Puff injection in the core samples, confirming that condensate recovery increase by 6% in the Huff-n-Puff injection operation. However, most studies focus on the primary depletion production; the physical mechanisms on the effectiveness of cyclic CH_4 injection are still limited.

Literatures have reported the evaluation of gas injection in shale oil reservoirs [15,17–19]. Sigmund et al. [20] and Brusilovsky [21] have conducted experiments in the porous media. They concluded that the phase behavior in the porous media deviates from the bulk phase. Recent studies have shown that nanopore confinement is an important factor since the nanopores cause high capillary pressure, affecting the properties of components as well as phase behavior further theoretically and experimentally [22–25]. Wang et al. [26] and Nojabaei et al. [23] modified the vapor-liquid phase equilibrium model based on Young-Laplace equation and Leverett J-function. They then incorporated the phase equilibrium model into the reservoir simulator to predict the well production in the tight oil reservoirs. Yang et al. [27] modified the Peng-Robison equation of state and introduced a new term representing the molecule-wall interaction. They reproduced the collected data with an overall error of 7.64% compared to the molecular simulation results. Nanofluidic devices were applied to investigate the nanopore effects. Luo et al. [28] and Alfi et al. [29] conducted the nanofluidic experiment and they all concluded that the bubble point shifts with the effect of confinement. Salahshoor et al. [1] reviewed the mathematical models and experimental studies to compare the phase behavior in conventional reservoirs and tiny pores.

Molecular diffusion is another key mechanism affecting the gas injection effectiveness. Yu et al. [30] has investigated that molecular diffusivity should be correctly included in the simulation model. In the process of CO_2-CH_4 displacement, diffusion is also an important mechanism [31]. Zhang et al. [32] compared the oil recovery of CO_2-EOR process and concluded that considering molecular diffusion is beneficial to improve the oil recovery. However, these investigations only focus on the CO_2 injection process; the impact on the CH_4 injection was not well understood. Recent studies have concluded that the diffusion coefficient of CH_4 is on the same order of CO_2 [33,34]; hence, the effect of molecular diffusion needs to be well examined.

Figure 1 shows the sketch of CH_4 injection process in the fractured horizontal well. As CH_4 is injected, the molecules will move into the fractures and diffuse into the matrix. The fluid phase behavior in nanopores should be determined. Due to the nanopore effects, the injected components will not distribute homogenous among different sizes of pores. Additionally, it will result in different swelling effect in the gas injection from conventional reservoirs because of the confined phase behavior in nanopores.

From the literature survey, there are still some limitations behind the previous studies. Although the EOR process is efficient in the tight oil reservoirs, few studies focus on the effect of confinement on the EOR effectiveness, especially for the CH_4 injection. Additionally, most of previous studies analyzed the operation parameters and the investigation of physical mechanisms affecting the CH_4 injection is limited. In order to fill this gap, we proposed a useful method incorporating the phase behavior model into the compositional simulator, which can accurately and efficiently evaluate the effect of key parameters on the CH_4 injection effectiveness. This work systematically analyzes the physical mechanisms and operation parameters; it can be easily used in the operations of EOR process.

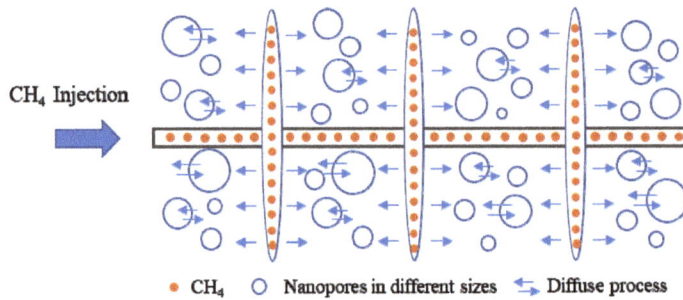

Figure 1. The sketch of CH$_4$ injection process in the fractured horizontal well (CH$_4$ molecules diffuse into different nanopores).

In this work, we evaluated the effect of confinement and CH$_4$ molecular diffusion on the cyclic CH$_4$ injection in the Eagle Ford shale reservoir. First, the methodology and detailed procedure were explained. Then, we built a reservoir model based on the fluid properties from the published Eagle Ford data [35]. The pore size distribution was obtained from the Eagle Ford rock samples [24]. Afterwards, a series of sensitivity analysis were performed to identify the impacts of the physical mechanisms on the effectiveness of cyclic CH$_4$ injection. Finally, we conducted the sensitivity analysis including operation parameters and physical mechanisms. This work provides a better analysis and optimization of CH$_4$ injection in the Eagle Ford shale reservoir.

2. Methodology

2.1. Shifts of Critical Properties

The nanopore effect on the critical temperatures and pressures has been reported in the literatures [24,36,37]. The interaction between the molecules and the pore walls is significant when the pore size is less than 10 nm [38,39]. In our study, the correlations published by Singh et al. [36] were applied to describe the critical property shifts [40]:

$$\Delta T_c^* = \frac{T_{cb} - T_{cp}}{T_{cb}} = 0.9409 \frac{\sigma_{LJ}}{r_p} - 0.2415 \left(\frac{\sigma_{LJ}}{r_p}\right)^2, \tag{1}$$

$$\Delta P_c^* = \frac{P_{cb} - P_{cp}}{P_{cb}} = 0.9409 \frac{\sigma_{LJ}}{r_p} - 0.2415 \left(\frac{\sigma_{LJ}}{r_p}\right)^2, \tag{2}$$

$$\sigma_{LJ} = 0.244 \sqrt[3]{\frac{T_{cb}}{P_{cb}}}, \tag{3}$$

where r_p represents the pore-throat radius, ΔT_c^* and ΔP_c^* express the relative critical temperature and pressure shift, respectively. T_{cb} and P_{cb} are the bulk critical temperature and pressure, respectively. T_{cp} and P_{cp} are the critical temperature and critical pressure in the confined space, respectively. σ_{LJ} is the Lennard-Jones size parameter (collision diameter).

2.2. Phase Equilibrium Calculation Considering Nanopore Confinement

In order to include the effect of confinement in the phase equilibrium model, the criterion of phase equilibrium can be rewritten as:

$$f_L^i(T, P_L, x_i) = f_V^i(T, P_V, y_i), \quad i = 1, \ldots, N_c, \tag{4}$$

$$P_V - P_L = P_{cap}, \tag{5}$$

where f_L^i and f_V^i express the fugacity of component i in the liquid and vapor phases, respectively. T is the reservoir temperature. P_V and P_L represent the pressures of the vapor and liquid phase, respectively. P_{cap} is the capillary pressure in the confined space, which is calculated using the Young-Laplace equation [41], defining as:

$$P_{cap} = \frac{2\sigma \cos \theta}{r_p}, \tag{6}$$

where θ represents the contact angle. In this model, the contact angle is assumed as zero and the angle between organic and inorganic pores was neglected. The interfacial tension, σ is calculated using the following equation:

$$\sigma = \left[\sum_i^{N_C} (\bar{\rho}_L[P]_i x_i - \bar{\rho}_V[P]_i y_i) \right]^4, \tag{7}$$

where $\bar{\rho}_L$ and $\bar{\rho}_V$ denote density of the liquid and vapor phases, respectively. $[P]_i$ is the parachor of component i. Parachor of pure component and mixture can be obtained from the work by Pedersen and Christensen [42].

The Peng-Robinson equation of state [43] is modified as Equation (8) considering the effect of confinement:

$$P = \frac{RT}{V_m - b} - \frac{a\alpha}{V_m^2 + 2bV_m - b^2}, \tag{8}$$

where V_m and R represent the mole volume of component i and the universal gas constant, respectively. a and b are the parameters obtained by van der Waals mixing rules.

When the confinement is included, Equation (8) should be solved separately for liquid and vapor phases:

$$(Z_L)^3 - (1 - B_L)(Z_L)^2 + (A_L - 2B_L - 3(B_L)^2)Z_L - (A_L B_L - (B_L)^2 - (B_L)^3) = 0, \tag{9}$$

$$(Z_V)^3 - (1 - B_V)(Z_V)^2 + (A_V - 2B_V - 3(B_V)^2)Z_V - (A_V B_V - (B_V)^2 - (B_V)^3) = 0, \tag{10}$$

where $A_L = \frac{a_L \alpha P_L}{R^2 T^2}$, $B_L = \frac{b_L P_L}{RT}$, $A_V = \frac{a_V \alpha P_V}{R^2 T^2}$, $B_V = \frac{b_V P_V}{RT}$. Z_L and Z_V are the compressibility of liquid and vapor phases, respectively. The non-linear equations are solved by Newton-Raphson method. The roots of Equations (9) and (10) are determined with the criterion of Gibbs free energy minimization in the liquid and vapor phases.

In the following section, we first built a reservoir model based on the typical fluid and fracture properties, and then performed sensitivity analysis of different parameters in the cyclic CH_4 injection. The fluid properties considering the nanopore effects were calculated through the phase equilibrium model. Afterwards, the properties were implemented into the reservoir simulator of CMG to evaluate the cyclic CH_4 injection effectiveness [44]. The detailed workflow of this work is presented in Figure 2.

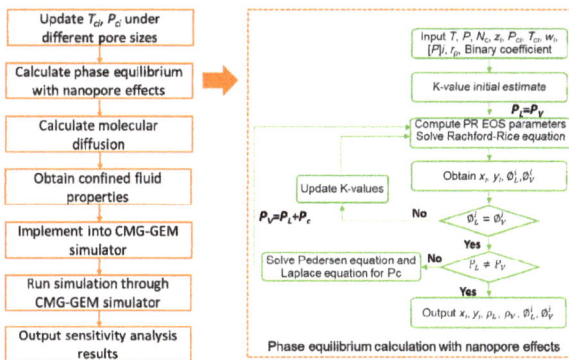

Figure 2. The workflow of evaluation of CH_4 injection effectiveness.

3. Base Case

In the simulation study, we set up the reservoir model using the CMG-GEM simulator [44]. The domain of the model is: 7785 ft in x direction, 1300 ft in y direction, and 40 ft in z direction. A horizontal well was set in the middle of the reservoir model, along with 76 hydraulic fractures. The fracture spacing is 80 ft and the fracture half-length is 210 ft. As reported, the reservoir temperature is 270 °F, the matrix porosity is 12%, and the initial reservoir pressure is 8125 psi. Table 1 summarizes the reasonable rock and fluid properties in the Eagle Ford shale reservoir [45]. The reservoir model is shown in Figure 3. Mohebbinia and Wong [46] have pointed out that molecular diffusion would be dominated in the low-permeability fractured reservoirs when gravitational drainage is inefficient. Hence, only diffusion mechanism was included in this work. The relative permeability curves are shown in Figure 4.

Table 1. Rock and fluid properties used in the reservoir model.

Properties	Value	Unit
Initial reservoir pressure	8125	psi
Reservoir temperature	270	°F
Reservoir thickness	100	ft
Water saturation	17%	-
Porosity	12%	-
Average matrix permeability	0.001	mD
Fracture half-length	210	ft
Fracture spacing	80	ft

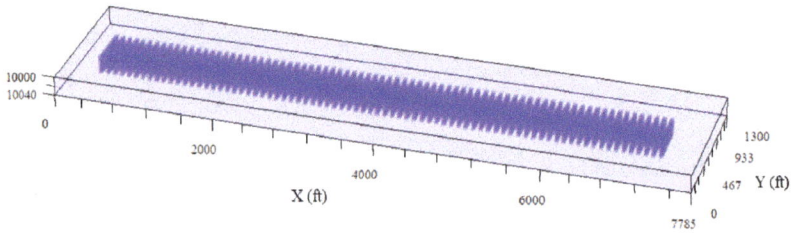

Figure 3. The reservoir simulation model in the cyclic CH_4 injection.

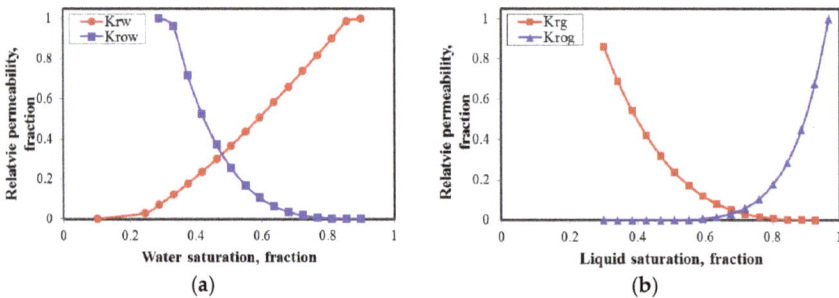

Figure 4. Relative permeability curves: (**a**) Water-oil relative permeability curve; (**b**) Liquid-gas relative permeability curve [38].

In this study, the fluid in the Eagle Ford reservoir is assumed containing six pseudo-components. Properties and binary interaction coefficients are listed in Tables 2 and 3, respectively. Oil gravity of 41 °API, gas-oil ratio of 1000 scf/stb, and formation volume factor of 1.65 rb/stb are obtained after tuning process. These properties have shown good agreements with the work by Orangi et al. [47].

Table 2. Properties of Eagle Ford oil modified from Orangi et al. [47].

Components	Mole Fraction (%)	Critical Temperature (K)	Critical Pressure (atm)	Acentric Factor	Molecular Weight (g/mol)
CO_2	1.18	304.20	72.8	0.225	44.01
N_2	0.16	126.20	33.5	0.040	28.01
CH_4	11.54	190.60	45.4	0.008	16.04
C_2-nC_5	26.44	274.74	36.5	0.172	52.02
C_6-C_{10}	38.09	438.68	25.1	0.284	103.01
C_{11+}	22.59	740.29	17.5	0.672	267.15

Table 3. Binary interaction coefficients for each component.

Component	CO_2	N_2	CH_4	C_2-nC_5	C_6-C_{10}	C_{11+}
CO_2	0	−0.020	0.1030	0.1299	0.1500	0.1500
N_2	−0.020	0	0.0310	0.0820	0.1200	0.1200
CH_4	0.1030	0.0310	0	0.0174	0.0462	0.1110
C_2-nC_5	0.1299	0.0820	0.0174	0	0.0073	0.0444
C_6-C_{10}	0.1500	0.1200	0.0462	0.0073	0	0.0162
C_{11+}	0.1500	0.1200	0.1110	0.0444	0.0162	0

In the base case, BHP of 1800 psi is the constraint for the production well at the beginning of the simulations. In the first three years, the well experiences a depletion production period. After that, it will be transferred to an injection well. The injection rate of CH_4 is set as 5000 Mscf/day. After 60 days of injection, the well will shut-in and begin a soaking period of 60 days. During the soaking period, the fluid is allowed to dissipate into the formation and mix with the fluid. Then the well is switched back on for another two-year production again. Thus, one cycle of CH_4 injection is finished. In this model, the production well experiences three cycles and the total production time is 15 years, as shown in Figure 5. A series of cases were conducted to simulate the cyclic CH_4 injection process. We compared the oil recovery factor in a 15-year period to investigate the effectiveness of CH_4 injection in the following sections.

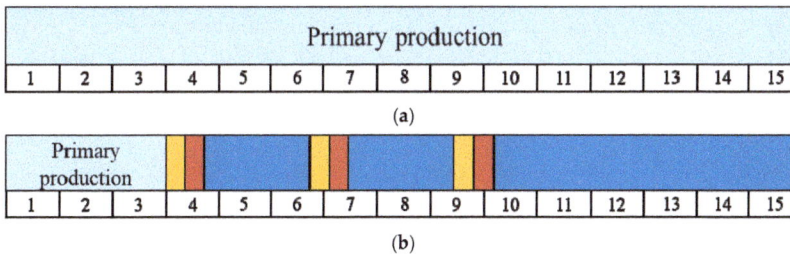

Figure 5. Production of different cases in total simulation time of 15 years: (**a**) Primary production; (**b**) Cyclic CH_4 injection (the yellow, red and dark blue bars represent CH_4 injection, soaking and EOR production period, respectively).

4. Results and Discussions

4.1. Effect of Confinement in Nanopores

The confinement is significant in the low permeability formations due to the nanopores. According to the experiment data from the Eagle Ford core samples [24], around 80% of the pores are 20 nm or less. The fluid properties under 5 nm, 10 nm, and 15 nm were calculated using the Equations (1) through (3), respectively, as listed in Table 4. The results show that critical temperatures and pressures suppress as the pore sizes reduce. With the procedure in Figure 2, we calculated the phase equilibrium and obtained the bubble point pressure under different pore sizes. As shown in Figure 6, the bubble point pressure significantly decreases, especially for the small pores, which further impacts the oil recovery in the tight oil reservoirs.

Table 4. Critical temperatures and pressures under different pore sizes.

Components	Critical Temperatures (K)				Critical Pressures (Bar)			
	Bulk	15 nm	10 nm	5 nm	Bulk	15 nm	10 nm	5 nm
CO_2	304.2	296.7	293.1	282.2	73.8	72.0	71.1	68.4
N_2	126.2	123.2	121.7	117.4	33.9	33.1	32.7	31.6
CH_4	190.6	185.9	183.6	176.8	46.0	44.9	44.3	42.7
C_2-nC_5	274.7	266.6	262.5	250.6	37.0	35.9	35.3	33.7
C_6-C_{10}	438.7	421.4	413.0	388.1	25.4	24.4	24.0	22.5
C_{11+}	740.3	701.4	682.4	627.0	17.7	16.8	16.3	15.0

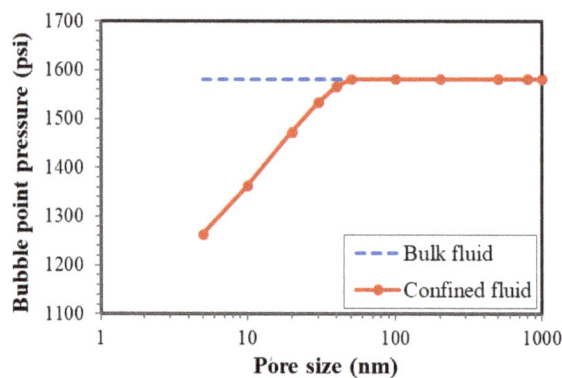

Figure 6. Bubble point pressure under different pore sizes.

Based on the pore size distribution of the formation, division of different pore regions are determined to represent the practical situation. Numbers of region ranging from 3 to 5 has been investigated and we finally decided 4 regions, which reduces computational cost and evaluates the confinement more reasonably. The contributions of each region are: less than 5 nm (42%), 5~10 nm (27%), 10~20 nm (13%), and larger than 20 nm (18%), respectively. The PVT properties of different pore sizes can be obtained in Section 2.1.

The effect of nanopore confinement on the well production was shown in Figure 7. It can be observed that the incremental oil recovery factor at the end of 15 years is 0.8% and 2.3% at the pore size of 10 nm and bulk, respectively, illustrating that the effect of confinement has a positive influence on the cyclic CH_4 injection effectiveness. Due to the confinement, the miscibility minimum pressure of the mixture and the oil viscosity decrease, leading to the improvement of well performance. Additionally, the bubble point pressure in the confined space is lower than the bulk phase, meaning that a longer time of single-phase production exist during the production. Hence, the confinement should be correctly included in the analysis of gas injection in the shale reservoirs. All the following cases include the confinement.

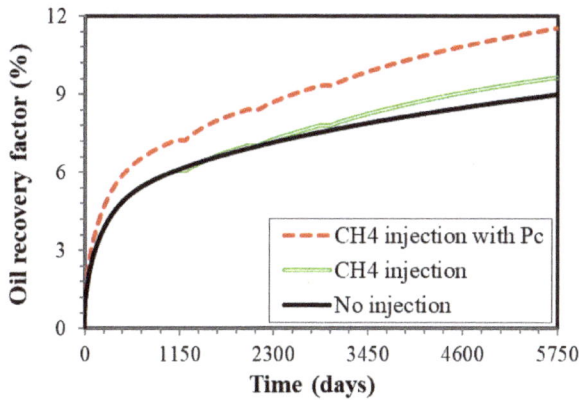

Figure 7. Effect of nanopore confinement on well performance in a 15-year period.

4.2. Molecular Diffusion

Molecular diffusion is another key parameter since CH_4 can move into the formation and mix with the oil during the soaking time. Neglecting diffusion coefficient will underestimate the ultimate oil recovery. In order to better analyze the mechanism of diffusion, we compared the oil recovery after 15 years with the coefficient ranging from 0.0001 cm^2/s to 0.01 cm^2/s. As shown in Figure 8, the incremental oil recovery is 1.92%, 2.36%, and 2.98%, with the coefficients of 0.0001, 0.001, and 0.01 cm^2/s, respectively. The results indicate that more CH_4 molecules will diffuse into the matrix instead of concentrating around the fractures with larger diffusion coefficient. Hence, more oil will be swept, resulting in larger oil production.

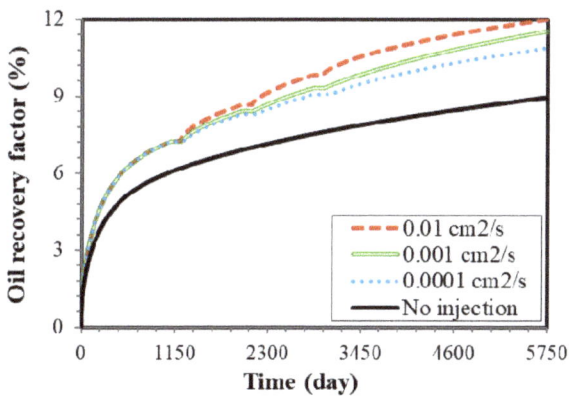

Figure 8. Effect of molecular diffusion on well production in a 15-year period.

4.3. Effect of Primary Depletion Period

The length of primary depletion period of 2, 3, and 5 years was studied, while other parameters were kept the same as the base case. As presented in Figure 9, the impact of primary depletion period is not noticeable since the increment is 2.28%, 2.34%, and 2.41%, respectively in this case. Delaying the start of gas injection is beneficial to improve the well production. However, if it starts too late, the production will decrease. Hence, decision of suitable primary depletion period is essential for the operations of CH_4 injection in the shale reservoirs.

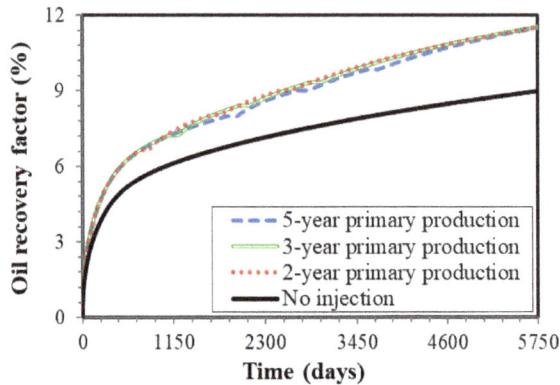

Figure 9. Comparison of oil recovery factor of different primary depletion period.

4.4. Effect of Injection Rate

Gas injection rate is directly related to the volume of CH_4 injected. A series of cases were conducted to investigate the effect of injection rate on the well performance of cyclic CH_4 injection. The rates were set as 2000, 5000, and 8000 Mscf/day, respectively. The results in Figure 10 show that the incremental oil recovery was 2.07%, 2.55%, and 2.82%, for the case of 2000, 5000, and 8000 Mscf/day, respectively, illustrating that higher injection rate leads to larger incremental oil recovery factor.

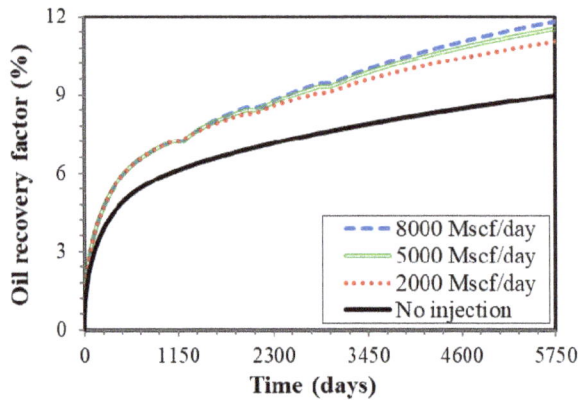

Figure 10. Comparison of oil recovery factor with different injection rates.

4.5. Effect of Injection Time

The length of injection time also impacts the CH_4 injection volume. We analyzed three cases with the injection time varying from 1 month to 3 months and kept other parameters as the same in the base case. As shown in Figure 11, oil recovery factor increases by 2.20%, 2.55%, and 2.76%, respectively. The results illustrate that longer injection time is beneficial to improve the efficiency of cyclic gas injection.

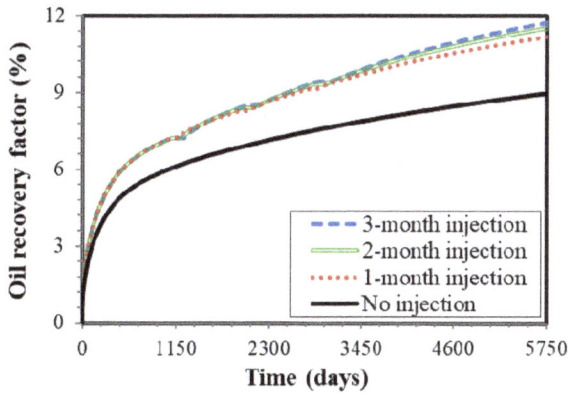

Figure 11. Comparison of oil recovery factor of different injection time.

4.6. Effect of Soaking Time

The soaking period affects the performance of Huff-n-Puff process as well. In this section, we conducted three cases with soaking time of 1 month, 2 months, and 3 months. As shown in Figure 12, the cases with the soaking time of 1 month, 2 months, and 3 months lead to the increment of 2.27%, 2.33%, and 2.37% after 15 years, respectively. As soaking period becomes longer, the CH_4 molecules will have more time to mix with oil phase adequately before its being produced back. Hence, the oil recovery factor improves with longer soaking time.

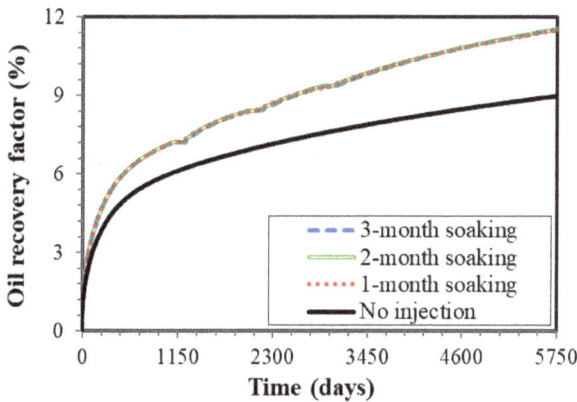

Figure 12. Comparison of oil recovery factor of different soaking time.

4.7. Effect of Number of Cycles

The number of cycles has significant impacts on the CH_4 injection effectiveness. We evaluated the well performance of the cases experiencing 1, 3, and 5 cycles, respectively. In Figure 13, compared the oil recovery factor of the scenarios with and without CH_4 injection for one-cycle treatment, the incremental oil recovery was boosted by 1.89%. Three-cycle processes yield the increment of 2.52%. For the five-cycle process, the incremental oil recovery is 2.72%. The increase of cycle numbers leads to the large incremental oil recovery; however, when it experiences 3 cycles or more, the impact on the well performance is diminished. Therefore, 3 cycles are suitable to reduce the operation cost and improve the oil production.

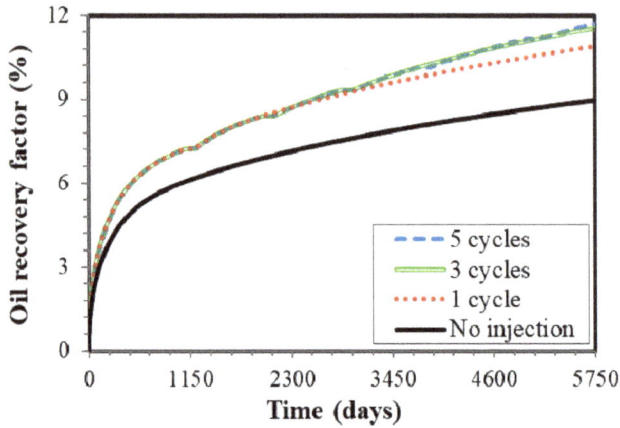

Figure 13. Comparison of oil recovery factor of different cycle treatments.

4.8. Effect of Matrix Permeability

In this section, we analyzed the effect of matrix permeability on the well performance. The matrix permeability was set ranging from 0.0001 mD to 0.01 mD. As shown in Figure 14, the oil recovery factor at the end of 15 years increases by 1.75%, 2.47% and 2.55%, corresponding to 0.0001, 0.001, and 0.01 mD, respectively. The matrix permeability influences the efficiency of CH_4 injection and higher permeability leads to larger incremental oil recovery. At the end of primary production, residual oil saturation is larger in the lower permeability formation and the diffusion mechanism is becoming more dominant. If the effect of confinement is included, more noticeable difference on the well performance will be observed in the higher permeability.

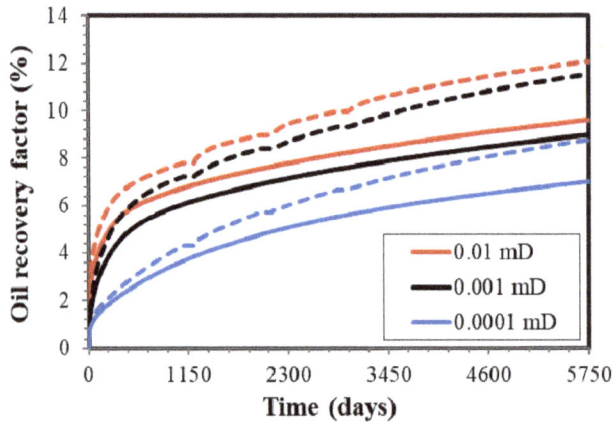

Figure 14. Effect of matrix permeability on oil recovery factor after 15 years (the solid and dash line represent the primary production and CH_4 injection, respectively).

We summarize the sensitivity parameters in Table 5 and represent by tornado plots shown in Figure 15. As shown, the most sensitive parameter for the cyclic CH_4 injection is the molecular diffusion, followed by matrix permeability, injection rate, injection time, and number of cycles. Primary depletion period and soaking time play the least important roles during the production time.

Table 5. Uncertain parameters and ranges for sensitivity analysis.

Parameters	Minimum	Medium	Maximum
Molecular diffusion/cm^2/s	0.0001	0.001	0.01
Primary depletion period/year	2	3	5
Injection time/day	30	60	90
Injection rate, Mscf/day	2000	5000	8000
Soaking time/day	30	60	90
Number of cycles	1	3	5
Matrix permeability, mD	0.001	0.01	0.1

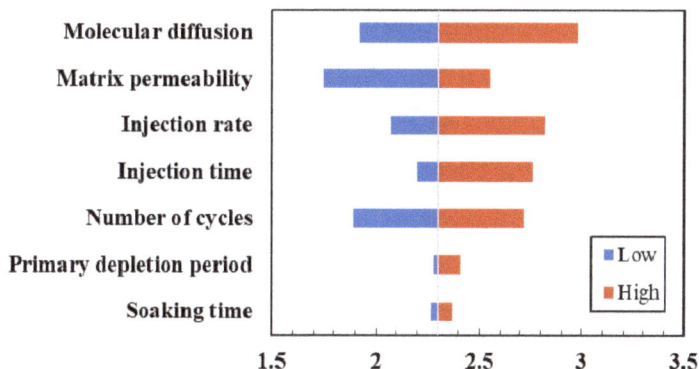

Figure 15. Tornado plot for the sensitivity analysis.

5. Conclusions

In this study, a numerical model is proposed to investigate the cyclic CH_4 injection in the Eagle Ford shale incorporating physical mechanisms such as molecular diffusion and confinement in nanopores. The following conclusions can be drawn:

(1) The effect of confinement in the nanopores is a significant factor in the simulation model to capture the real mechanism during the cyclic CH_4 injection;

(2) A series of simulations were performed to evaluate the impacts of key parameters on the process of enhanced oil recovery, concluding that molecular diffusion is the most sensitive, followed by matrix permeability, injection rate, injection time, and number of cycles;

(3) The impacts of primary depletion period and soaking time are less favorable for the cyclic CH_4 injection process;

(4) This work provides a better understanding of factors affecting the efficiency of cyclic CH_4 injection, which can hopefully guide the operations in the shale reservoir.

Author Contributions: Conceptualization, Y.Z. and Y.D.; methodology and investigation Y.Z.; writing—original draft preparation, Y.Z.; writing—review and editing, J.H. and Y.S.; project administration, Y.D.; funding acquisition, Y.Z., Y.D. and J.H.

Funding: This research was funded by "National Natural Science Foundation of China, grant number 51804282 and 51674010", "National Science and Technology Major Project of China, grant number 2017ZX05009-005", "Fundamental Research Funds for the Central Universities, grant number 59/53200759016", and "PetroChina Innovation Foundation (2017D-5007-0312)".

Acknowledgments: We would like to appreciate Computer Modeling Group Ltd. for providing the CMG software for this investigation.

Conflicts of Interest: The authors declare no conflict of interest.

References

1. Salahshoor, S.; Fahes, M.; Teodoriu, C. A review on the effect of confinement on phase behavior in tight formations. *J. Nat. Gas Sci. Eng.* **2018**, *51*, 89–103. [CrossRef]

2. Mullen, J. Petrophysical characterization of the Eagle Ford shale in South Texas. In Proceedings of the Canadian Unconventional Resources & International Petroleum Conference, Calgary, AB, Canada, 19–21 October 2010.

3. Shelley, R.F.; Saugier, L.D.; Al-Tailji, W.; Guliyev, N.; Shah, K. Understanding hydraulic fracture stimulated horizontal Eagle Ford completions. In Proceedings of the SPE/EAGE European Unconventional Resources Conference and Exhibition, Vienna, Austria, 20–22 March 2012.

4. Zhang, L.; Kou, Z.; Wang, H.; Zhao, Y.; Dejam, M.; Guo, J.; Du, J. Performance analysis for a model of a multi-wing hydraulically fractured vertical well in a coalbed methane gas reservoir. *J. Pet. Sci. Eng.* **2018**, *166*, 104–120. [CrossRef]

5. Dejam, M.; Hassanzadeh, H.; Chen, Z. Semi-analytical solution for pressure transient analysis of a hydraulically fractured vertical well in a bounded dual-porosity reservoir. *J. Hydrol.* **2018**, *565*, 289–301. [CrossRef]

6. Xu, Z.H.; Zhao, P.Q.; Wang, Z.L.; Ostadhassan, M.; Pan, Z.H. Characterization and consecutive prediction of pore structures in tight oil reservoirs. *Energies* **2018**, *11*, 2705. [CrossRef]

7. Qian, K.; Yang, S.L.; Dou, H.E.; Wang, Q.; Wang, L.; Huang, Y. Experimental investigation on microscopic residual oil distribution during CO_2 Huff-and-Puff process in tight oil reservoirs. *Energies* **2018**, *11*, 2843. [CrossRef]

8. Dejam, M.; Hassanzadeh, H.; Chen, Z. Pre-Darcy flow in porous media. *Water Resour. Res.* **2017**, *53*, 8187–8210. [CrossRef]

9. Dejam, M.; Hassanzadeh, H.; Chen, Z. Pre-Darcy flow in tight and shale formations. In Proceedings of the 70th Annual Meeting of the APS (American Physical Society) Division of Fluid Dynamics, Denver, CO, USA, 19–21 November 2017.

10. Hawthorne, S.B.; Gorecki, C.D.; Sorensen, J.A.; Steadman, E.N.; Harju, A.J.; Melzer, S. Hydrocarbon mobilization mechanisms from upper, middle, and lower Bakken reservoir rocks exposed to CO_2. In Proceedings of the SPE Unconventional Resources Conference, Calgary, AB, Canada, 5–7 November 2013.

11. Ren, B.; Ren, S.; Zhang, L.; Chen, G.; Zhang, H. Monitoring on CO_2 migration in a tight oil reservoir during CCS-EOR in Jilin Oilfield China. *Energy* **2016**, *98*, 108–121. [CrossRef]

12. Sheng, J.J. Critical review of field EOR projects in shale and tight reservoirs. *J. Pet. Sci. Eng.* **2017**, *159*, 654–665. [CrossRef]

13. Wang, X.; Luo, P.; Er, V.; Huang, S.S. Assessment of CO_2 flooding potential for Bakken Formation, Saskatchewan. In Proceedings of the Canadian Unconventional Resources & International Petroleum Conference, Calgary, AB, Canada, 19–21 October 2010.

14. Sanchez-Rivera, D.; Mohanty, K.; Balhoff, M. Reservoir simulation and optimization of Huff-n-Puff operations in the Bakken shale. *Fuel* **2015**, *147*, 82–94. [CrossRef]

15. Alfarge, D.; Wei, M.; Bai, B. Feasibility of CO_2-EOR in shale-oil reservoirs: Numerical simulation study and pilot tests. In Proceedings of the Carbon Management Technology Conference, Houston, TX, USA, 17–20 July 2017.

16. Meng, X.; Yu, Y.; Sheng, J.J.; Watson, M.; Mody, F. An experimental study on huff-n-puff gas injection to enhance condensate recovery in shale gas reservoirs. In Proceedings of the Unconventional Resources Technology Conference, San Antonio, TX, USA, 20–22 July 2015.

17. Shi, R.; Kantzas, A. Enhanced heavy oil recovery on depleted long core system by CH_4 and CO_2. In Proceedings of the SPE International Thermal Operations and Heavy Oil Symposium, Calgary, AB, Canada, 20–23 October 2008.

18. Gamadi, T.D.; Elldakli, F.; Sheng, J.J. Compositional simulation evaluation of EOR potential in shale oil reservoirs by cyclic natural gas injection. In Proceedings of the Unconventional Resources Technology Conference, Denver, CO, USA, 25–27 August 2014.

19. Shen, Z.; Sheng, J.J. Experimental study of asphaltene aggregation during CO_2 and CH_4 injection in shale oil reservoirs. In Proceedings of the SPE Improved Oil Recovery Conference, Tulsa, OK, USA, 11–13 April 2016.

20. Sigmund, P.M.; Dranchuk, P.M.; Morrow, N.R.; Purvis, R.A. Retrograde condensation in porous media. *SPE J.* **1973**, *13*, 93–104. [CrossRef]

21. Brusilovsky, A.I. Mathematical simulation of phase behavior of natural multicomponent systems at high pressures with an equation of state. *SPE Res. Eng.* **1992**, *7*, 117–122. [CrossRef]

22. Du, L.; Chu, L. Understanding anomalous phase behavior in unconventional oil reservoirs. In Proceedings of the SPE Canadian Unconventional Resources Conference, Calgary, AB, Canada, 30 October–1 November 2012.

23. Nojabaei, B.; Johns, R.T.; Chu, L. Effect of capillary pressure on phase behavior in tight rocks and shales. *SPE Res. Eval. Eng.* **2013**, *16*, 281–289. [CrossRef]

24. Sanaei, A.; Jamili, A.; Callard, J.; Mathur, A. Production modeling in the Eagle Ford gas condensate window: Integrating new relationships between core permeability, pore size, and confined PVT properties. In Proceedings of the SPE Western North American and Rocky Mountain Joint Meeting, Denver, CO, USA, 16–18 April 2014.

25. Zhang, Y.; Lashgari, H.R.; Di, Y.; Sepehrnoori, K. Capillary pressure effect on hydrocarbon phase behavior in unconventional reservoirs. In Proceedings of the SPE Low Perm Symposium, Denver, CO, USA, 5–6 May 2016.

26. Wang, L.; Parsa, E.; Gao, Y.; Ok, J.T.; Neeves, K.; Yin, X.; Ozkan, E. Experimental study and modeling of the effect of nanoconfinement on hydrocarbon phase behavior in unconventional reservoirs. In Proceedings of the SPE Western North American and Rocky Mountain Joint Meeting, Denver, CO, USA, 16–18 April 2014.

27. Yang, G.; Fan, Z.; Li, X. Determination of confined fluid phase behavior using modified Peng-Robinson equation of state. In Proceedings of the Unconventional Resources Technology Conference, Houston, TX, USA, 23–25 July 2018.

28. Luo, S.; Nasrabadi, H.; Lutkenhaus, J.L. Effect of confinement on the bubble points of hydrocarbons in nanoporous media. *AIChE J.* **2016**, *62*, 1772–1780. [CrossRef]

29. Alfi, M.; Nasrabadi, H.; Banerjee, D. Effect of confinement on bubble point temperature shift of hydrocarbon mixtures: Experimental investigation using nanofluidic devices. In Proceedings of the SPE Annual Technical Conference and Exhibition, San Antonio, TX, USA, 9–11 October 2017.

30. Yu, W.; Lashgari, H.R.; Wu, K.; Sepehrnoori, K. CO_2 injection for enhanced oil recovery in Bakken tight oil reservoirs. *Fuel* **2015**, *159*, 354–363. [CrossRef]

31. Du, X.D.; Gu, M.; Duan, S.; Xian, X.F. Investigation of CO_2–CH_4 displacement and transport in shale for enhanced shale gas recovery and CO_2 sequestration. *J. Energy Resour. Technol.* **2017**, *139*, 012909. [CrossRef]

32. Zhang, Y.; Yu, W.; Li, Z.; Sepehrnoori, K. Simulation study of factors affecting CO_2 Huff-n-Puff process in tight oil reservoirs. *J. Pet. Sci. Eng.* **2018**, *163*, 264–269. [CrossRef]

33. Wu, K.; Li, X.; Guo, C.; Wang, C.; Chen, Z. A unified model for gas transfer in nanopores of shale-gas reservoirs: Coupling pore diffusion and surface diffusion. *SPE J.* **2016**, *21*, 1583–1611. [CrossRef]

34. Chen, C.; Gu, M. Investigation of cyclic CO_2 huff-and-puff recovery in shale oil reservoirs using reservoir simulation and sensitivity analysis. *Fuel* **2017**, *188*, 102–111. [CrossRef]

35. Simpson, M.D.; Patterson, R.; Wu, K. Study of stress shadow effects in Eagle Ford Shale: Insight from field data analysis. In Proceedings of the 50th U.S. Rock Mechanics/Geomechanics Symposium, American Rock Mechanics Association, Houston, TX, USA, 26–29 June 2016.

36. Singh, S.K.; Sinha, A.; Deo, G.; Singh, J.K. Vapor-liquid phase coexistence, critical properties, and surface tension of confined alkanes. *J. Phys. Chem. C* **2009**, *113*, 7170–7180. [CrossRef]

37. Jin, L.C.; Ma, Y.X.; Jamili, A. Investigating the effect of pore proximity on phase behavior and fluid properties in shale formations. In Proceedings of the SPE Annual Technical Conference and Exhibition, New Orleans, LA, USA, 30 September–2 October 2013.

38. Campos, M.D.; Akkutlu, I.Y.; Sigal, R.F. A molecular dynamics study on natural gas solubility enhancement in water confined to small pores. In Proceedings of the SPE Annual Technical Conference and Exhibition, New Orleans, LA, USA, 4–7 October 2009.

39. Devegowda, D.; Sapmanee, K.; Civan, F.; Sigal, R.F. Phase behavior of gas condensates in shales due to pore proximity effects: Implications for transport, reserves and well productivity. In Proceedings of the SPE Annual Technical Conference and Exhibition, San Antonio, TX, USA, 8–10 October 2012.

40. Zarragoicoechea, G.J.; Kuz, V.A. Critical shift of a confined fluid in a nanopore. *Fluid Phase Equilib.* **2004**, *220*, 7–9. [CrossRef]

41.   Adamson, A.W. Capillarity. In *Physical Chemistry of Surfaces*, 5th ed.; John Wiley & Sons: New York City, NY, USA, 1990.

42.   Pedersen, K.S.; Christensen, P.L. Flash and phase envelope calculations. In *Phase Behavior of Petroleum Reservoir Fluids*, 1st ed.; CRC Press Taylor & Francis Group: Boca Raton, FL, USA, 2006.

43.   Peng, D.Y.; Robinson, D.B. A new two-constant equation of state. *Ind. Eng. Chem. Fundam.* **1976**, *15*, 59–64. [CrossRef]

44.   CMG. *GEM-GEM User's Guide*; Computer Modeling Group Ltd.: Calgary, AB, Canada, 2012.

45.   Condon, S.M.; Dyman, T.S. *Geologic Assessment of Undiscovered Conventional Oil and Gas Resources in the Upper Cretaceous Navarro and Taylor Groups, Western Gul Province*; Texas: USGS Digital Data Series; USGU: Reston, VG, USA, 2006.

46.   Mohebbinia, S.; Wong, T. Molecular diffusion calculations in simulation of gasfloods in fractured reservoirs. In Proceedings of the SPE Reservoir Simulation Conference, Montgomery, AL, USA, 20–22 February 2017.

47.   Orangi, A.; Nagarajan, N.R.; Honarpour, M.M.; Rosenzweig, J.J. Unconventional shale oil and gas-condensate reservoir production, impact of rock, fluid, and hydraulic fractures. In Proceedings of the SPE Hydraulic Fracturing Technology Conference and Exhibition, The Woodlands, TX, USA, 24–26 January 2011.

MDPI

St. Alban-Anlage 66

4052 Basel

Switzerland

Tel. +41 61 683 77 34

Fax +41 61 302 89 18

www.mdpi.com

Energies Editorial Office

E-mail: energies@mdpi.com

www.mdpi.com/journal/energies

www.ingramcontent.com/pod-product-compliance
Lightning Source LLC
Chambersburg PA
CBHW051837210326
41597CB00033B/5680